Springer Series in Materials Science

Volume 235

The Springer Series in Materials Science covers the complete spectrum of materials physics, including fundamental principles, physical properties, materials theory and design. Recognizing the increasing importance of materials science in future device technologies, the book titles in this series reflect the state-of-the-art in understanding and controlling the structure and properties of all important classes of materials.

More information about this series at http://www.springer.com/series/856

Michelle J.S. Spencer · Tetsuya Morishita
Editors

Silicene

Structure, Properties and Applications

 Springer

Editors
Michelle J.S. Spencer
School of Science
RMIT University
Melbourne, VIC
Australia

Tetsuya Morishita
National Institute of Advanced Industrial
 Science and Technology (AIST)
Tsukuba
Japan

ISSN 0933-033X   ISSN 2196-2812   (electronic)
Springer Series in Materials Science
ISBN 978-3-319-28342-5   ISBN 978-3-319-28344-9   (eBook)
DOI 10.1007/978-3-319-28344-9

Library of Congress Control Number: 2015950444

This Springer imprint is published by SpringerNature
The registered company is Springer International Publishing AG Switzerland

Preface

The field of nanomaterials has become one of the most quickly growing areas in science due to the unique properties and potential applications of these materials in electronics, medicine, consumer goods, defence, amongst others. Nanomaterials come in different shapes including zero-dimensional (0D), one-dimensional (1D) and two-dimensional (2D) forms. They are classified by having at least one of their dimensions less than 100 nm in size.

The class of 2D nanomaterials are characterised by large lateral dimensions and small thicknesses of the order of less than, typically, several nanometres. They are often referred to as nanosheets and can be considered akin to extremely thin sheets of paper. A wide variety of materials can be grown as 2D nanomaterials and can be composed of one or multiple elements. The elemental 2D nanomaterials usually have names ending in 'ene' and include the material that is the focus of this book, namely silicene. Silicene is a crystalline two-dimensional (2D) nanomaterial composed entirely of silicon (Si) atoms. The atoms in this single layer are arranged in a hexagonal honeycomb structure which, when viewed from the side, are buckled.

Other 2D elemental nanomaterials include the well-known structure graphene (composed of carbon), as well as phosphorene (composed of P), germanene (composed of Ge) and stanene (composed of Sn). There are also 2D nanomaterials composed of different compounds (usually consisting of two elements covalently bonded together). Examples include the transition metal di-chalcogenides having the composition of MX_2, with M being a transition metal atom (Mo, W, etc.) and X a chalcogen atom (S, Se or Te). Common examples include MoS_2, WS_2, $MoSe_2$, WSe_2 and $MoTe_2$.

This book is dedicated to discussing the structure, properties and applications of silicene. While there are a number of excellent review articles that have been published on this material, we have brought together top scientist working on different aspects of silicene to provide a state-of-the-art and up-to-date account of a wide variety of aspects of the material.

At the time of publication of this book >700 papers have been published on silicene (based on a search of Web of Science using the term 'silicene'). More than 450 of these were published in 2014 and 2015 alone. Of course, this does not find all articles published on this material, particular the early work performed in the area before the term silicene was used or recognised. Indeed, some of our early papers used the term 'Si nanosheets' to refer to multilayer or functionalised silicene. The numbers, however, do give an indication of the popularity of the material and how fast the field is growing.

The name silicene was first coined by G.G. Guzman-Verri and L.C. Lew Yan Voon in 2007 in their paper on the electronic structure of silicon-based nanostructures in Physical Review B. Thicker variants of silicene are referred to as nanosheets, or more recently, multilayer silicene.

Silicene has been studied using a variety of methods including different experimental techniques as well as different theoretical or computational approaches. The information provided by the different techniques has helped to generate complimentary information to advance our understanding of this material.

This book is divided into 3 sections. The first section of the book (Part I) is dedicated to unmodified free-standing silicene and multilayer silicene. Chapter 1 discusses the physical properties of free-standing silicene, including its structure, mechanical, electronic, electric field, topological, optical, transport, magnetic and thermal properties. A brief discussion on defects and doping is also included. As silicene has not been synthesised unmodified in the free-standing form, the results are primarily theoretical in nature. Chapter 2 reviews the topological properties of 2D elemental honeycomb nanosheets of the Group 14 elements, which includes silicene. Chapter 3 discusses the formation, structural and electronic properties of unmodified free-standing multilayer silicene that has been studied using molecular dynamics and density functional theory (DFT) approaches.

The second section of the book (Part II) reviews the properties of silicene that are modified by molecular chemisorption or when interfaced with non-metallic surfaces. Specifically, Chap. 4 discusses the soft chemical synthesis of silicene functionalised with different chemical groups and the potential applications of such materials. Chapter 5 reviews the theoretical studies and in particular the DFT calculations of functionalised silicene and the associated structural, electronic and dynamic properties. Chapter 6 covers theoretical work examining the interaction between silicene and non-metallic surfaces such as aluminium nitride, layered metal (di)chalcogenides, ZnS surfaces as well as possible growth of silicene on MoS_2. The findings are compared to experimental results where available.

The third section of the book (Part III) covers the studies of silicene on different substrates and presents both the theoretical and experimental results. There is a significant body of work that has focused on the deposition of silicene onto single crystal silver (Ag) surfaces and we dedicate Chaps. 7–12 to the different aspects of work in this area. In Chap. 7, the experimental studies of silicene grown on the Ag (111) single crystal surface are reviewed. There are many different overlays that form on this surface and the structures and their associated electronic properties are presented. Chapter 8 covers the growth of silicene on the Ag(111) and Au(110) surfaces. While there is some overlap with Chap. 7 we believe that this is warranted due to the volume of work in this area, and in particular, the inclusion of multilayer silicene on the Ag(111) surface is emphasised. Chapter 9 discusses the growth of silicon nanoribbons on the Ag(110) surface. While we have mainly focused the book on extended silicene, we believe that having a chapter focused on nanoribbons (similar to strips, or 'ribbons' of silicene) will be of interest to the reader, as there is a school of thought as to whether the nanoribbons can actually be distinguished from pure silicene. Chapter 10 gives a comprehensive review of the theoretical studies of silicene on different silver surfaces. Chapter 11 presents work on the adsorption and interaction of different molecules on silicene. This aspect of the field is particularly important for determining the potential application of silicene in nanoelectronics or other devices, where they need to operate under environmental conditions. Chapter 12 concludes the book with a review of the studies that have examined deposition of silicene on substrates other than silver and include several conductive substrates such as $ZrB_2(0001)$, $ZrC(111)$, $Ir(111)$ and $Au(110)$, in particular.

As editors and authors of this book, we are very pleased to have been involved in publishing what we believe is the first book wholly dedicated to this exciting new material. The chapters presented here will provide a solid foundation for those starting out in the field, including undergraduate and postgraduate students studying chemistry, physics, engineering and nanotechnology. For the experts already working on silicene and 2D nanomaterials as experimentalists and/or theoreticians, we believe that this will provide a source of invaluable information and a useful reference.

Of course, the work on silicene is still a growing area of research and there are many questions yet to be answered about the potential of this material. We are very excited to be part of the discoveries in this area and look forward to seeing where this material will go in the future. We hope that you as the reader will share in our passion for science and 2D nanomaterials!

Melbourne
Tsukuba

Michelle J.S. Spencer
Tetsuya Morishita

Contents

Contributors

Valeri V. Afanas'ev Department of Physics and Astronomy, University of Leuven, Leuven, Belgium

Ryuichi Arafune International Center for Materials Nanoarchitectonics (WPI-MANA), National Institute for Materials Science, Tsukuba, Ibaraki, Japan

Bernard Aufray Aix-Marseille Université, CNRS, Marseille, France

Azzedine Bendounan TEMPO Beamline, Synchrotron Soleil, Gif-sur-Yvette Cedex, France

Jean-Paul Biberian Aix-Marseille Université, CNRS, Marseille, France

Yi Du Institute for Superconducting and Electronic Materials (ISEM), Australian Institute for Innovative Materials (AIIM), University of Wollongong, Wollongong, NSW, Australia

Gérald Dujardin Institut des Sciences Moléculaires d'Orsay, ISMO-CNRS, Bât. 210, Université Paris-Sud, Orsay, France

Bénédicte Ealet Aix-Marseille Université, CNRS, Marseille, France

Hanna Enriquez Institut des Sciences Moléculaires d'Orsay, ISMO-CNRS, Bât. 210, Université Paris-Sud, Orsay, France

Motohiko Ezawa Department of Applied Physics, University of Tokyo, Tokyo, Japan

Antoine Fleurence Japan Advanced Institute of Science and Technology, Nomi, Ishikawa, Japan

Jean-Yves Hoarau Aix-Marseille Université, CNRS, Marseille, France

M. Houssa Department of Physics and Astronomy, University of Leuven, Leuven, Belgium

Haik Jamgotchian Aix Marseille Université, CNRS, Marseille, France

Abdelkader Kara Department of Physics, University of Central Florida, Orlando, FL, USA

Lok C. Lew Yan Voon School of Science and Mathematics, The Citadel, Charleston, SC, USA

Chun Liang Lin Department of Advanced Materials Science, Graduate School of Frontier Science, University of Tokyo, Chiba, Japan

Hichem Maradj Laboratoire LSMC, Université d'Oran es-sénia, Oran, Algeria

Andrew J. Mayne Institut des Sciences Moléculaires d'Orsay, ISMO-CNRS, Bât. 210, Université Paris-Sud, Orsay, France

Tetsuya Morishita Research Center for Computational Design of Advanced Functional Materials, National Institute of Advanced Industrial Science and Technology (AIST), Tsukuba, Ibaraki, Japan

Hideyuki Nakano Toyota Central R&D Labs., Inc., Nagakute, Aichi, Japan

Masataka Ohashi Toyota Central R&D Labs., Inc., Nagakute, Aichi, Japan

Hamid Oughaddou Institut des Sciences Moléculaires d'Orsay, ISMO-CNRS, Bât. 210, Université Paris-Sud, Orsay, France; Département de Physique, Université de Cergy-Pontoise, Cergy-Pontoise Cedex, France

Fausto Sirroti TEMPO Beamline, Synchrotron Soleil, Gif-sur-Yvette Cedex, France

Michelle J.S. Spencer School of Science, RMIT University, Melbourne, VIC, Australia

André Stesmans Department of Physics and Astronomy, University of Leuven, Leuven, Belgium

Noriaki Takagi Department of Advanced Materials Science, Graduate School of Frontier Science, University of Tokyo, Chiba, Japan

Mohammed Rachid Tchalala Institut des Sciences Moléculaires d'Orsay, ISMO-CNRS, Bât. 210, Université Paris-Sud, Orsay, France

Xun Xu Institute for Superconducting and Electronic Materials (ISEM), Australian Institute for Innovative Materials (AIIM), University of Wollongong, Wollongong, NSW, Australia

Handan Yildirim School of Chemical Engineering, Purdue University, Lafayette, IN, USA

Part I
Free Standing Silicene

Chapter 1
Physical Properties of Silicene

Lok C. Lew Yan Voon

Abstract In this chapter, we discuss the physical properties of free-standing silicene. Silicene is a single atomic layer of silicon much like graphene. The interest in silicene is exactly the same as that for graphene, in being two-dimensional and possessing a Dirac cone. One advantage relies on its possible application in electronics, whereby its natural compatibility with the current Si technology might make fabrication much more of a commercial reality. Since free-standing has not yet been made, all of the results are theoretical in nature, though most properties are not expected to differ significantly for silicene on a substrate.

1.1 Introduction

Silicene [1] is a single atomic layer of silicon (Si) much like graphene [2]. Early work, both theoretical [3–7] and experimental [8, 9], went mostly unnoticed until silicene nanoribbons were reported to have been fabricated on a silver substrate by Kara et al. in [10]. Since then, silicene sheets have been grown mainly on Ag(111) starting in 2012 [11–15]; these were achieved under ultrahigh vacuum conditions by evaporation of silicon wafer and slow deposition onto a substrate at 220–260 °C.

The interest in silicene is exactly the same as that for graphene, in being two-dimensional (2D) and possessing a Dirac cone [1]. One advantage relies on its possible application in electronics, whereby its natural compatibility with the current Si technology might make fabrication much more of a commercial reality. Indeed, a field-effect transistor made out of silicene has finally been demonstrated in 2015 [16].

L.C. Lew Yan Voon (✉)
School of Science and Mathematics, The Citadel, 171 Moultrie Street,
Charleston, SC 29409, USA
e-mail: llewyanv@citadel.edu

© Springer International Publishing Switzerland 2016
M.J.S. Spencer and T. Morishita (eds.), *Silicene*, Springer Series
in Materials Science 235, DOI 10.1007/978-3-319-28344-9_1

In this chapter, we discuss the physical properties of free-standing silicene. Since the latter has not yet been made, all of the results are theoretical in nature, though most properties are not expected to differ significantly for silicene on a substrate. Thus, the results on free-standing silicene are relevant for studies on silicene on substrates.

1.2   Structure

Numerous theoretical studies, almost all based on first-principles calculations, of the structural properties, have been published and they are all consistent. First-principles calculations are invariably based upon density-functional theory (DFT), whether using the local-density approximation (LDA) or the generalized-gradient approximation (GGA) to the exchange-correlation potential. The local density approximation is known to lead to overbinding and, thus, a slightly smaller lattice constant.

In a DFT calculation, an initial geometry is assumed and the atoms are moved so as to minimize the total energy while preserving the lattice symmetry. Takeda and Shiraishi [4] first carried out this process for a single layer of Si. Given the already known existence of graphite, they assumed a hexagonal lattice for Si as well (with a superiodicity perpendicular to the plane with a large vacuum layer, typically at least ~ 10 Å) and allowed the in-plane lattice constant a to vary, as well as the position of the basis atom (B in Fig. 1.1) within the unit cell while preserving the imposed D_{3d} symmetry. Realizing the fact that Si is not known to form the flat sp^2 bonding, they allowed the B atom to move out of the A atom plane. They found what they called the corrugated structure to have a lower total energy than the flat one, and a local minimum for $a = 3.855$ Å and a deformation angle of 9.9°. A more recent GGA calculation puts the lowering in energy at 30 meV/atom and a binding energy of 4.9 eV/atom, which is lower than that for bulk silicon (diamond structure) by 0.6 eV/atom [7]. As Takeda and Shiraishi pointed out [4], the corrugated structure makes sense since it resembles closely the (111) plane of bulk cubic silicon. Two ways of explaining the buckled structure instead of the flat structure of graphene are via the weakening of the π double bond due to the larger separation of the Si atoms,

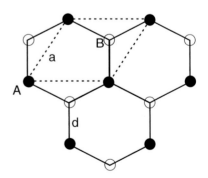

Fig. 1.1 Crystal structure of silicene. The lattice is hexagonal, a unit cell is the *dashed one*, and the basis consists of two Si atoms labelled *A* and *B*

	d (Å)	a (Å)	Δz (Å)
Silicene [20]	2.248	3.820	0.44
Graphene [20]	1.414	2.456	
Germanene [20]	2.382	4.0	0.64
Silicon [33]	2.35		
g-Si [33]	2.35	4.07	
T-Si [33]		2.427	

Table 1.1 Structural parameters of various 2D sheets

and via the pseudo Jahn-Teller effect with coupling of the electronic ground state to the next one by a vibrational mode [17].

All DFT calculations since then have reproduced very similar structural parameters. Typical structural parameters are given in Table 1.1. The out-of-plane height ΔZ of the Si atom was found to be 0.53 Å by Ding and Ni [18], and 0.44 Å by Cahangirov et al. [19] and 0.45 Å by us [20]. In bulk Si, the out-of-plane Si atom is 0.78 Å from the (111) plane. Thus, the bonding in silicene can be viewed as in between sp^2 and sp^3. The bond length d is much larger than for graphene because of the larger size of silicon compared to carbon.

The fact that the above structure is a local minimum in the total energy does not guarantee stability. In fact, Cahangirov et al. [19] found that silicene has another local minimum with a larger binding energy at a higher buckling, with $\Delta Z \approx 2$ Å (referred to as the high buckled or HB structure compared to the low buckled or LB one). However, a variety of tests showed the LB structure to be the stable one. First, they found that, on a (2×2) supercell, the HB structure developed clustering upon structural optimization; the same group had also earlier obtained the same LB structure on a (2×2) supercell [7]. Second, both the HB and planar (PL) structures have phonon spectra with imaginary frequency modes, a signature of lattice instability. Third, the LB structure was found to be preserved upon performing ab initio molecular dynamics (MD) on a (4×4) supercell with temperature as high as 1000 K.

More recent work has questioned whether the silicene structure is, in fact, the most stable one. Thus, Kaltsas and Tsetseris [21] started with different configurations by taking the surface layer of various Si surface reconstructions and optimizing the structure. In the process, they found that structures based on the $\sqrt{3} \times \sqrt{3}$, 5×5, and 7×7 reconstructions are actually all more stable than the perfect silicene structure, by 48, 17 and 6 meV per atom, respectively (Fig. 1.2).

In spite of the above discussions, a number of authors have explored the possibility of silicene being flat like graphene. A graphitic form of silicon, g-Si, with presumably flat silicon layers, were studied by Yin and Cohen in [22] and found to be metastable compared to the diamond structure. Based upon earlier ideas, they suggested that a pressure of −69 kbar could lead to a stable graphitic silicon. Wang et al. [5] were, on the other hand, interested in the nature of the bonding and predicted that the interlayer bonding would be stronger than the van der Waals one. As already indicated, a number of authors have compared the stability of a single

(a)　　　　　　　**(b)**　　　　　　　**(c)**　　　　　　　**(d)**

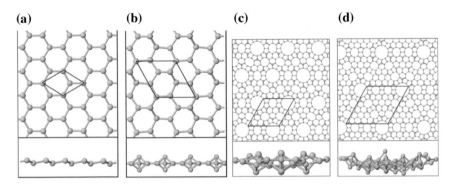

Fig. 1.2 Other possible stable forms of silicene: **a** silicene, **b** $\sqrt{3} \times \sqrt{3}$, **c** 5×5, and **d** 7×7. Reproduced from [21] with permission from the PCCP Owner Societies

sheet of flat silicene to that of corrugated silicene [4, 6, 7, 19]. However, a few studies assumed a flat silicene structure [23, 24], with no apparent check on the stability of such. Unfortunately, this early confusion about the structure of silicene has led to some unwarranted criticisms [25].

Nevertheless, others have looked at ways a flat sheet could be coerced to form. Three general ideas have been proposed. The first is based upon using a host to confine the Si layer [26]. Thus, an MD study using the Tersoff potential showed that liquid Si atoms confined between two parallel-plane walls separated by less than 7.5 Å can assume a flat hexagonal lattice structure when the liquid Si is quenched down from 2400 K down to 0 K [26]. The Si atoms were assumed to interact with the walls using a 9–3 Lennard-Jones potential. The stability of the structure was checked using first-principles MD up to 300 K. The second approach proposes using externally applied biaxial tensile strain [27–29]. It is very much expected that the buckling would be reduced by a biaxial tensile strain; Liu et al. [27] found it to be reduced to 0.34 Å with a 12.5 % strain. Wang et al. [29] reported that they obtained a stable flat silicene with a biaxial strain of 20 %. However, this disagrees with other studies. Thus, two groups found the buckling to reach a low (about 0.3 Å for a 10 % strain in [28] and about 0.23 Å for a 7 % strain in [30]) and then increase again (Fig. 1.3). Furthermore, a number of workers found some instability settling in around 17 % [28, 30, 31]. Yang et al. [30] proposed that a ZZ uniaxial strain of 16 % could reduce the buckling to zero but that the resulting structure is unstable from phonon calculations. The third approach for "flattening" the silicene was based on the hope that adsorption of other atoms or of a substrate might lead to a planar structure. However, functionalization studies show that the opposite happens, whereby the bonding becomes more sp^3-like [20]. For completeness, we mention that a recent Monte Carlo calculation using Tersoff potential did not find any buckling [32], contrary to all the DFT calculations.

Fig. 1.3 Buckling parameter of silicene under external tensile strain. *AC* and *ZZ* refer to uniaxial strains (armchair and zigzag directions, respectively) while *EQ* refers to biaxial. Reprinted from [30], with permission from Elsevier [30]

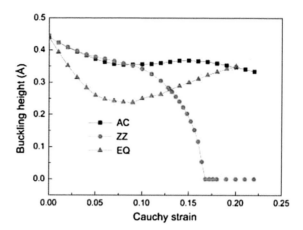

1.3 Mechanical Properties

Two-dimensional materials are predicted to have much higher mechanical strengths than bulk materials [47]. Graphene is already known to have excellent mechanical strength due to its sp^2 bonding (Table 1.2). A 2D sheet should also have transversal flexibility. Additionally, the ability to support large strain means the nonlinear elastic regime can easily be attained. Thus, much of the work on mechanical properties have investigated the four regimes of linear elasticity, nonlinear elasticity, plasticity and fracture.

The elastic properties of a hexagonal 2D sheet can be characterized by elastic constants C_{11} and C_{12}, the Poisson ratio v, and an in-plane stiffness C. If the strain energy (difference in total energy with and without strain) is assumed to depend quadratically on an external applied strain ε_{ij} (harmonic approximation),

$$E_s(\epsilon_{xx}, \epsilon_{yy}) = a_1 \epsilon_{xx}^2 + a_2 \epsilon_{yy}^2 + a_3 \epsilon_{xx} \epsilon_{yy}, \tag{1.1}$$

then one can show that the elastic parameters are related [34] as follows:

$$a_1 = a_2 = \frac{hA_0}{2} C_{11}, \tag{1.2}$$

$$a_3 = hA_0 C_{12}, \tag{1.3}$$

$$v = -\frac{\epsilon_{\text{trans}}}{\epsilon_{\text{axial}}} = \frac{C_{12}}{C_{11}} = \frac{a_3}{2a_1}, \tag{1.4}$$

$$C = hC_{11}\left[1 - \left(\frac{C_{11}}{C_{12}}\right)^2\right] = \frac{[2a_1 - (a_3)^2/2a_1]}{A_0}, \tag{1.5}$$

Table 1.2 Physical properties of silicene compared to graphene

Property	Silicene	Graphene
Lattice constant (Å)	3.86 [20]	2.46 [34]
Buckling (Å)	0.44 [20]	0
Elastic constant C (N/m)	62 [34]	335 [34]
	60.06 (ZZ), 63.51 (AC) [35]	
	50.44 (ZZ), 62.31 (AC) [36]	
	86.48, 85.99 [37]	328.02, 328.30 [37]
Poisson ratio ν	0.30 [34]	0.16 [34]
	0.41 (ZZ), 0.37 (AC) [35]	
Young's modulus (T Pa nm)	0.178 [27]	0.420 [27]
Bending modulus D (eV)	38.63 [36]	1.5 [38]
Phonon D-mode (K point) (cm^{-1})	545 [28]	~1350 [39]
Phonon G-mode (K point) (cm^{-1})	550 [28]	~1580 [39]
Fermi velocity (m/s)	~10^5	~10^6
Deformation potential E_1 (eV)	2.13 [37]	5.14, 5.00 [37]
Intrinsic mobility (10^5 cm^2 V^{-1} s^{-1})		
μ_h	2.23, 2.22 [37]	3.22, 3.51 [37]
μ_e	2.58, 2.57 [37]	3.39, 3.20 [37]
Work function (eV)	4.8 [31]	4.56 [40]
Thermal conductivity κ at 300 K (W/m K)	9.4 [41]	3000–5000 [42]
Thermopower S (μV/K)	87 [43]	−32 [44]
Thermoelectric figure of merit ZT	<0.5 [45]	0.17–1.02 [46]

where h and A_0 are the effective thickness and equilibrium area of the supercell, respectively. An isotropic model is appropriate to describe the response to a biaxial strain for a hexagonal crystal. Peng et al. [47] found linear elasticity to only apply for strains smaller than 3 % in magnitude, beyond which higher-order elastic coefficients would be needed.

Sahin et al. [34] computed the Poisson ratio ν and in-plane stiffness constant C by stretching the sheets in the plane, computing the change in the total energy with the deformation, and fitting to a quadratic function. They obtained ν = 0.3 and $C = 62$ J/m^2 for silicene, compared to 0.16 and 335 J/m^2, respectively, for graphene. The bulk modulus of silicene is reported to be 3.5 times smaller than for graphene [48]. Liu et al. [27] define the modified Young's modulus as

$$Y = \frac{1}{A_0} \frac{\partial^2 E}{\partial \epsilon^2}\bigg|_{\epsilon=0}.$$ (1.6)

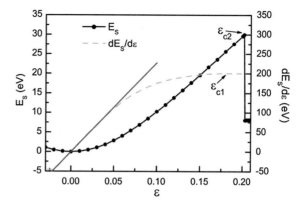

Fig. 1.4 Strain energy and its first differential as a function of biaxial tensile strain [31]

This is the same definition as the in-plane stiffness [31]. They obtained $Y = 0.178$ T Pa nm, about 40 % of the graphene value.

Qin et al. [31] showed that the anharmonic regime is reached for a biaxial strain in the range -2 to 4 % (Fig. 1.4), where the negative sign is for compressive strain. Beyond the harmonic regime, they identified two critical points. Beyond the first one ($\varepsilon_{c1} \approx 18$ % which they also call the "ultimate strain" [49]), $dE_s/d\varepsilon$ reaches a maximum value, i.e., it takes less tension to stretch the structure and it is unstable under certain acoustic waves (imaginary phonon frequencies), a phenomenon known as phonon instability. On the other hand, from an analysis of the phonon frequencies, Liu et al. determined that silicene would be unstable under a compressive biaxial strain larger than 5 %. Beyond the second critical point ($\varepsilon_{c2} \approx 20$ %), the strain energy decreases sharply and this corresponds to the yield point. Kaloni et al. [28] found this to occur for $\varepsilon \approx 17$ %. The change in the buckling parameter was reported by Kaloni et al. [28] and Peng et al. [47]. While an initial decrease with expansion is expected, both groups found the buckling to increase again for a strain larger than 10 %.

It has been pointed out that the strength of 2D materials can only be captured with hexagonal rings [47], leading to the need for a 6-atom unit cell for silicene. This unit cell allows one to model possible soft modes correctly. Most of the above calculations, though, have used smaller unit cells, which could be acceptable for smaller strains.

Mechanical response under a uniaxial strain has also been carried out [29, 30, 35, 47, 49]. Due to the difference in bond orientations, strain along perpendicular directions (conventionally taken to be x and y and chosen to be either the zigzag (ZZ) or armchair (AC) directions) could lead to different responses. For example, Qin et al. [49] found that the Poisson ratio is isotropic and constant for low strain (below 2 %) but then decreases (increases) for AC (ZZ) strain (Fig. 1.5a); similar results were obtained by Yang et al. [30]. Indeed, Wang et al. [29] obtained a Poisson ratio as high as 0.62 for ZZ strain. The response also becomes anisotropic beyond the harmonic regime (Fig. 1.5b). The ultimate strain was computed to be 0.17 (0.15) for AC (ZZ) strain. The buckling parameter, on the other hand, was

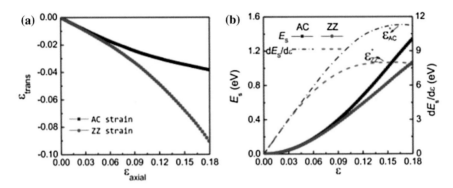

Fig. 1.5 **a** Transverse strain as a function of axial strain. **b** Strain energy and its derivative as a function of axial strain. From [49]

found to decrease in a linear fashion without chirality effects by Zhao [35], whereas Peng et al. [47] only found this result to be true until a strain of 0.16. Wang et al. [29] found the planar structure to be stable for a tensile strain of 0.2 while others found the buckling parameter to remain nonzero for uniaxial strains between −0.1 and 0.4.

Zhao [35] defines the in-plane strength by

$$f_u = \sigma_u h, \tag{1.7}$$

where σ_u is the ultimate stress. She obtained f_u to be 5.66 N/m (ZZ) and 7.07 N/m (AC). Failure was identified for a tensile strain of 0.33 (ZZ) and 0.23 (AC). Yang et al. [30] used DFT-LDA to calculate the stress-strain relationships of low-buckled silicene. In addition to elastic instability, they also considered the possibility of phonon instability by computing the phonon frequencies; the latter was done using finite differences to compute the dynamical matrices for a $10 \times 10 \times 1$ supercell. Anharmonicity and anisotropy was found to set in for a strain larger than 0.03. The ideal strengths for equiaxial tension and AC uniaxial tension were found to be 7.59 N/m and 6.76 N/m, respectively. For ZZ uniaxial tension, there exists two ideal strengths (5.26 N/m and 5.29 N/m) due to the phase transition of silicene from the original low-buckled structure to a planar structure at a strain of 0.16. Phonon instability was found to occur after elastic instability. Instability for a biaxially strained structure at 0.22 strain occured due to phonon modes perpendicular to the plane; this differs from graphene.

Classical MD (using the LAMMPS code) studies of the mechanical properties for large but finite nanosheets have been carried out [36, 50, 51]. Roman and Crawford [36] used a simulation region of 10×10 nm, the ReaxFF potential to describe Si–Si interactions, a microcanonical ensemble, and a nominal temperature of 10 K to limit temperature fluctuations but also to observe failure events. The obtained structural parameters were consistent with DFT values, though the buckling parameter was somewhat higher at 0.69 Å. The stress-strain curves provide the Young's modulus (from the slope at small strain), tensile strength, and fracture strain (when the peak stress is reached). They also computed the out-of-plane bending stiffness by deforming the sheets into partial cylindrical tubes and minimizing the strain energy with respect to the curvature:

$$U = \frac{1}{2}D\kappa^2, \tag{1.8}$$

where κ, the curvature, was in the range of 0.05–0.3 nm^{-1}, and D is the bending modulus per unit width. D was obtained to be 38.63 ± 0.501 eV. Silicene is more flexible but harder to break than silicon [51]. The fracture strength and strain were also found to be smaller at higher temperature; plasticity was not observed.

Summarizing the mechanical properties, the Young's modulus, bulk modulus and ultimate stress of silicene are predicted to be lower than for graphene. On the other hand, the Poisson ratio of silicene is higher than for graphene.

1.4 Electronic Properties

1.4.1 Band Structure

The silicene lattice is hexagonal leading to a hexagonal Brillouin zone. The unique band structure feature of silicene and graphene (in the absence of spin-orbit interaction) is the presence of valence and conduction bands with linear dispersions, so-called Dirac cones, crossing at the Fermi energy and at the K and K' points in the Brillouin zone. The Dirac cones form valleys in the Brillouin zone and the two degenerate bands at a given point originate from the A and B sublattices of the silicene structure. For graphene, these Dirac electrons originate from the p_z states on each sublattice; thus, in a tight-binding (TB) calculation, the Hamiltonian is of order 2 and the Dirac cones can be related to pseudospins. For silicene, due to the lack of mirror symmetry, p_z states are coupled to p_x and p_y states, as well as s states.

The band structure of buckled and flat silicene have been compared in a number of papers [1, 6, 19, 52] whereas Lebègue and Eriksson [23] compared silicene to planar Ge; similar calculations were performed by Houssa et al. [53–55], Wang [56] and Suzuki and Yokomizo [24]. Both forms of silicene have been shown to have electronic properties very similar to graphene [1, 6] in that they both have a zero π–π^* gap at the K point. While the Dirac cones have been observed for graphene, it remains a prediction for free-standing silicene [1] and whether it is present for silicene on silver remains a controversy [57–65] in spite of early claims of observation [15, 66, 67].

All the above calculations were done by ignoring the spin-orbit coupling. Inclusion of the latter effect has shown that silicene would open a small gap of 1.55 meV and, therefore, might be better than graphene at displaying the quantum

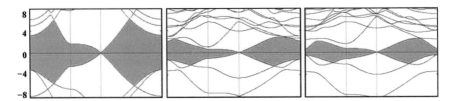

Fig. 1.6 DFT band structures of graphene, silicene and germanene. Reprinted figure with permission from [34]. Copyright (2009) by the American Physical Society

spin Hall effect (QSHE) [68]. The band structures of graphene, silicene and germanene, as obtained by Sahin et al. [34] are reproduced in Fig. 1.6.

While most calculations of the band structure have been carried out using ab initio methods, empirical methods such as TB [1, 68] and $k \cdot p$ [69, 70] have been used as well as they provide semi-analytical results, are much more efficient, and provide a physical picture. Beyond the linear approximation, the most general Hamiltonian for silicene allowed by symmetry can be written down as, to leading orders and in the presence of strain , electric field E and magnetic field B, (but without spin-orbit coupling) [70]

$$\mathcal{H} = \mathcal{H}_i + \mathcal{H}_e, \tag{1.9}$$

$$\mathcal{H}_e = \mathcal{H}^\epsilon + \mathcal{H}^E + \mathcal{H}^B, \tag{1.10}$$

$$\mathcal{H}_i = a_1 \left(k_y \mathbf{J}_x - k_x \mathbf{J}_y \right) + a_2 \left(k_x^2 + k_y^2 \right)$$
$$+ a_4 k_y \left(3k_x^2 - k_y^2 \right) \mathbf{J}_z + a_5 \left(k_x^2 + k_y^2 \right) \left(k_y \mathbf{J}_x - k_x \mathbf{J}_y \right) + \cdots, \tag{1.11}$$

$$\mathcal{H}^\epsilon = e_1 \left(\epsilon_{xx} + \epsilon_{yy} \right) + e_2 \epsilon_{zz} + e_3 \left(\epsilon_{xx} + \epsilon_{yy} \right) \left(k_y \mathbf{J}_x - k_x \mathbf{J}_y \right)$$
$$+ e_4 \left[\left(\epsilon_{xx} - \epsilon_{yy} \right) k_y + 2\epsilon_{xy} k_x \right] \mathbf{J}_z + e_5 \epsilon_{zz} \left(k_y \mathbf{J}_x - k_x \mathbf{J}_y \right) + \cdots, \tag{1.12}$$

$$\mathcal{H}^E = c_1 E_z \mathbf{J}_z + c_2 \left(E_x^2 + E_y^2 \right) + c_3 E_z^2 + c_4 \left(E_x k_y - E_y k_x \right)$$
$$+ c_5 \left[\left(k_x^2 - k_y^2 \right) E_x - 2k_x k_y E_y \right] + c_6 \left(E_x^2 + E_y^2 \right) \left(k_y \mathbf{J}_x - k_x \mathbf{J}_y \right)$$
$$+ c_8 \left[\left(E_x^2 - E_y^2 \right) k_y + 2E_x E_y k_x \right] \mathbf{J}_z + \cdots, \tag{1.13}$$

$$\mathcal{H}^B = b_1 \left(B_x \mathbf{J}_x + B_y \mathbf{J}_y \right) + b_2 B_z \mathbf{J}_z + b_3 \left(k_y B_x - k_x B_y \right) + \cdots, \tag{1.14}$$

where the dots refer to higher-order terms. The a_i's, b_i's, c_i's and e_i's are $k \cdot p$ parameters. The H_i terms are intrinsic band-structure Hamiltonians while the other ones exist in the presence of external fields. In the above Hamiltonian, the $\mathbf{J}_i$ matrices represent the pseudospin degree of freedom and are the (2 × 2) Pauli spin matrices. In the band structure, the a_2 and a_3 terms provide quadratic in

k contributions while the a_4 and a_5 are of cubic order but only the a_4 term gives rise to an anisotropic term. Hence, one can readily see that the band structure of silicene to linear and quadratic orders is isotropic and anisotropic effects only manifest themselves if cubic terms become important. For comparison, the corresponding $\mathcal{H}$ for graphene [71] is

$$\mathcal{H} = a_{61}\left(k_y \mathbf{J}_x + k_x \mathbf{J}_y\right) + a_{11}\left(k_x^2 + k_y^2\right) + a_{62}\left[(k_y^2 - k_x^2)\mathbf{J}_x + 2k_x k_y \mathbf{J}_y\right], \quad (1.15)$$

and the anisotropic term is quadratic in the wave vector.

In the presence of spin-orbit coupling, Ezawa [72] gave

$$H_\eta = \hbar v_F \left(\eta k_x \tau_x + k_y \tau_y\right) + \lambda_{SO} \eta \tau_z \sigma_z + a\lambda_{R2} \eta \tau_z (k_y \sigma_x - k_x \sigma_y), \quad (1.16)$$

where $\eta = \pm 1$ for the two valleys K and K', σ_z are the Pauli matrices associated with the electron spin, τ_z are the Pauli matrices associated with pseudospin, and the Rashba term is now present due to inversion asymmetry.

The Fermi velocity of the linear bands is an important parameter. Using DFT, Cahangirov et al. [19] estimated them to be $\sim 10^6$ m/s for silicene, basically the same value as for graphene. However, Guzmán-Verri and Lew Yan Voon [1], using TB models, evaluated them to be $\sim 10^5$ m/s for silicene. Dzade et al. [48] also obtained a smaller Fermi velocity for silicene. More recent calculations also find a slightly smaller value for silicene than for graphene. This can be easily understood from the reduced hopping in silicene since the Si atoms are more distant from each other.

1.4.2   Strain

A few calculations of a free-standing silicene sheet under strain have been carried out [27–31, 35, 49, 73–76]. The standard approach is to first obtain the unstrained relaxed structure and then distort the unit cell in the appropriate direction such that the strain in that direction is given by

$$\epsilon = \frac{a - a_0}{a_0}, \quad (1.17)$$

where a_0 (a) is the equilibrium (strained) lattice parameter. Since the symmetry is reduced for a uniaxial strain, it is convenient to use a rectangular 1×2 supercell with four atoms. However, a larger unit cell is needed in order to correctly simulate the strength of the nanosheet [47].

A relatively large strain can be applied to the nanosheets and a biaxial strain is a natural expectation if the sheets are deposited on a substrate. A biaxial tensile strain was found to lead to a semimetal-metal transition when the strain is larger than 7 % [27, 28, 31, 73]. This is due to the lowering of the conduction band at the Γ-point;

the Dirac point was also found to increase in energy (but remaining degenerate) [31], leading to the possibility of *p*-type self-doping [28]. Similarly, biaxial compressive strain leads to a lowering of the Dirac point below the Fermi level, leading to *n*-type doping [73]. These changes have been correlated with the changing character of the bonding between sp^3 and sp^2. The above change in character does not occur for graphene since the bonding is pure sp^2 and the atoms remain in a plane. The Fermi velocity is found to decrease slowly with strain, decreasing to 94 % of the unstrained value for strain up to 7 % [31]. Biaxial strain is also found to lead to superconductivity [74]. In particular, for an electron doping of 3.5×10^{14} cm^{-2} and a tension of 5 %, the critical temperature was calculated to be 18.7 K using the Eliashberg theory.

On the other hand, a uniaxial strain is expected to lead to a gap opening due to the symmetry lowering. Indeed, a gap was reported to open up for uniaxial tensile strain, up to 0.08 eV for strain along the zigzag (ZZ) direction and up to 0.04 eV for strain along the armchair (AC) direction [35], at about 8 and 5 %, respectively (Fig 1.7). Mohan et al. [75] similarly obtained a small band gap for tensile strain but found that a direct band gap of 389 meV is formed for 6 % uniaxial compression. An indirect band gap of 379 meV is found for 6 % biaxial compression. They computed the corresponding deformation potentials,

$$\frac{dE}{d\epsilon}, \tag{1.18}$$

and found them to be fairly constant for strains below around 6 %. However, Qin et al. [49] and Yang et al. [30] did not obtain a gap and only obtained the Dirac point to shift. The latter interpreted the disagreement of Zhao to the latter not using sufficient *k* points near the crossing. The lack of band-gap opening has been confirmed using $k \cdot p$ theory [70]; from (1.12), it can be seen that both bands at the Dirac point have the same deformation potential, which means they would both be affected equally by the strain. Qin et al. did obtain a dependence of the Fermi velocity with the type of uniaxial strain as well as a wave-vector dependence. Yang et al. also included spin-orbit coupling in their calculations. The spin-orbit coupling

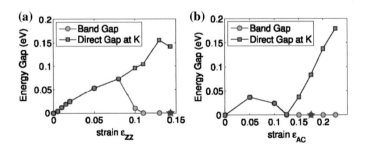

Fig. 1.7 DFT-PBE band gap with uniaxial tensile strain for silicene. Reprinted from [35]. Copyright (2012), with permission from Elsevier [35]

gap was found to initially decrease with increasing strain and then increase again after a strain of about 0.05.

Qin et al. computed the work function as a function of strain. The work function is defined as the minimum energy to remove an electron and is given by

$$\Phi = E_{\text{vac}} - E_F. \tag{1.19}$$

The vacuum energy E_{vac} was determined as the average electrostatic potential energy in a plane parallel to the silicene layer and asymptotically away. For a biaxial strain, the work function initially increased from the unstrained value of 4.8 eV, then saturates to around 5.1 eV for a strain above 15 % [31]. For a uniaxial strain [49], the change is isotropic up to 3 % beyond which the ZZ strain leads to a larger work function but no saturation is observed. The strain dependence has been interpreted in terms of the change in the Fermi level.

1.5 Electric Field

Electric field effects on silicene have been investigated [77–79]. The field has an effect on the buckling parameter. The vertical electric field was found to increase the buckling parameter quadratically with the field [77].

Ni et al. [77] found that a vertical electric field opened a band gap in single layer silicene, contrary to graphene, and the gap increased linearly with the field up to about 1 V/Å. The reason for the gap opening is because the two atoms in the unit cell experience different electric potentials due to the different heights. Geometry optimization and electronic structure was performed using Dmol^3. Ni et al. obtained a rate of 0.157 eÅ while Drummond et al. [78] got 0.0742 eÅ. The latter also indicated that the gap actually starts closing for $E_\perp \approx 0.5$ VÅ^{-1} due to the overlap of the conduction band at Γ and the valence band at K. The electric field also leads to an almost linear increase in the effective mass; for example, for a field of 0.4 V/Å, the hole mass was found to be $0.015m_0$ ($0.033m_0$) along the $K\Gamma$ (MK) direction and about 2 % different for the electron mass.

1.6 Topological Properties

Silicene is similar to graphene in being a Kane-Mele type 2D topological insulator (TI) [80]. A TI has a bulk energy gap but gapless edge states that allow correlated charge and spin transport (Fig. 1.8). They can be distinguished from the more common band insulators because the charge transport is protected from disorder (due to the correlation with spin) and, mathematically, this can be represented by a different topological order or quantum number, the Z_2 invariant [82]. There are two topological quantum numbers: the Chern number C and the Z_2 index. The latter is

(a)   **(b)**

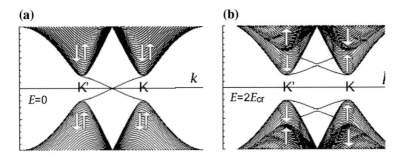

Fig. 1.8 One-dimensional energy bands for a silicene nanoribbon. **a** The bands crossing the gap represent edge states, demonstrating that it is a topological insulator. **b** All states are gapped, demonstrating that it is a band insulator. Reproduced from [81] with kind permission from Springer Science and Business Media

also the same as the spin-Chern number C_s when the spin σ_z is a good quantum number. Qualitatively, this is so far no different from graphene. Indeed, Liu et al. [68] used the fact that the Hamiltonian of the buckled structure can be continuously obtained from the flat one [1] to demonstrate the QSH state for silicene. The QSH of the TI is generated with the assistance of the spin-orbit coupling. The larger spin-orbit coupling gap of 1.55 meV [68] for silicene makes it more practical than graphene for realizing a TI.

It can be expected that the topological properties for silicene would be different from graphene in the presence of an external electric field since the buckled structure of silicene leads to a gap opening, contrary to the case for graphene. In fact, the interplay of the spin-orbit gap and an electric field induced gap allows for a transition between a TI and a band insulator (where the gapless edge states are not protected by topology). This quantum phase transition was computed to occur for a vertical field of 20 mV/Å [78]. Additionally, this is accompanied by a transition from the QSHE to the quantum valley Hall effect (QVHE) [83, 84]. In the presence of an exchange field M and an electric field E_z, one can plot a phase diagram (Fig. 1.9). Ezawa computed four principal phases [81]: band insulator (BI), quantum anomalous Hall (QAH), quantum spin Hall (QSH), valley-polarized metal (VPM) and spin valley-polarized metal (SVPM). The QAH is characterized by an insulating bulk gap and chiral gapless edges; it is induced by the internal magnetization and by the spin-orbit coupling, i.e., it displays quantized Hall conductance in the absence of an external magnetic field. The valley-polarized metal refers to silicene with electrons moved from the conduction band at K to the valence band at K' in a perpendicular electric field. Ezawa has further exploited the buckled structure of silicene to postulate aditional topological phases not encountered for graphene. Thus, an inhomogeneous electric field is shown to generate a helical zero mode away from the edges by closing the band gap spatially [85], while circularly polarized light was used to trigger a topological phase transition from one TI to another TI as a result of the photon dressing of the bands [86]. It was also postulated to break the valley degeneracy by introducing different exchange fields on

Fig. 1.9 Phase diagram for silicene in exchange field (M) and electric field E_z. The insulator phases are indexed by the Chern and spin-Chern numbers (C, C_s). Reprinted figure with permission from [154]. Copyright (2012) by the American Physical Society

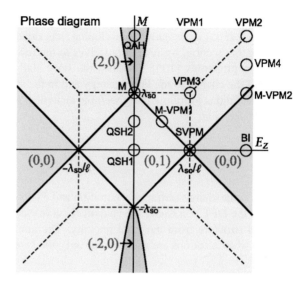

the two sublattices [72] leading to such states as a QSQAH one (with one valley being a QSH state and the other being a QAH one) and single-valley semimetals (one valley is closed, the other is open). This ability to control the K and K' valleys independently is termed valleytronics. The different exchange field could be generated by adsorbing different tranition metal atoms to the two sublattices or by sandwiching the silicene between two different ferromagnets. The valley-selective band structure can be probed by using circularly-polarized light leading to circular dichroism. A complete tabulation of the topological indices for the various phases in silicene is provided in [87].

Silicene with a topological domain wall in the presence of a perpendicular electric field has been shown to display the quantum valley Hall effect [88]. Generating multiple topological states using a spatially modulated electric field has also been considered [89].

1.7 Doping

Doping refers to the addition of a small concentration of impurities into a host lattice and has been particularly well-studied for semiconductors due to the substantial impact on the electrical properties. For two reasons, this has been less well studied for silicene. First, silicene does not have a band gap. Instead, one sees calculations of doping for silicene nanoribbons [90–101] and for silicane (hydrogen-terminated silicene) [102] instead as they both have band gaps. Second, the first-principles calculations undertaken cannot handle large unit cells and, therefore, even an impurity calculation ends up in a large concentration regime.

Thus, a B-substituted silicene structure with the stoichiometry B:Si = 1:3 has been considered [103] or a study of Ni-doping was done with a stoichiometry of 1:2 [104]. In this section, we restrict ourselves to the few studies of low doping on the electronic properties [105]. Possibly the only study of low doping on silicene looked at the change in the Fermi energy and in the structure [105]. They found the DFT calculations to follow fairly well the analytical relation resulting from a linear energy dispersion:

$$\sigma = \text{sign}\,(E_F)\frac{E_F^2}{(\text{eV})^2}C \times 10^{13}\text{ cm}^{-2},\qquad(1.20)$$

where σ is the charge carrier concentration and E_F the Fermi energy. Any deviation between the DFT points and the analytical curve was attributed to the deviation of the band structure from the ideal linearity. The lattice was also found to expand (shrink) when electrons are added (removed); the latter result is similar to graphene [105].

1.8 Optical Properties

A fascinating property of graphene is the universal low-frequency optical absorbance predicted by a non-interacting Dirac fermion theory in 2D and equal to $\pi\alpha$, where $\alpha = 1/137.076$ is the Sommerfeld fine structure constant; this has been observed experimentally [106]. Since this result is independent of the atomic species, buckling, and orbital hybridization, it should apply to silicene as well. A proof based upon the independent-particle approximation has been provided [107].

Optical properties have been computed using DFT [107–111]. Figure 1.10 compares the results for graphene, silicene and germanene obtained in the absence

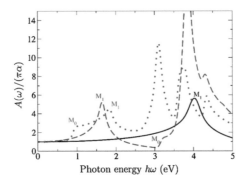

Fig. 1.10 Spectral absorbance of graphene (*black solid line*), silicene (*red dashed line*) and germanene (*red dotted line*). Reprinted figure with permission from [108]. Copyright (2013) by the American Physical Society

of self-energy and excitonic corrections and at normal incidence. The optical properties at higher energies do differ due to differences in the band structure. Inclusion of quasiparticle many-body effects was found to have no influence on the low-frequency absorbance but led to a blueshift of the peaks at higher energies, mainly because of the increase in the interband energies [109]. However, inclusion of spin-orbit coupling leads to important changes at low frequency [109]. First, a SO gap is opened leading to transparency. Second, the near-gap absorbance is enhanced by a factor of 2; this is due to the fact that we now have the equivalent of a massive Dirac particle in 2D. Third, the absorbance decays back down to the universal value between the gap and the higher-energy peaks.

The dielectric functions also reveal the nature of the plasmons (collective excitations of the electron gas) since the latter are probed using electron energy loss spectroscopy (EELS) and peaks in the latter correspond to dips in the imaginary part of the dielectric function, $\epsilon_2(\omega)$. Two features in EELS of silicene for in-plane polarization are below 5 eV and above 5 eV, corresponding to the π and $\pi + \sigma$ peaks, respectively. The π plasmon results from the collective π–π^* transitions while $\pi + \sigma$ plasmon results from the π–σ^* and σ–σ^* transitions [112]. The energies are lower than for graphene [110]. Calculations of the dielectric properties [75] show that the π plasmon in silicene disappears with tensile and asymmetric strains; the $\pi + \sigma$ plasmons are red-shifted for tensile strains; and the $\pi + \sigma$ plasmons are blue-shifted for compressive strains. Plasmons have been proposed as probes of topological phase transitions [113, 114].

Temperature effects on plasmons have been studied [115]. At $T = 0$ and in the absence of the spin-orbit gap, only interband transitions are possible. At finite temperature, intraband transitions become possible as well. The low density-of-state (DOS) at the Dirac point reduces the likelihood of low-frequency plasmons. However, the SO gap for silicene increases the DOS and, at finite temperatures, allows intraband plasmons as well as interband ones. When the zero of ε_1 is in the gap of ε_2, the plasmon is undamped. At a critical temperature, the interband transition is negligible and the intraband one dominates, leading to damped plasmons.

The effect of aluminium (Al) and phosphorus (P) substitutional doping on the optical properties has been studied [116] and found to differ from graphene. For EELS with parallel polarization, similar to the results for graphene, no new EELS peak occurs. However, for perpendicular polarization, two new EELS peaks emerge for P doping and was attributed to the buckling in silicene.

1.9 Transport

Charge carrier transport is one of the motivations for studying silicene given the potential for transistor action and compatibility with current Si electronics. While the carrier mobility in silicene will be strongly dependent upon the environment it is in, understanding the intrinsic mobility for a free-standing film is still important. A recent calculation [37] obtained an electron (a hole) mobility of

2.57×10^5 cm^2 V^{-1} s^{-1} (2.22×10^5 cm^2 V^{-1} s^{-1}) at room temperature, slightly lower than for graphene but much larger than for bulk Si. They used a deformation potential theory and the Boltzmann transport equation, with acoustic phonons as the scattering source. The deformation potential was calculated by stretching and compressing the lattice by up to 1.5 % along the zigzag and armchair directions. A kinetic equation approach within the relaxation time approximation, treating a finite-temperature screened electron-impurity interaction, has been used to investigate transport in the presence of impurities and a gap [117]. A residual conductivity is found when the chemical potential is in the gap and arises from interband correlation.

1.10 Magnetic Properties

We here consider the effect of an external magnetic field on the electronic properties. For standard semiconductors, this is typically observed by performing a magneto-optical experiment, whereby the Landau levels formed are probed by optical transitions. This is a very sensitive approach to characterizing the band structure. For free-standing silicene, magneto-optical properties have been studied theoretically revealing interesting spin-valley effects [118–121]. For example, the low-energy dispersion is obtained from (1.16) with the neglect of the Rashba SO term as

$$E_n = sgn(n)\sqrt{\frac{1}{4}\Delta_{SO}^2 + 2|n|v_F^2\hbar eB}, \qquad (1.21)$$

for $n \neq 0$ and $E_n = \sigma\eta\Delta_{SO}/2$ for $n = 0$, where n is the subband index, and when starting with the Dirac-Kane-Mele Hamiltonian (Fig. 1.11). The selection rules for interband transitions are found to be $\Delta n = \pm 1$.

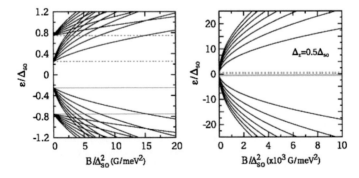

Fig. 1.11 Landau levels as a function of magnetic field for small (*left*) and large (*right*) field. Reprinted figure with permission from [119]. Copyright (2013) by the American Physical Society

For graphene, Chang [122], using a simple effective-mass approach and the Luttinger-Kohn formalism [123], showed that the Landau levels are given by

$$E_n = sgn(n)\sqrt{2}\hbar v_F \sqrt{|n|B}. \qquad (1.22)$$

The basic result is that the positions and intensities of the absorption lines scale with $\sqrt{B}$ and that the line corresponding to the transition from the lowest $n = 0$ to the $n = 1$ Landau level is anomalous due to the Dirac nature [124]. This behaviour has been observed in infrared spectroscopy [125–127].

The effect on the band structure in the presence of an external electric field and a perpendicular magnetic field has also been studied [128]. They found that, for finite E_z, the spin and valley degeneracies of the Landau levels are lifted and this leads to additional plateaus in the Hall conductivity, at half-integer values of $4e^2/h$, due to spin intra-Landau-level transitions that are absent in graphene.

1.11 Phonons

A number of papers have included a calculation of phonon modes in order to establish the stability of structures obtained by energy minimization [19, 28, 48]. The dispersion relations for graphene and silicene are compared in Fig. 1.12. The highest G- and D-modes are 550 and 545 cm^{-1} [28].

Silicene has a flexural mode, as does graphene. However, because of the symmetry reduction due to the buckling, this mode for silicene has both a z and an xy componemt [41].

Dzade et al. [48] reported significant violation of the acoustic sum rule when using density functional perturbation theory within the Born-Oppenheimer approximation; hence, they reported results using the frozen phonon method. They found the flexural modes to be much smaller in energy than for graphene leading to the conclusion that silicene would more likely form ripples.

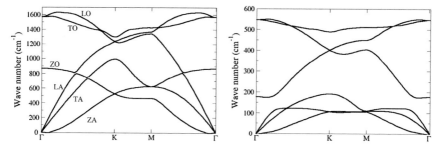

Fig. 1.12 Phonon spectrum for graphene (*left*) and silicene (*right*). Reprinted from [155], Copyright (2014), with permission from Elsevier

Kaloni et al. [28] studied the change in the phonon spectrum under tensile biaxial strain. All the modes soften with increasing strain due to the weakening of the Si–Si bond. They obtained negative frequency modes for a strain larger than 17 %. They also computed the Grüneisen parameter for the G-mode:

$$\gamma_G = -\frac{\Delta\omega_G}{2\omega_G^0 \epsilon}, \tag{1.23}$$

where ω_G^0 is the unstrained frequency and $\Delta\omega_G$ is the change in the frequency under the biaxial strain. The Grüneisen parameter was found to decrease from 1.64 ($\epsilon = 5$ %) to 1.34 ($\epsilon = 20$ %) and then increase again, following the trend with the buckling parameter.

Electron-optical phonon coupling matrix elements are calculated to be small [129], being about a factor of 25 times smaller than in graphene. A consequence is that long momentum relaxation lengths and high carrier mobilities are predicted for silicene as carrier relaxation via phonon scattering is inhibited.

1.12 Thermal Properties

A few thermal properties have been studied.

1.12.1 Thermal Stability

The thermal stability has been studied using reactive MD [130] and pristine silicene was found to be stable up to 1500 K, after which it formed a three-dimensional amorphous structure. Defects have been found to reduce the thermal stability of silicene by as much as 30 % [130]. Passivating the defects improves the stability.

1.12.2 Lattice Thermal Conductivity

As a semimetal, the thermal conductivity of silicene has been assumed to be mainly due to phonon transport. Initial calculations have used classical MD and gave values in the range 5–65 W/mK at 300 K. Thus, an equilibrium MD simulation using Tersoff bond-order potential and the Green-Kubo approach for κ on a variety of supercells ranging from 5×3 to 17×10 gave an in-plane thermal conductivity κ of 20 W/mK [131], compared to a value of ~ 3000–5000 W/mK for suspended graphene [42] and 150–200 W/mK for silicon. It should be noted that they observed a small anisotropy in the value of the conductivity of the order of 1–4 W/mK and a general trend of increasing κ with increasing supercell size. From an analysis of the partial density of states (PDOS), they associated the reduction in κ to two reasons. One is the lowering of the PDOS for frequencies below 20 THz (phonon softening), and the other is due to the blueshift in the PDOS to higher frequencies (phonon stiffening). The result is a reduction in the phonon modes present for heat conduction at low temperatures.

Nonequilibrium MD simulations have also been carried out [132–137]. Wang and Sun, Hu et al., Ng et al., Liu et al. (who also used the Stillinger-Weber potential), and Yeo and Liu used the Tersoff potential, while Pei et al. used the MEAM potential. Hu et al. used a simulation region with 1000 unit cells in one direction and the atomic positions were relaxed. Thermal conductivity was then calculated using non-equilibrium MD and extracted from the Fourier law by computing the heat flux [133]. They found that κ (~ 40 W/mK) is significantly influenced by the out-of-plane flexural modes. However, studies have shown that the Tersoff potential for bulk Si does not reproduce the buckling [133]; the MEAM potential used by the latter gave a buckling of 0.85 nm at 10 K, which is doubled accepted values. This work (as well as Wang and Sun) also used the reverse non-equilibrium MD whereby the heat flux is imposed and the temperature gradient obtained; there is a length dependence to κ and they studied in the range of 33–210 nm. They got a similar κ to the work by Pei et al. [134]. On the other hand, Wang and Sun (Ng et al.) got a κ of about 55(65) W/mK and Liu et al. [136] got 55–65 W/mK. Studies as a function of length have also been carried out using MD [137]. Graphene was found to possess significantly higher thermal conductivities than silicene at every length scale and chirality, and this was attributed to the higher phonon group velocities of the dominant acoustic modes in graphene.

Zhang et al. [138] has used a modified Stillinger-Weber potential in their study and the LAMMPS code. They were able to reproduce the buckling of silicene and phonon dispersions in agreement with DFT. In their case, they did both equilibrium and non-equilibrium MD as well as using anharmonic lattice dynamics (ALD). The MD methods gave a κ below 12 W/mK while the ALD results could be even lower. They used the ALD method to obtain the phonon mode contributions. They deduced that more than 50 % of the thermal conductivity is due to phonon modes with wavelengths less than 10 nm. Their conclusion, though, differs from the other work cited above. They deduced that as much as 80 % of κ arises from in-plane longitudinal modes, with most of the rest due to the transverse modes. Another calculation using the Tersoff potential instead also found the in-plane phonons to contribute over 85 % to κ. Nevertheless, they also associate the difference to phonon softening and phonon stiffening.

The temperature dependence of the thermal conductivity of free-standing and SiO_2 supported silicene has been investigated using both equilibrium and non-equilibrium MD from 300 to 900 K [139], inclusive of quantum corrections. The substrate leads to a 78 % reduction at 300 K. However, Zhang et al. [140] found that the thermal conductivity could either increase or decrease depending on the substrate. Thus, they found an increase in the thermal conductivity of silicene supported on the 6H-SiC substrate and attributed this increase to the augmented lifetime of the majority of the acoustic phonons, while they found a significant decrease in the thermal conductivity of silicene supported on the 3C-SiC substrate results, due to the reduction in the lifetime of almost the entire phonon spectrum. This is in contrast to graphene where substrates have been found to always lead to a decrease in the phonon transport [140]. For suspended silicence, κ has been computed over $0 < T < 400$ K [141]. They obtained a maximum value of about 100 W/mK near 100 K, compared to around 1000 W/mK for bulk Si [141] and 3000–5000 W/mK for graphene.

Xie et al. [41] reported a value of 9.4 W/mK at 300 K using first-principles calculations; out-of-plane vibrations was only found to contribute less than 10 % of the overall thermal conductivity. The difference compared to graphene was related to the buckling which breaks the reflection symmetry and leads to scattering of the flexural modes with other modes as the former are no longer pure out-of-plane vibrations.

The effect of defects [131], strain [133, 134], and isotopic doping [134, 136] on κ has been studied. Not surprisingly, defects were found to reduce κ due to the increased defect-phonon scattering. In fact, Li and Zhang [131] found that removing a single atom out of 448 atoms from the domain led to a 78 % reduction in κ. Under tensile strain, κ was found to initially increase at small strains (below 4 %) and then decrease with larger strains [134]; this was done on a 33 nm × 33 nm silicene sheet. The initial increase could be attributed to the reduced buckling under strain and, therefore, an increase in in-plane stiffness. Higher strains lead to stretching of the Si–Si bonds and, therefore, a decrease in in-plane stiffness. Isotopic doping is found to decrease κ as well; for example, 50 % of ^{30}Si in a ^{28}Si lattice led to a 20 % reduction.

Thermal conductance G and rectification at an interface with graphene has been studied [142]. They found that G not only increases with the temperature but also with the monolayer length at a given temperature until it reaches a saturated value. At 300 K, the saturated value is 250 MW/m^2 2 K. In contrast, they found that R decreased with increasing monolayer length and temperature. Both G and R were significantly affected by tensile strain applied on silicene along the interface direction, but both were almost independent of the heat flux J. They determined a critical value $J = 42$ GW/m^2 above which low-frequency kinetic waves were excited and provided an additional channel for heat transport.

1.12.3 Thermopower and Thermoelectricity

The lower thermal conductivity of silicene compared to graphene could be advantageous for thermoelectric properties since the latter are improved by lower thermal conductivity and higher thermopower. Thermopower is defined by the voltage drop that appears across a material due to an imposed temperature gradient, and a measure is via the Seebeck coefficient,

$$S = \frac{1}{eT}\frac{L_1}{L_0},$$ (1.24)

where T is the temperature and

$$L_m(\mu) = \frac{2}{h} \int_{-\infty}^{\infty} dE\, T(E)(E - \mu)^m \left(-\frac{\partial f(E, \mu)}{\partial E} \right),$$ (1.25)

where $T(E)$ is the transmission at energy E, $F(E, \mu)$ is the Fermi-Dirac distribution, and μ is the chemical potential. The thermopower was calculated with and without an external field [43]. A four-band TB model was used for the electronic band structure, transport properties were calculated using the non-equilibrium Green's function method, and the simulation length was 1000 unit cells. They found the thermopower to be somewhat insensitive to temperature changes between 100 and 500 K and to have a peak value of 87 μV/K. In the presence of an electric field, and with the chemical potential in the band gap, there is an increase in κ up to 300 μV/K at 300 K.

Thermoelectricity is measured by the figure of merit,

$$ZT = \frac{\sigma S^2 T}{\kappa},$$ (1.26)

where σ is the electrical conductance. The thermoelectric coefficients of silicene at room temperature have been computed [45] using the BOLTZTRAP and VASP codes. They were found to have a ZT_e less than 0.5. On the other hand, for graphene, the figure of merit has been estimated to be 0.17–1.02 for isotope-doped graphene with a sheet carrier concentration of 10^{16}–10^{17} m^{-2} and a temperature of 300–450 K [46].

1.13 Stability to Oxidation

A DFT study of the stability to oxygen exposure has been carried out [143, 144]. They found that the molecule dissociates easily on silicene without any energy barrier and the oxygen atoms form strong bonds with Si atoms. Coupled with the low diffusivity, oxides are formed. Hence, silicene is predicted to be unstable in air. In a subsequent work, they have also studied the stability of silicane [144]. There are now two minor energy barriers of O_2 molecule adsorption and dissociation on silicane, thus silicane has higher stability than free-standing silicene in oxygen. However, they also argue that, once the O_2 molecule dissociates into two O atoms, desorption of O atoms will be very difficult due to its high energy barrier. An experimental study of oxidation showed the availability to create a band gap [145].

1.14   Defects

Given that defects are always present in real samples, it is useful to briefly discuss the variety of defects than can exist in silicene and the impact on certain physical properties.

The simplest type of defect would be a point defect such as a vacancy [146, 147]. For graphene, the ground state is a nonmagnetic reconstructed monovacancy (MV) [147]. For silicene, on the other hand, the three Si atoms around the symmetric MV core move to the core center and eliminate their dangling bonds by forming a fourth bond, becoming nonmagnetic [146]. However, this has subsequently been found not to be the lowest energy MV. A vacancy type, monovacancy MV-1, not previously found in graphene, has been predicted in silicene using first-principles calculations [147] on a 8 × 8 unit cell. It has a planelike sp^3 hybridization at its defect core. The diffusion coefficient of MV-1 is calculated to be 2.3×10 cm^2/s, much higher than that of the MV in graphene. Silicene with MV-1 is metallic.

Vacancy clusters, extended line defects (ELDs), and adatoms have been studied using first-principles calculations [148, 149]. Divacancies have been found to have lower mobilities than single vacancies [148] and divacancies induce small gaps in silicene. Si adatoms induce long-range spin polarization and a band gap, thus achieving an all-silicon magnetic semiconductor. Small defects were found to have a tendency to coalescence forming highly stable vacancy clusters [149]. The 5|8|5 ELD was found to be easier to form in the silicene than graphene because of the mixed sp2/sp3 hybridization of silicene. Stone-Wales defects have also been studied [148, 150, 151]. The Stone-Wales defect is a topological defect formed by the 90° rotation of a dimer bond which results in four hexagons turning into two pairs of pentagon-heptagon rings (Fig. 1.13). In particular, it was found that the

(a)   **(b)**

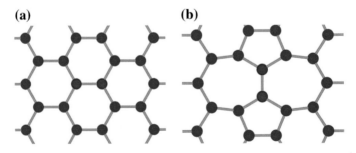

Fig. 1.13 Representation of a Stone-Wales defect in silicene. Reprinted from [150], Copyright (2014), with permission from Elsevier

formation energy and kinetic barrier are lower in silicene than in graphene. A band gap of 0.01 eV is created. The effect of vacancies and Stone-Wales defects on the mechanical properties have been studied using molecular dynamic finite element method with Tersoff potential [152]. They found that pristine and lowly defective silicene sheets exhibit almost the same elastic nature up to the fracture points. However, a single defect significantly weakened the silicene sheet, leading to a considerable reduction in the fracture strength. Thus, one 2-atom vacancy in the reduced the fracture stress by 1820 % and the fracture strain by 3335 %. The weakening effects of Stone-Wales defects varied with the tensile direction and the orientation of these defects.

A defect that is particular to 2D materials but absent for graphene is the buckling interface formed between two pieces of silicene with oppositely oriented buckling [153]. This leads to a line defect that has a low formation energy but has a higher reactivity than the pristine silicene itself. The latter was deduced by studying the adsorption of a single gold atom and they turn out to have a binding energy of −3.50 eV.

1.15 Summary

Silicene is another example of the novel materials belonging to the class of 2D materials beyond graphene. Yet, it has some differences since the sheet is puckered, has a different point-group symmetry and a different pseudospin. Thus, it is predicted to have Dirac cones just as for graphene but with a larger spin-orbit gap and with a band gap opening under a perpendicular electric field. Physical properties are, generally, less extreme than for graphene (e.g., slightly lower mechanical strengths and electron Fermi velocity) but it can also have "better" physical properties such as better thermoelectric figure of merit and a richer topological phase diagram.

Currently, free-standing silicene (without functional groups) have not been realized experimentally. Nevertheless, a proper understanding of the properties of free-standing silicene is a good starting point for understanding the properties of silicene on a substrate or of functionalization.

References

1. G.G. Guzmán-Verri, L.C. Lew Yan Voon, Phys. Rev. B (Condens. Matter Mater. Phys.) **76**(7), 075131 (2007). doi:10.1103/PhysRevB.76.075131. URL http://link.aps.org/abstract/PRB/v76/e075131
2. K.S. Novoselov, A.K. Geim, S.V. Morozov, D. Jiang, Y. Zhang, S.V. Dubonos, I.V. Grigorieva, A.A. Firsov, Science **306**(5696), 666 (2004)
3. K. Takeda, K. Shiraishi, Phys. Rev. B **39**(15), 11028 (1989). doi:10.1103/PhysRevB.39.11028

4. K. Takeda, K. Shiraishi, Phys. Rev. B **50**(20), 14916 (1994). doi:10.1103/PhysRevB.50. 14916
5. Y. Wang, K. Scheerschmidt, U. Gösele, Phys. Rev. B **61**, 12864 (2000). doi:10.1103/PhysRevB.61.12864. URL http://link.aps.org/doi/10.1103/PhysRevB.61.12864
6. X. Yang, J. Ni, Phys. Rev. B **72**(19), 195426 (2005). doi:10.1103/PhysRevB.72.195426
7. E. Durgun, S. Tongay, S. Ciraci, Phys. Rev. B **72**(7), 075420 (2005)
8. H. Nakano, M. Ishii, H. Nakamura, Chem. Commun. 2945–2947 (2005). doi:10.1039/B500758E. URL http://dx.doi.org/10.1039/B500758E
9. H. Nakano, T. Mitsuoka, M. Harada, K. Horibuchi, H. Nozaki, N. Takahashi, T. Nonaka, Y. Seno, H. Nakamura, Angew. Chem. **118**(38), 6451 (2006)
10. A. Kara, C. Léandri, M.E. Dávila, P. de Padova, B. Ealet, H. Oughaddou, B. Aufray, G.L. Lay, J. Supercond. Novel Magn. **22**, 259 (2009)
11. D. Chiappe, C. Grazianetti, G. Tallarida, M. Fanciulli, A. Molle, Adv. Mat. **24**, 5088 (2012)
12. B. Feng, Z. Ding, S. Meng, Y. Yao, X. He, P. Cheng, L. Chen, K. Wu, Nano Lett. **12**(7), 3507 (2012). doi:10.1021/nl301047g. URL http://pubs.acs.org/doi/abs/10.1021/nl301047g
13. H. Jamgotchian, Y. Colignon, N. Hamzaoui, B. Ealet, J.Y. Hoarau, B. Aufray, J.P. Bibérian, J. Phys. Condensed Mat. **24**(17), 172001 (2012)
14. C.L. Lin, R. Arafune, K. Kawahara, N. Tsukahara, E. Minamitani, Y. Kim, N. Takagi, M. Kawai, Appl. Phys. Express **5**, 045802 (2012)
15. P. Vogt, P. De Padova, C. Quaresima, J. Avila, E. Frantzeskakis, M.C. Asensio, A. Resta, B. Ealet, G. Le Lay, Phys. Rev. Lett. **108**, 155501 (2012). Doi:10.1103/PhysRevLett.108. 155501. URL http://link.aps.org/doi/10.1103/PhysRevLett.108.155501
16. L. Tao, E. Cinquanta, D. Chiappe, C. Grazianetti, M. Fanciulli, M. Dubey, A. Molle, D. Akinwande, Nat. Nano. **10**, 227 (2015)
17. J.R. Soto, B. Molina, J.J. Castro, Phys. Chem. Chem. Phys. **17**, 7624 (2015). doi:10.1039/C4CP05912C. URL http://dx.doi.org/10.1039/C4CP05912C
18. Y. Ding, J. Ni, Appl. Phys. Lett. **95**(8), 083115 (2009)
19. S. Cahangirov, M. Topsakal, E. Aktürk, H. Sahin, S. Ciraci, Phys. Rev. Lett. **102**(23), 236804 (2009)
20. L.C. Lew Yan Voon, E. Sandberg, R.S. Aga, A.A. Farajian, Appl. Phys. Lett. **97**, 163114 (2010)
21. D. Kaltsas, L. Tsetseris, Phys. Chem. Chem. Phys. **15**(24), 9710 (2013). doi:10.1039/C3CP50944C. URL http://dx.doi.org/10.1039/C3CP50944C
22. M.T. Yin, M.L. Cohen, Phys. Rev. B **29**, 6996 (1984). doi:10.1103/PhysRevB.29.6996. URL http://link.aps.org/doi/10.1103/PhysRevB.29.6996
23. S. Lebègue, O. Eriksson, Phys. Rev. B (Condens. Matter Mater. Phys.) **79**(11), 115409 (2009). doi:10.1103/PhysRevB.79.115409. URL http://link.aps.org/abstract/PRB/v79/e115409
24. T. Suzuki, Y. Yokomizo, Physica E **42**, 2820 (2010)
25. R. Hoffmann, Angewandte Chemie International Edition **52**(1), 93 (2013). doi:10.1002/anie. 201206678. URL http://dx.doi.org/10.1002/anie.201206678
26. T. Morishita, K. Nishio, M. Mikami, Phys. Rev. B **77**(8), 081401 (2008). doi:10.1103/PhysRevB.77.081401
27. G. Liu, M.S. Wu, C.Y. Ouyang, B. Xu, EPL (Europhys. Lett.) **99**(1), 17010 (2012). URL: http://stacks.iop.org/0295-5075/99/i=1/a=17010
28. T.P. Kaloni, Y.C. Cheng, U. Schwingenschlögl, J. App. Phys. **113**, 104305 (2013)
29. B. Wang, J. Wu, X. Gu, H. Yin, Y. Wei, R. Yang, M. Dresselhaus, Appl. Phys. Lett. **104**(8), 081902 (2014). Doi:http://dx.doi.org/10.1063/1.4866415. URL http://scitation.aip.org/content/aip/journal/apl/104/8/10.1063/1.4866415
30. C. Yang, Z. Yu, P. Lu, Y. Liu, H. Ye, T. Gao, Phonon instability and ideal strength of silicene under tension. Comput. Mater. Sci. **95**(0), 420 (2014). doi:http://dx.doi.org/10.1016/j.commatsci.2014.07.046. URL http://www.sciencedirect.com/science/article/pii/S0927025614005291
31. R. Qin, C.H. Wang, W. Zhu, Y. Zhang, AIP Adv. **2**(2), 022159 (2012)

32. V. Bocchetti, H.T. Diep, H. Enriquez, H. Oughaddou, A. Kara, J. Phys. Conf. Ser. **491**(1), 012008 (2014). URL http://stacks.iop.org/1742-6596/491/i=1/a=012008
33. Y. Zhang, R. Tsu, Nanoscale Res. Lett. **5**, 805 (2010)
34. H.S. Sahin, S. Cahangirov, M. Topsakal, E. Bekaroglu, E. Akturk, R.T. Senger, S. Ciraci, Phys. Rev. B **80**(15), 155453 (2009). doi:10.1103/PhysRevB.80.155453
35. H. Zhao, Phys. Lett. A **376**(46), 3546 (2012). Doi:10.1016/j.physleta.2012.10.024. URL http://www.sciencedirect.com/science/article/pii/S0375960112010523
36. R.E. Roman, S.W. Cranford, Comput. Mater. Sci. **82**(0), 50 (2014). Doi:http://dx.doi.org/10.1016/j.commatsci.2013.09.030. URL http://www.sciencedirect.com/science/article/pii/S092702561300565X
37. Z.G. Shao, X.S. Ye, L. Yang, C.L. Wang, J. Appl. Phys. **114**(9), 093712 (2013)
38. Q. Lu, M. Arroyo, R. Huang, J. Phys. D Appl. Phys. **42**(10), 102002 (2009). URL http://stacks.iop.org/0022-3727/42/i=10/a=102002
39. A.C. Ferrari, J.C. Meyer, V. Scardaci, C. Casiraghi, M. Lazzeri, F. Mauri, S. Piscanec, D. Jiang, K.S. Novoselov, S. Roth, A.K. Geim, Phys. Rev. Lett. **97**(18), 187401 (2006). doi:10.1103/PhysRevLett.97.187401. URL http://link.aps.org/abstract/PRL/v97/e187401
40. R. Yan, Q. Zhang, W. Li, I. Calizo, T. Shen, C.A. Richter, A.R. Hight-Walker, X. Liang, A. Seabaugh, D. Jena, H. Grace Xing, D.J. Gundlach, N.V. Nguyen, Applied Physics Letters **101**(2), 022105 (2012). doi:http://dx.doi.org/10.1063/1.4734955. URL http://scitation.aip.org/content/aip/journal/apl/101/2/10.1063/1.4734955
41. H. Xie, M. Hu, H. Bao, Appl. Phys. Lett. **104**(13), 131906 (2014). doi:http://dx.doi.org/10.1063/1.4870586. URL http://scitation.aip.org/content/aip/journal/apl/104/13/10.1063/1.4870586
42. A.A. Balandin, S. Ghosh, W. Bao, I. Calizo, D. Teweldebrhan, F. Miao, C.N. Lau, Nano Lett. **8**(3), 902 (2008)
43. Y. Yan, H. Wu, F. Jiang, H. Zhao, Eur. Phys. J. B **86**(11), 1 (2013). doi:10.1140/epjb/e2013-40818-3. URL http://dx.doi.org/10.1140/epjb/e2013-40818-3
44. A.N. Sidorov, A. Sherehiy, R. Jayasinghe, R. Stallard, D.K. Benjamin, Q. Yu, Z. Liu, W. Wu, H. Cao, Y.P. Chen, Z. Jiang, G.U. Sumanasekera, Appl. Phys. Lett. **99**(1), 013115 (2011). doi:http://dx.doi.org/10.1063/1.3609858. URL http://scitation.aip.org/content/aip/journal/apl/99/1/10.1063/1.3609858
45. K. Yang, S. Cahangirov, A. Cantarero, A. Rubio, R. D'Agosta, Phys. Rev. B **89**, 125403 (2014). Doi:10.1103/PhysRevB.89.125403. URL http://link.aps.org/doi/10.1103/PhysRevB.89.125403
46. R. Verma, S. Bhattacharya, S. Mahapatra, IEEE Trans. Electr. Dev. **60**(6), 2064 (2013). doi:10.1109/TED.2013.2258159
47. Q. Peng, X. Wen, S. De, RSC Adv. **3**, 13772 (2013). doi:10.1039/C3RA41347K. URL http://dx.doi.org/10.1039/C3RA41347K
48. N.Y. Dzade, K.O. Obodo, S.K. Adjokatse, A.C. Ashu, E. Amankwah, C.D. Atiso, A.A. Bello, E. Igumbor, S.B. Nzabarinda, J.T. Obodo, A.O. Ogbuu, O.E. Femi, J.O. Udeigwe, U. V. Waghmare, J. Phys. Condens. Matter **22**, 375502 (2010)
49. R. Qin, W. Zhu, Y. Zhang, X. Deng, Nano. Res. Lett. **9**(1), 521 (2014)
50. R. Ansari, S. Rouhi, S. Ajori, Superlattices Microstruct. **65**(0), 64 (2014). doi:http://dx.doi.org/10.1016/j.spmi.2013.10.039. URL http://www.sciencedirect.com/science/article/pii/S0749603613003765
51. Q.X. Pei, Z.D. Sha, Y.Y. Zhang, Y.W. Zhang, J. Appl. Phys. **115**(2), 023519 (2014). doi:http://dx.doi.org/10.1063/1.4861736. URL http://scitation.aip.org/content/aip/journal/jap/115/2/10.1063/1.4861736
52. S. Wang, L. Zhu, Q. Chen, J. Wang, F. Ding, J. Appl. Phys. **109**(5), 053516 (2011)
53. M. Houssa, G. Pourtois, V.V. Afanas'ev, A. Stesmans, Appl. Phys. Lett. **96**, 082111 (2010)
54. M. Houssa, G. Pourtois, V.V. Afanas'ev, A. Stesmans, Appl. Phys. Lett. **97**, 112106 (2010)
55. M. Houssa, G. Pourtois, M.M. Heyns, V.V. Afanas'ev, A. Stesmans, J. Electrochem. Soc. **158**(2), H107 (2011)
56. S. Wang, J. Phys. Soc. Jpn. **79**, 064602 (2010)

57. R. Arafune, C.L. Lin, R. Nagao, M. Kawai, N. Takagi, Phys. Rev. Lett. **110**, 229701 (2013). doi:10.1103/PhysRevLett.110.229701. URL http://link.aps.org/doi/10.1103/PhysRevLett.110.229701
58. P. Gori, O. Pulci, F. Ronci, S. Colonna, F. Bechsted, J. Appl. Phys. **114**(11), 113710 (2013)
59. C.L. Lin, R. Arafune, K. Kawahara, M. Kanno, N. Tsukahara, E. Minamitani, Y. Kim, M. Kawai, N. Takagi, Phys. Rev. Lett. **110**, 076801 (2013). doi:10.1103/PhysRevLett.110.076801. URL http://link.aps.org/doi/10.1103/PhysRevLett.110.076801
60. D. Tsoutsou, E. Xenogiannopoulou, E. Golias, P. Tsipas, A. Dimoulas, Appl. Phys. Lett. **103** (23), 231604 (2013). doi:http://dx.doi.org/10.1063/1.4841335. URL http://scitation.aip.org/content/aip/journal/apl/103/23/10.1063/1.4841335
61. M.X. Chen, M. Weinert, Nano Lett. doi:10.1021/nl502107v. URL http://pubs.acs.org/doi/abs/10.1021/nl502107v
62. N.W. Johnson, P. Vogt, A. Resta, P. De Padova, I. Perez, D. Muir, E.Z. Kurmaev, G. Le Lay, A. Moewes, Adv. Funct. Mater. **24**(33), 5253 (2014). doi:10.1002/adfm.201400769. URL http://dx.doi.org/10.1002/adfm.201400769
63. S.K. Mahatha, P. Moras, V. Bellini, P.M. Sheverdyaeva, C. Struzzi, L. Petaccia, C. Carbone, Phys. Rev. B **89**, 201416 (2014). DOI 10.1103/PhysRevB.89.201416. URL http://link.aps.org/doi/10.1103/PhysRevB.89.201416
64. E. Scalise, E. Cinquanta, M. Houssa, B. van den Broek, D. Chiappe, C. Grazianetti, G. Pourtois, B. Ealet, A. Molle, M. Fanciulli, V. Afanas ev, A. Stesmans, Appl. Surf. Sci. **291** (1), 113 (2013). doi:http://dx.doi.org/10.1016/j.apsusc.2013.08.113. URL http://www.sciencedirect.com/science/article/pii/S0169433213016048
65. H. Ishida, Y. Hamamoto, Y. Morikawa, E. Minamitani, R. Arafune, N. Takagi, New J. Phys. **17**(1), 015013 (2015). URL http://stacks.iop.org/1367-2630/17/i=1/a=015013
66. L. Chen, C.C. Liu, B. Feng, X. He, P. Cheng, Z. Ding, S. Meng, Y. Yao, K. Wu, Phys. Rev. Lett. **109**, 056804 (2012). doi:10.1103/PhysRevLett.109.056804. URL http://link.aps.org/doi/10.1103/PhysRevLett.109.056804
67. B. Feng, H. Li, C.C. Liu, T.N. Shao, P. Cheng, Y. Yao, S. Meng, L. Chen, K. Wu, ACS Nano **7**(10), 9049 (2013). doi:10.1021/nn403661h. URL http://pubs.acs.org/doi/abs/10.1021/nn403661h
68. C.C. Liu, W. Feng, Y. Yao, Phys. Rev. Lett. **107**, 076802 (2011). doi:10.1103/PhysRevLett.107.076802. URL http://link.aps.org/doi/10.1103/PhysRevLett.107.076802
69. F. Geissler, J.C. Budich, B. Trauzettel, New J. Phys. **15**(8), 085030 (2013). URL http://stacks.iop.org/1367-2630/15/i=8/a=085030
70. L.C.L.Y. Voon, A. Lopez-Bezanilla, J. Wang, Y. Zhang, M. Willatzen, New J. Phys. **17**(2), 025004 (2015). URL http://stacks.iop.org/1367-2630/17/i=2/a=025004
71. R. Winkler, U. Zülicke, Phys. Rev. B **82**, 245313 (2010). doi:10.1103/PhysRevB.82.245313. URL http://link.aps.org/doi/10.1103/PhysRevB.82.245313
72. M. Ezawa, Phys. Rev. B **87**, 155415 (2013). doi:10.1103/PhysRevB.87.155415. URL http://link.aps.org/doi/10.1103/PhysRevB.87.155415
73. Y. Wang, Y. Ding, Solid State Commun. **155**(2), 6 (2013)
74. A. Durajski, D. Szczes'niak, R. Szczes'niak, Solid State Commun. **200**, 17 (2014). doi:http://dx.doi.org/10.1016/j.ssc.2014.09.007. URL http://www.sciencedirect.com/science/article/pii/S0038109814003664
75. B. Mohan, A. Kumar, P. Ahluwalia, Physica E: Low-dimensional systems and nanostructures **61**(0), 40 (2014). doi:http://dx.doi.org/10.1016/j.physe.2014.03.013. URL http://www.sciencedirect.com/science/article/pii/S1386947714001003
76. C.H. Yang, Z.Y. Yu, P.F. Lu, Y.m. Liu, S. Manzoor, M. Li, S. Zhou, Proc. SPIE **8975**, 89750K (2014). doi:10.1117/12.2038401. URL http://dx.doi.org/10.1117/12.2038401
77. Z. Ni, Q. Liu, K. Tang, J. Zheng, J. Zhou, R. Qin, Z. Gao, D. Yu, J. Lu, Nano Lett. **12**(1), 113 (2012). doi:10.1021/nl203065e. URL http://pubs.acs.org/doi/abs/10.1021/nl203065e
78. N.D. Drummond, V. Zólyomi, V.I. Fal'ko, Phys. Rev. B **85**, 075423 (2012). doi:10.1103/PhysRevB.85.075423. URL http://link.aps.org/doi/10.1103/PhysRevB.85.075423

79. V. Vargiamidis, P. Vasilopoulos, G.Q. Hai, J. Phys. Condens. Matter **26**(34), 345303 (2014). URL http://stacks.iop.org/0953-8984/26/i=34/a=345303
80. C.L. Kane, E.J. Mele, Phys. Rev. Lett. **95**, 226801 (2005). doi:10.1103/PhysRevLett.95.226801. URL http://link.aps.org/doi/10.1103/PhysRevLett.95.226801
81. M. Ezawa, Eur. J. Phys. B **85**(11), 1 (2012)
82. C.L. Kane, E.J. Mele, Phys. Rev. Lett. **95**, 146802 (2005). doi:10.1103/PhysRevLett.95.146802. URL http://link.aps.org/doi/10.1103/PhysRevLett.95.146802
83. M. Tahir, U. Schwingenschlogl, Sci. Rep. **3**(1075), 1 (2013)
84. M. Tahir, A. Manchon, K. Sabeeh, U. Schwingenschogl, Appl. Phys. Lett. **102**(16), 162412 (2013)
85. M. Ezawa, New J. Phys. **14**(3), 033003 (2012). URL http://stacks.iop.org/1367-2630/14/i=3/a=033003
86. M. Ezawa, Phys. Rev. Lett. **110**, 026603 (2013). doi:10.1103/PhysRevLett.110.026603. URL http://link.aps.org/doi/10.1103/PhysRevLett.110.026603
87. M. Ezawa, JPS Conf. Proc. **1**(1), 012003 (2014)
88. Y. Kim, K. Choi, J. Ihm, H. Jin, Phys. Rev. B **89**, 085429 (2014). doi:10.1103/PhysRevB.89.085429. URL http://link.aps.org/doi/10.1103/PhysRevB.89.085429
89. S.K. Wang, J. Wang, K.S. Chan, New J. Phys. **16**(4), 045015 (2014). URL http://stacks.iop.org/1367-2630/16/i=4/a=045015
90. D.Q. Fang, S.L. Zhang, H. Xu, RSC Adv. **3**, 24075 (2013). doi:10.1039/C3RA42720J. URL http://dx.doi.org/10.1039/C3RA42720J
91. H.X. Luan, C.W. Zhang, F.B. Zheng, P.J. Wang, J. Phys. Chem. C. doi:10.1021/jp4005357. URL http://pubs.acs.org/doi/abs/10.1021/jp4005357
92. L. Ma, J.M. Zhang, K.W. Xu, V. Ji, Phys. B **425**, 66 (2013)
93. F. Zheng, C. Zhang, P. Wang, S. Li, J. Appl. Phys. **113**(15), 154302 (2013)
94. F. Zheng, C. Zhang, S. Yan, F. Li, J. Mater. Chem. C **1**, 2735 (2013). doi:10.1039/C3TC30097H. URL http://dx.doi.org/10.1039/C3TC30097H
95. A.B. Chen, X.F. Wang, P. Vasilopoulos, M.X. Zhai, Y.S. Liu, Phys. Chem. Chem. Phys. **16**, 5113 (2014). doi:10.1039/C3CP55447C. URL http://dx.doi.org/10.1039/C3CP55447C
96. J. Chen, X.F. Wang, P. Vasilopoulos, A.B. Chen, J.C. Wu, ChemPhysChem **15**(13), 2701 (2014). doi:10.1002/cphc.201402171. URL http://dx.doi.org/10.1002/cphc.201402171
97. Y.J. Dong, X.F. Wang, P. Vasilopoulos, M.X. Zhai, X.M. Wu, J. Phys. D Appl. Phys. **47**(10), 105304 (2014). URL http://stacks.iop.org/0022-3727/47/i=10/a=105304
98. Y. Liu, X. Yang, X. Zhang, X. Hong, X.F. Wang, J. Feng, C. Zhang, RSC Adv. (2014). doi:10.1039/C4RA07791A. URL http://dx.doi.org/10.1039/C4RA07791A
99. A. Lopez-Bezanilla, J. Phys. Chem. C **118**(32), 18788 (2014). doi:10.1021/jp5060809. URL http://dx.doi.org/10.1021/jp5060809
100. L. Ma, J.M. Zhang, K.W. Xu, V. Ji, Phys. E Low-Dimension. Syst. Nanostruct. **60**(0), 112 (2014). doi:http://dx.doi.org/10.1016/j.physe.2014.02.013. URL http://www.sciencedirect.com/science/article/pii/S1386947714000642
101. J.M. Zhang, W.T. Song, K.W. Xu, V. Ji, Comput. Mater. Sci. **95**(0), 429 (2014). doi:http://dx.doi.org/10.1016/j.commatsci.2014.08.019. URL http://www.sciencedirect.com/science/article/pii/S0927025614005618
102. X. Pi, Z. Ni, Y. Liu, Z. Ruan, M. Xu, D. Yang, Phys. Chem. Chem. Phys. **17**, 4146 (2015). doi:10.1039/C4CP05196C. URL http://dx.doi.org/10.1039/C4CP05196C
103. X. Tan, F. Li, Z. Chen, J. Phys. Chem. C **118**(45), 25825 (2014). doi:10.1021/jp507011p. URL http://dx.doi.org/10.1021/jp507011p
104. A. Manjanath, V. Kumar, A.K. Singh, Phys. Chem. Chem. Phys. **16**, 1667 (2014). doi:10.1039/C3CP54655A. URL http://dx.doi.org/10.1039/C3CP54655A
105. Y.C. Cheng, Z.Y. Zhu, U. Schwingenschlögl, Europhys. Lett. **95**, 17005 (2011)
106. R.R. Nair, P. Blake, A.N. Grigorenko, K.S. Novoselov, T.J. Booth, T. Stauber, N.M.R. Peres, A.K. Geim, Science **320**(5881), 1308 (2008)
107. F. Bechstedt, L. Matthes, P. Gori, O. Pulci, Appl. Phys. Lett. **100**, 261906 (2012)

108. L. Matthes, P. Gori, O. Pulci, F. Bechstedt, Phys. Rev. B **87**, 035438 (2013). doi:10.1103/PhysRevB.87.035438. URL http://link.aps.org/doi/10.1103/PhysRevB.87.035438
109. L. Matthes, O. Pulci, F. Bechstedt, J. Phys. Cond. Matter **25**, 395305 (2013)
110. L. Matthes, O. Pulci, F. Bechstedt, New J. Phys. **16**(10), 105007 (2014). URL http://stacks.iop.org/1367-2630/16/i=10/a=105007
111. H. Bao, J. Guo, W. Liao, H. Zhao, Appl. Phys. A, 1–5 (2014). doi:10.1007/s00339-014-8837-x. URL http://dx.doi.org/10.1007/s00339-014-8837-x
112. B. Mohan, A. Kumar, P. Ahluwalia, Phys. E: Low-Dimension. Syst. Nanostruct. **53**, 233 (2013). doi:10.1016/j.physe.2013.05.014. URL http://www.sciencedirect.com/science/article/pii/S1386947713001872
113. C.J. Tabert, E.J. Nicol, Phys. Rev. B **89**, 195410 (2014). doi:10.1103/PhysRevB.89.195410. URL http://link.aps.org/doi/10.1103/PhysRevB.89.195410
114. H.R. Chang, J. Zhou, H. Zhang, Y. Yao, Phys. Rev. B **89**, 201411 (2014). doi:10.1103/PhysRevB.89.201411. URL http://link.aps.org/doi/10.1103/PhysRevB.89.201411
115. J.Y. Wu, S.C. Chen, M.F. Lin, New J. Phys. **16**(12), 125002 (2014). URL http://stacks.iop.org/1367-2630/16/i=12/a=125002
116. R. Das, S. Chowdhury, A. Majumdar, D. Jana, RSC Adv. **5**, 41 (2015). doi:10.1039/C4RA07976K. URL http://dx.doi.org/10.1039/C4RA07976K
117. Y. Yao, S.Y. Liu, X.L. Lei, Phys. Rev. B **91**, 115411 (2015). doi:10.1103/PhysRevB.91.115411. URL http://link.aps.org/doi/10.1103/PhysRevB.91.115411
118. C.J. Tabert, E.J. Nicol, Phys. Rev. Lett. **110**, 197402 (2013). doi:10.1103/PhysRevLett.110.197402. URL http://link.aps.org/doi/10.1103/PhysRevLett.110.197402
119. C.J. Tabert, E.J. Nicol, Phys. Rev. B **88**, 085434 (2013). doi:10.1103/PhysRevB.88.085434. URL http://link.aps.org/doi/10.1103/PhysRevB.88.085434
120. N. Singh, U. Schwingenschlgl, Phys. Status Solidi (RRL) Rapid Res. Lett. **8**(4) (2014). doi:10.1002/pssr.201409025. URL http://dx.doi.org/10.1002/pssr.201409025
121. V.Y. Tsaran, S.G. Sharapov, Phys. Rev. B **90**, 205417 (2014). doi:10.1103/PhysRevB.90.205417. URL http://link.aps.org/doi/10.1103/PhysRevB.90.205417
122. C.P. Chang, J. Appl. Phys. **110**(1), 013725 (2011). doi:http://dx.doi.org/10.1063/1.3603040. URL http://scitation.aip.org/content/aip/journal/jap/110/1/10.1063/1.3603040
123. J.M. Luttinger, W. Kohn, Phys. Rev. **97**(4), 869 (1955)
124. V.P. Gusynin, S.G. Sharapov, J.P. Carbotte, Phys. Rev. Lett. **98**, 157402 (2007). doi:10.1103/PhysRevLett.98.157402. URL http://link.aps.org/doi/10.1103/PhysRevLett.98.157402
125. Y. Zhang, Y.W. Tan, H.L. Stormer, P. Kim, Nature **438**, 201 (2005)
126. Y. Zhang, Z. Jiang, J.P. Small, M.S. Purewal, Y.W. Tan, M. Fazlollahi, J.D. Chudow, J.A. Jaszczak, H.L. Stormer, P. Kim, Phys. Rev. Lett. **96**, 136806 (2006). doi:10.1103/PhysRevLett.96.136806. URL http://link.aps.org/doi/10.1103/PhysRevLett.96.136806
127. Z. Jiang, E.A. Henriksen, L.C. Tung, Y.J. Wang, M.E. Schwartz, M.Y. Han, P. Kim, H.L. Stormer, Phys. Rev. Lett. **98**, 197403 (2007). doi:10.1103/PhysRevLett.98.197403. URL http://link.aps.org/doi/10.1103/PhysRevLett.98.197403
128. K. Shakouri, P. Vasilopoulos, V. Vargiamidis, F.M. Peeters, Phys. Rev. B **90**, 235423 (2014). doi:10.1103/PhysRevB.90.235423. URL http://link.aps.org/doi/10.1103/PhysRevB.90.235423
129. N.J. Roome, J.D. Carey, ACS Appl. Mater. Interfaces. doi:10.1021/am501022x. URL http://pubs.acs.org/doi/abs/10.1021/am501022x
130. G. Berdiyorov, F. Peeters, RSC Adv. **4**(3), 1133 (2014)
131. H. peng Li, R. qin Zhang, EPL (Europhy. Lett.) **99**(3), 36001 (2012). URL http://stacks.iop.org/0295-5075/99/i=3/a=36001
132. L. Wang, H. Sun, J. Mol. Model. **18**, 4811 (2012). URL http://dx.doi.org/10.1007/s00894-012-1482-4. 10.1007/s00894-012-1482-4
133. M. Hu, X. Zhang, D. Poulikakos, Phys. Rev. B **87**, 195417 (2013). doi:10.1103/PhysRevB.87.195417. URL http://link.aps.org/doi/10.1103/PhysRevB.87.195417
134. Q.X. Pei, Y.W. Zhang, Z.D. Sha, V.B. Shenoy, J. Appl. Phys. **114**(3), 033526 (2013). doi:10.1063/1.4815960

135. T. Ng, J. Yeo, Z. Liu, Int.J. Mech. Mater. Des. **9**, 105 (2013). doi:10.1007/s10999-013-9215-0. URL http://dx.doi.org/10.1007/s10999-013-9215-0
136. B. Liu, C.D. Reddy, J. Jiang, H. Zhu, J.A. Baimova, S.V. Dmitriev, K. Zhou, J. Phys. D Appl. Phys. **47**(16), 165301 (2014). URL http://stacks.iop.org/0022-3727/47/i=16/a=165301
137. J.J. Yeo, Z.S. Liu, J. Comput. Theor. Nanosci. **11**(8), 1790 (2014-08-01T00:00:00). doi:10.1166/jctn.2014.3568. URL http://www.ingentaconnect.com/content/asp/jctn/2014/00000011/00000008/art00011
138. X. Zhang, H. Xie, M. Hu, H. Bao, S. Yue, G. Qin, G. Su, Phys. Rev. B **89**, 054310 (2014). doi:10.1103/PhysRevB.89.054310. URL http://link.aps.org/doi/10.1103/PhysRevB.89.054310
139. Z. Wang, T. Feng, X. Ruan, J. Appl. Phys. **117**(8), 084317 (2015). doi:http://dx.doi.org/10.1063/1.4913600. URL http://scitation.aip.org/content/aip/journal/jap/117/8/10.1063/1.4913600
140. X. Zhang, H. Bao, M. Hu, Nanoscale **7**, 6014 (2015). Doi:10.1039/C4NR06523A. URL http://dx.doi.org/10.1039/C4NR06523A
141. M. Kamatagi, J. Elliott, N. Sankeshwar, A. Lindsay Greer, in *Physics of Semiconductor Devices*, ed. by V.K. Jain, A. Verma, Environmental science and engineering (Springer International Publishing, 2014), pp. 617–619. doi:10.1007/978-3-319-03002-9157. URL http://dx.doi.org/10.1007/978-3-319-03002-9157
142. B. Liu, J.A. Baimova, C.D. Reddy, S.V. Dmitriev, W.K. Law, X.Q. Feng, K. Zhou, Carbon **79**(0), 236 (2014). doi:http://dx.doi.org/10.1016/j.carbon.2014.07.064. URL http://www.sciencedirect.com/science/article/pii/S0008622314007027
143. G. Liu, X.L. Lei, M.S. Wu, B. Xu, C.Y. Ouyang, EPL (Europhys. Lett.) **106**(4), 47001 (2014). URL http://stacks.iop.org/0295-5075/106/i=4/a=47001
144. G. Liu, X.L. Lei, M.S. Wu, B. Xu, C.Y. Ouyang, J. Phys. Condens. Matter **26**(35), 355007 (2014). URL http://stacks.iop.org/0953-8984/26/i=35/a=355007
145. Y. Du, J. Zhuang, H. Liu, X. Xu, S. Eilers, K. Wu, P. Cheng, J. Zhao, X. Pi, K.W. See, G. Peleckis, X. Wang, S.X. Dou, ACS Nano **8**(10), 10019 (2014). doi:10.1021/nn504451t. URL http://dx.doi.org/10.1021/nn504451t. PMID: 25248135
146. V.O. Özçelik, H.H. Gurel, S. Ciraci, Phys. Rev. B **88**, 045440 (2013). doi:10.1103/PhysRevB.88.045440. URL http://link.aps.org/doi/10.1103/PhysRevB.88.045440
147. R. Li, Y. Han, T. Hu, J. Dong, Y. Kawazoe, Phys. Rev. B **90**, 045425 (2014). doi:10.1103/PhysRevB.90.045425. URL http://link.aps.org/doi/10.1103/PhysRevB.90.045425
148. J. Gao, J. Zhang, H. Liu, Q. Zhang, J. Zhao, Nanoscale **5**, 9785 (2013). doi:10.1039/C3NR02826G. URL http://dx.doi.org/10.1039/C3NR02826G
149. S. Li, Y. Wu, Y. Tu, Y. Wang, T. Jiang, W. Liu, Y. Zhao, Sci. Rep. **5**(7881) (2015)
150. A. Manjanath, A.K. Singh, Chem. Phys. Lett. **592**(0), 52 (2014). doi:10.1016/j.cplett.2013.12.010. URL http://www.sciencedirect.com/science/article/pii/S0009261413014905
151. V.V. Hoang, H.T.C. Mi, J. Phys. D Appl. Phys. **47**(49), 495303 (2014). URL http://stacks.iop.org/0022-3727/47/i=49/a=495303
152. M.Q. Le, D.T. Nguyen, Appl. Phys. A, 1–9 (2014). doi:10.1007/s00339-014-8904-3. URL http://dx.doi.org/10.1007/s00339-014-8904-3
153. M.P. Lima, A. Fazzio, A.J.R. da Silva, Phys. Rev. B **88**, 235413 (2013). doi:10.1103/PhysRevB.88.235413. URL http://link.aps.org/doi/10.1103/PhysRevB.88.235413
154. M. Ezawa, Phys. Rev. Lett. **109**, 055502 (2012)
155. P. Gori et al., Thermophysical properties of the novel 2D materials graphene and silicene: insights from Ab initio calculations. Energy Proc. **45**, 512 (2014)

Chapter 2
Topological Physics of Honeycomb Dirac Systems

Motohiko Ezawa

Abstract In this chapter, we review the topological properties of honeycomb Dirac systems. Such systems are materialized in silicene, germanene and stanene, which are the monolayer honeycomb structures of silicon, germanium and tin, respectively. We first analyze the Hall conductance in terms of the Chern number. Then we show that the Hall effects are understood by the Chern-Simons action, which is the topological field theory of the massive Dirac system. We also introduce the entanglement entropy, the entanglement spectrum and the quantum distance in view of the bulk-edge correspondence.

2.1 Introduction

The success of graphene has excited intensive studies of monolayer atomic systems [1, 2]. The dispersion of graphene is linear around the Fermi energy and hence is described by the Dirac theory. Graphene is described by the massless Dirac theory since graphene is a semi-metal. Recently a new class of honeycomb monolayer systems have been experimentally synthesized and have attracted much attention. They are silicene, germanene and stanene, which are monolayer honeycomb structures made of silicon, germanium and tin, respectively [3–6]. There is the spin-orbit (SO) interaction in these materials, which leads to the quantum spin Hall (QSH) effect [7], as is the typical signature of topological insulators indexed by the Chern number. They are described by the massive Dirac fermion. A prominent feature is that the band gap can be controlled by applying an electric field since the structure is buckled [8–10]. Various topological phases are predicted including the quantum anomalous Hall (QAH) effect [11, 12] and quantum-spin-quantum-anomalous Hall effects [13].

M. Ezawa (✉)
Department of Applied Physics, University of Tokyo, Tokyo Hongo 7-3-1, Japan
e-mail: ezawa@ap.t.u-tokyo.ac.jp

© Springer International Publishing Switzerland 2016
M.J.S. Spencer and T. Morishita (eds.), *Silicene*, Springer Series
in Materials Science 235, DOI 10.1007/978-3-319-28344-9_2

These novel monolayer materials can be experimentally grown on substrates. Silicene is grown on Ag [14–16], ZrB$_2$ [17], Ir [18], MoS$_2$ [19], germanene is grown on Au [20], Pt [21], Al [22], Ge$_2$Pt crystal [23] and stanene is grown on Bi$_2$Te$_2$ [24]. However, in general, the effect of the substrates is strong enough to destroy the Dirac nature of monolayer materials. Recently, this problem has been overcome by transferring silicene from the Ag substrate to an alumina substrate, where it has been shown that silicene acts as a field-effect transistor at room temperature [25].

In this chapter, we review the recent progress of topological monolayer materials based on the Dirac theory. We first derive the Dirac theory for the honeycomb lattice from the tight-binding model. Second, we derive the Chern-Simons action by taking the functional integration in the Dirac theory, where the Chern number is introduced. Third, we derive the Thouless-Kohmoto-Nightingale-Nijs (TKNN) formula [26] for the Hall conductance from the Chern-Simons theory. The Hall conductance is quantized inside the bulk band gap for the massive Dirac fermions. Then, we explore the bulk-edge correspondence [27, 28], which is the easiest way to know if the bulk is a topological insulator. The emergence of topological edge states in a nanoribbon is a strong indication that the system is topological. Finally, we introduce the notion of the entanglement spectrum, the entanglement entropy and the quantum distance, which provide other methods to demonstrate the bulk-edge correspondence.

2.2 Graphene and Silicene

The basic structure of graphene and silicene is a honeycomb lattice. It consists of two triangular sublattices made of inequivalent lattice sites A and B (Fig. 2.1a). The reciprocal lattice is also a honeycomb lattice in the momentum space (Fig. 2.1b), which constitutes the Brillouin zone.

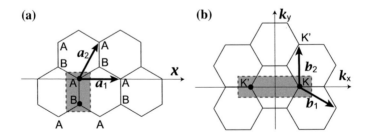

Fig. 2.1 a The honeycomb structure, made of two fundamental vectors $\mathbf{a}_1$ and $\mathbf{a}_2$, consists of two sublattices made of A and B sites. A *dotted rectangular* represents a unit cell. **b** The reciprocal lattice is also a honeycomb lattice. A *dotted rectangular* represents a unit cell, which contains two inequivalent points K and K'

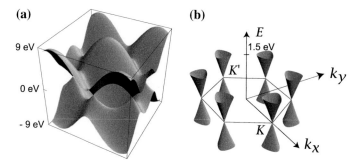

Fig. 2.2 Band structure of graphene. **a** Six valleys are seen in this figure. **b** The gap is closed at the K and K' points, where the band structure looks like a cone. It is called the Dirac cone because the dispersion is linear

Graphene: Graphene is described by the simplest tight-binding model for a honeycomb lattice [1, 2],

$$\hat{H}_0 = -t \sum_{\langle i,j \rangle s} c_{is}^\dagger c_{js}, \tag{2.1}$$

where $c_{is}^\dagger$ creates an electron with spin polarization $s = \uparrow\downarrow$ at site i, $\langle i,j \rangle$ runs over all the nearest neighbor hopping sites, and t is the transfer energy.

The Hamiltonian (2.1) is rewritten as

$$\hat{H}_0 = t \sum_s \int d^2 k' \left(c_{As}^\dagger, c_{Bs}^\dagger \right) \begin{pmatrix} 0 & f(\mathbf{k}') \\ f^*(\mathbf{k}') & 0 \end{pmatrix} \begin{pmatrix} c_{As} \\ c_{Bs} \end{pmatrix} \tag{2.2}$$

in the momentum space, with $s = \pm$ being the spin index, and

$$f(\mathbf{k}') = e^{-iak'_y/\sqrt{3}} + 2e^{iak'_y/2\sqrt{3}} \cos \frac{ak'_x}{2}. \tag{2.3}$$

The energy spectrum is obtained as

$$E(\mathbf{k}') = t \sqrt{1 + 4 \cos \frac{ak'_x}{2} \cos \frac{\sqrt{3}ak'_y}{2} + 4 \cos^2 \frac{ak'_x}{2}}. \tag{2.4}$$

We illustrate the band structure (2.4) in Fig. 2.2a. It consists of valleys or cones near the Fermi surface as shown in Fig. 2.2b. The gap closes at $\mathbf{k}' = \mathbf{K}_\eta$ with

$$\mathbf{K}_\eta = \frac{1}{a} \left(\eta \frac{4\pi}{3}, 0 \right) \quad \text{with } \eta = \pm. \tag{2.5}$$

These points are the K and K' points.

We are interested in physics near the Fermi energy. We make the Taylor expansion of (2.3) around $\mathbf{k}' = \mathbf{K}_\eta$. By setting $\mathbf{k}' = \mathbf{K}_\eta + \mathbf{k}$, we obtain

$$f\left(\mathbf{k} + \mathbf{K}_\eta\right) = \eta k_x - i k_y \quad \text{for } |\mathbf{k}| \ll a^{-1}, \tag{2.6}$$

and the dispersion is linear,

$$E\left(\mathbf{k} + \mathbf{K}_\eta\right) = \pm \hbar v_{\mathrm{F}} \sqrt{k_x^2 + k_y^2}. \tag{2.7}$$

Hence, the low-energy physics near the Fermi energy is described by the Dirac theory,

$$H_s^\eta = \begin{pmatrix} 0 & \hbar v_{\mathrm{F}}(\eta k_x - i k_y) \\ \hbar v_{\mathrm{F}}(\eta k_x + i k_y) & 0 \end{pmatrix} = \hbar v_{\mathrm{F}}(\eta k_x \tau_x + k_y \tau_y), \tag{2.8}$$

where $v_{\mathrm{F}} = \frac{\sqrt{3}}{2\hbar} at$ is the Fermi velocity.

The K and K' points are also referred to as the K_η points or the Dirac points. The cone-shaped parts of the energy spectrum are referred to as the Dirac cones (Figs. 2.2 and 2.3a2). We note that the number of the Dirac cones is always even in the tight-binding theory, which is known as the Nielsen-Ninomiya theorem [29].

Silicene and tunable band gap: The basic nature of silicene is described also by the tight-binding model (2.1). There are two additional features. One is the presence of the SO interaction, which makes silicene a topological insulator. The other is its buckled structure with a layer separation between the two sublattices (Fig. 2.3b1). This freedom allows us to tune the gap by introducing a potential difference between the two sublattices. When we apply an electric field E_z perpendicular to silicene, the tight-binding Hamiltonian reads

$$\hat{H} = -t \sum_{\langle i,j \rangle s} c_{is}^\dagger c_{js} + i \frac{\lambda_{\mathrm{SO}}}{3\sqrt{3}} \sum_{\langle\langle i,j \rangle\rangle s} s v_{ij} c_{is}^\dagger c_{js} - \ell \sum_{is} \mu_i E_z c_{is}^\dagger c_{is}, \tag{2.9}$$

where $\langle\langle i,j \rangle\rangle$ run over all the next-nearest neighbor hopping sites. The spin index stands for $s = \uparrow\downarrow$ for indices and for $s = \pm$ within equations. It describes germanene and stanene as well.

We explain each term. (i) The first term represents the usual nearest-neighbor hopping with the transfer energy t. (ii) The second term represents the effective SO coupling with λ_{SO}, where $v_{ij} = +1$ if the next-nearest-neighboring hopping is anticlockwise and $v_{ij} = -1$ if it is clockwise with respect to the positive z axis [31].

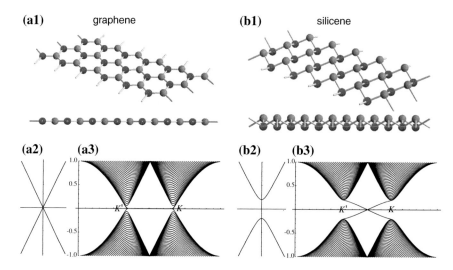

Fig. 2.3 (**a1**) The lattice structure of graphene is planar, but (**b1**) that of silicene is buckled. *Red* and *blue balls* represent A and B sites. (**a2**) The band gap of graphene is closed, where the dispersion is linear near the Fermi energy. (**b2**) The band gap of silicene is open. (**a3**) A flat line connecting the K and K' points represents gapless flat edge modes in a nanoribbon. It contains 4-fold degenerate edge states for up/down-spin and left/right movers. (**b3**) Two lines connecting the tips of the Dirac cones represents edge modes of a nanoribbon. Each line contains 2-fold degenerate edge states (Color figure online)

(iii) The third term represents the staggered sublattice potential with $\mu_i = +1\,(-1)$ for the A (B) site [8]. Explicit values of these parameters are summarized in Table 2.1. By diagonalizing the Hamiltonian by setting $E_z = 0$, we obtain the band structure illustrated as in Fig. 2.3b2. The prominent feature is that the gap is open due to the SO interaction, and hence silicene is an insulator. A large SO interaction with $\lambda_{SO} = 0.3$ eV is materialized in functionalized stanene [32], which will be a topological insulator at room temperature.

The low-energy physic near the Fermi energy can be derived from the tight-binding model. It is given by the Dirac Hamiltonian (2.8) together with the diagonal mass term around each Dirac point indexed by s and η,

Table 2.1 The parameters charactering graphene, silicene and germanene

	t (eV)	v	a (Å)	λ_{SO}	ℓ	θ
Graphene	2.8	9.8	2.46	10^{-3}	0	90
Silicene	1.6	5.5	3.86	3.9	0.23	101.7
Germanene	1.3	4.6	4.02	43	0.33	106.5
Stanene	1.3	4.9	4.70	100	0.40	107.1

Here, v_F is in the unit of 10^5 m/s, and λ_{SO} in the unit of meV. ℓ is the buckle height, while θ is the bond angle. Taken from [30]

$$H_s^\eta = \begin{pmatrix} \Delta_s^\eta & \hbar v_F(\eta k_x - i k_y) \\ \hbar v_F(\eta k_x + i k_y) & -\Delta_s^\eta \end{pmatrix} = \hbar v_F(\eta k_x \tau_x + k_y \tau_y) + \Delta_s^\eta \tau_z, \quad (2.10)$$

where

$$\Delta_s^\eta = \eta s \lambda_{SO} - \ell E_z \equiv -\ell(E_z - \eta s E_{cr}), \quad (2.11)$$

and

$$E_{cr} \equiv \lambda_{SO}/\ell. \quad (2.12)$$

Here, λ_{SO} is the SO interaction strength, ℓ is the separation between the A and B sublattices, and E_z is the applied perpendicular electric field.

The energy spectrum reads

$$E(\mathbf{k}) = \pm\sqrt{(\hbar v_F k)^2 + (\Delta_s^\eta)^2}. \quad (2.13)$$

The gap is given by $2|\Delta_s^\eta| = 2\ell|E_z - \eta s E_{cr}|$.

It is important that the band gap is tunable by controlling the external electric field E_z [8]. The gap is open when $E_z = 0$. As $|E_z|$ increases, the gap become narrower (Fig. 2.4), and it closes at $E_z = \eta s E_{cr}$, where silicene is semi-metallic just as in graphene. As $|E_z|$ increases further, the gap opens again.

Generalized Dirac mass terms: There are actually other ways to control the band gap by introducing other interactions to silicene [13]. Since each Dirac cone is indexed by two parameters $\eta = \pm$ and $s = \pm$, the most general Dirac mass must have the following expression,

$$\Delta_s^\eta = \eta s \lambda_{SO} - \lambda_V + \eta \lambda_H + s \lambda_{SX}, \quad (2.14)$$

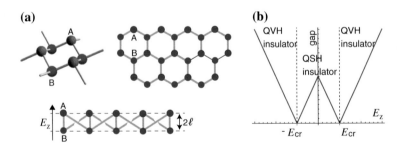

(a)   **(b)**

Fig. 2.4 Electrically tunable band gap of silicene. **a** The buckled structure is illustrated. Electric potential difference may be induced between the A and B sites. **b** Silicene is in the QSH phase as it is, but it is transformed into the quantum-valley Hall (QVH) phase by applying the electric field more than the critical value. See Table 2.2 for the QVH phase. It becomes a semi-metal at the critical electric field ($\pm E_{cr}$), where the gap closes

so that it has four independent parameters, λ_{SO}, λ_V, λ_H and λ_{SX}. We have already discussed the first two terms representing the SO interaction and the sublattice staggered potential with $\lambda_V = \ell E_z$. The third term describes the Haldane interaction induced by the photo-irradiation, where $\lambda_\Omega = \hbar v_F^2 \mathcal{A}^2 \Omega^{-1}$ with Ω the frequency and $\mathcal{A}$ the dimensionless intensity [12]. The fourth term describes the antiferromagnetic exchange magnetization. We may write down the tight-binding terms that yield the fourth and fifth terms as [13],

$$i\frac{\lambda_H}{3\sqrt{3}} \sum_{\langle\langle i,j\rangle\rangle s} v_{ij} c_{is}^\dagger c_{js}, \quad \lambda_{SX} \sum_{is} s\mu_i c_{is}^\dagger c_{is}. \tag{2.15}$$

There are a variety of 2D materials whose low-energy physics is described by the Dirac Hamiltonian (2.10) with the Dirac mass (2.14). We call them general honeycomb systems. Examples are monolayer antiferromagnetic manganese chalcogenophosphates ($MnPX_3$, X = S, Se) [33] and perovskite G-type antiferromagnetic insulators grown along [111] direction [34].

2.3 Topological Field Theory

The well-known example of topological insulators is the integer quantum Hall system. When v Landau levels are filled, the electric field E_j drives the Hall current,

$$J_i = \sigma_{ij} E_j, \quad (i, j = x, y), \tag{2.16}$$

with

$$\sigma_{xy} = \frac{e^2}{h} v, \tag{2.17}$$

where $v = 1, 2, 3, \ldots$. Thus the Hall conductivity is quantized. The quantized Hall conductance is robust against impurities and randomness. This is because of the topological nature of the Hall conductance. Namely, the quantized value cannot change its value continuously even if we increase external perturbations continuously from zero. This is the characteristic feature common to all topological insulators.

There exists a gap between the valence and conduction bands in the massive Dirac theory (2.10). We demonstrate that, when the Fermi level is present within the band gap, the system becomes a topological insulator by deriving similar formulas to the Hall current (2.16) with (2.17).

2.3.1 Chern-Simons Theory

We start with the generic formula for the current in the presence of the electric potential A_i,

$$J^i = \frac{\partial S_{\text{eff}}}{\partial A_i}, \tag{2.18}$$

where S_{eff} is the effective action. We calculate the effective action for the Dirac fermion coupled with the gauge potential A_i.

The Hamiltonian reads

$$H = \hbar v_F \left[\eta (k_x - eA_x)\sigma_x + (k_y - eA_y)\sigma_y \right] + m\sigma_z. \tag{2.19}$$

We separate it into the unperturbed part H_0 and the perturbation term V,

$$H_0 = \hbar v_F \left[\eta k_x \sigma_x + k_y \sigma_y \right] + m\sigma_z, \tag{2.20}$$

$$V = -\hbar v_F \left[eA_x \sigma_x + eA_y \sigma_y \right]. \tag{2.21}$$

The effective action is given by

$$S_{\text{eff}} = \text{Tr} \ln \left(-G_0^{-1} + V \right) = \text{Tr} \ln \left(-G_0^{-1} \right) + \text{Tr} \ln (1 - G_0 V), \tag{2.22}$$

with the Green function

$$\hat{G}_0(\mathbf{k}, i\hbar\omega_n) = \frac{1}{i\hbar\omega_n - H_0} = \frac{i\hbar\omega \mathbb{I} + \hbar v_F (\eta k_x \sigma_x + k_y \sigma_y) + m\sigma_z}{(i\hbar\omega_n)^2 - v_F^2 k^2 - m^2}. \tag{2.23}$$

The effective action is expanded in the power series of eA_i. The nontrivial lowest order term is given by the second order term in eA_i,

$$S_{\text{eff}} = \sum_{\mathbf{q}} \Pi^{\mu\nu}(\mathbf{q}, \omega) A_\mu(\mathbf{q}) A_\nu(-\mathbf{q}). \tag{2.24}$$

It is enough to keep the expansion to this order in the linear response theory. We note that the term $\Pi^{\mu\nu}$ is the self-energy of the photon in the ordinary $3 + 1$ dimensional quantum electrodynamics. However, as we now see, it yields the Chern-Simons action in the $2 + 1$ dimensions. In what follows the two-point correlation function

$$\Pi^{\mu\nu}(\mathbf{x} - \mathbf{y}, \omega) = \frac{1}{i} \left. \frac{\partial^2 S_{\text{eff}}[A_\mu]}{\partial A_\mu(\mathbf{x})\partial A_\nu(\mathbf{y})} \right|_{A_\mu = 0} \tag{2.25}$$

plays the key role.

We first evaluate the effective action in the one-loop approximation,

$$S_{\mathrm{eff}}^{(1)} \simeq \mathrm{Tr}(G_0 V G_0 V) = \sum_{\mathbf{q}} \Pi_{(1)}^{\mu\nu}(\mathbf{q}, \omega) A_\mu(\mathbf{q}) A_\nu(-\mathbf{q}), \qquad (2.26)$$

where

$$\Pi_{(1)}^{\mu\nu}(\mathbf{q}, \omega) = -e^2 \sum_{\mathbf{k}} \sum_{i\omega_n} \mathrm{Tr}\left[\widehat{G}_0(\mathbf{k}, i\hbar\omega_n)\, \sigma^\mu \widehat{G}_0(\mathbf{k}+\mathbf{q}, i\hbar(\omega_n+\omega_m))\sigma^\nu\right]. \quad (2.27)$$

With the use of the Green function (2.23), the susceptibility (2.27) is given by

$$\Pi_{\mu\nu}^{(1)}(\mathbf{q}, \omega) = -e^2 \sum_{\mathbf{k}\in 2D} \int_{-\infty}^{\infty} \frac{d\omega}{2\pi} \frac{1}{(i\hbar\omega_n)^2 - v_F^2 k^2 - m^2}$$
$$\times \frac{\varrho_{\mu\nu}(\mathbf{q}, \omega)}{(i\hbar(\omega_n+\omega_m) - \mu)^2 - v_F^2(\mathbf{k}+\mathbf{q})^2 - m^2}, \qquad (2.28)$$

with

$$\varrho^{\mu\nu}(\mathbf{q}, \omega) = \mathrm{Tr}[((i\hbar\omega)\,\mathbb{I} + \hbar v_F(\mathbf{k}\cdot\sigma) + m\sigma_z)\sigma^\mu$$
$$\times (i\hbar(\omega+\omega')\mathbb{I} + \hbar v_F((\mathbf{k}+\mathbf{q})\cdot\sigma) + m\sigma_z)\,\sigma^\nu]. \qquad (2.29)$$

Here, μ, ν run over x, y, z. By using the relation

$$\mathrm{Tr}[\sigma^\mu \sigma^\nu \sigma^\rho] = 2i\varepsilon^{\mu\nu\rho} \qquad (2.30)$$

it is written as

$$\Pi_{(1)}^{\mu\nu}(\mathbf{q}, \omega) = \varepsilon^{\mu\nu\rho} q_\rho \Pi_{(1)}(q^2, m). \qquad (2.31)$$

It is estimated as [35, 36]

$$\Pi_{(1)}(q^2, m) = -\eta e^2 m \int \frac{d^3 k}{(2\pi)^3} \frac{1}{[\hbar^2(k+q)^2 + m^2][\hbar^2 k^2 + m^2]}$$
$$= -\eta e^2 \frac{m}{4\pi\hbar|q|} \arcsin\left(\frac{\hbar|q|}{\sqrt{\hbar^2 q^2 + 4m^2}}\right). \qquad (2.32)$$

In the long wave-length limit ($q/m \to 0$), we find

$$\Pi_{(1)}^{\mu\nu}(\mathbf{q}, \omega) = -\frac{e^2}{8\pi\hbar} \eta \frac{m}{|m|} \varepsilon^{\mu\nu\rho} q_\rho. \qquad (2.33)$$

Substituting this into the effective action (2.24) we rewrite it as

$$S_{\text{eff}}^{(1)} = -\frac{e^2}{8\pi\hbar}\eta\frac{m}{|m|}\varepsilon^{\mu\nu\rho}\int d^3x A_\mu \partial_\nu A_\rho.$$

(2.34)

This is known as the Chern-Simons action.
Consequently, from (2.18) we obtain

$$J^i = \frac{\partial S_{\text{eff}}^{(1)}}{\partial A_i} = -\frac{e^2}{h}\frac{\eta}{2}\frac{m}{|m|}\varepsilon^{ij}E_j = \sigma_{ij}^\eta E_j,$$

(2.35)

or

$$\sigma_{ij}^\eta = \frac{e^2}{h}\varepsilon^{ij}C^\eta,$$

(2.36)

with

$$C^\eta = -\frac{\eta}{2}\frac{m}{|m|}.$$

(2.37)

This is the Chern number [37, 38], which appears in the effective action (2.34) as

$$S_{\text{eff}}^{(1)} = \frac{e^2}{4\pi\hbar}C^\eta\varepsilon^{\mu\nu\rho}\int d^3x A_\mu \partial_\nu A_\rho.$$

(2.38)

The Chern number is a half integer instead of an integer, $C^\eta = \pm\frac{1}{2}$. This is because we have only consider the massive Dirac theory at the Dirac point K_η.

2.3.2 Chern-Simons Action and Chern Number

We have calculated the Hall conductance in the one-loop approximation. There are actually no higher-order correction terms due to the self-energy correction and the vertex correction, which is known as the Coleman-Hill theorem [39].

We reformulate the Hall conductance or the Chern number in such a way that they do not depend on the explicit form of the Hamiltonian. The Chern number is derived by calculating the self-energy diagram of a photon. It is in general given by

$$\Pi_{\mu\nu}(q) = i\int\frac{d^3p}{(2\pi)^3}\text{Tr}[G(p)\Gamma_\mu(p,p-q)G(p-q)\Gamma_\nu(p-q,p)],$$

(2.39)

where $G(p)$ and $i\Gamma_\mu(p,p')$ denote the fermion propagator and the vertex function, respectively. They include higher-order corrections. In the long wave-length limit $(q/m \to 0)$, the linear term is given by

$$
\begin{aligned}
C &= \varepsilon^{\alpha\mu\nu} \lim_{q\to 0} \frac{\partial}{\partial q^\alpha} \Pi_{\mu\nu}(q) \\
&= i\varepsilon^{\alpha\mu\nu} \lim_{q\to 0} \frac{\partial}{\partial q^\alpha} \int \frac{d^3 p}{(2\pi)^3} \mathrm{Tr}[G(p)\Gamma_\mu(p,p-q)G(p-q)\Gamma_\nu(p-q,p)] \\
&= i\varepsilon^{\alpha\mu\nu} \lim_{q\to 0} \int \frac{d^3 p}{(2\pi)^3} \mathrm{Tr}\left[G(p) \frac{\Gamma_\mu(p,p-q)}{\partial^\alpha q} G(p)\Gamma_\nu(p,p) \right. \\
&\quad \left. + G(p)\Gamma_\mu(p,p)G(p)\frac{\partial\Gamma_\nu(p-q,p)}{\partial q^\alpha} + G(p)\Gamma_\mu(p,p)\frac{\partial G(p-q)}{\partial q^\alpha}\Gamma_\nu(p,p) \right].
\end{aligned}
$$
(2.40)

The first two terms vanish due to the symmetry property. The gauge invariance implies that the vertex function satisfies the Ward-Takahashi identity,

$$
\Gamma_\mu(p,p) = \frac{\partial G^{-1}(p)}{\partial p_\mu}.
$$
(2.41)

By inserting this relation to (2.40), we find

$$
C = \frac{i\varepsilon^{\mu\nu\rho}}{24\pi^2} \lim_{q\to 0} \int d^3 p\, \mathrm{Tr}\left[G(p) \frac{\partial G^{-1}(p)}{\partial p^\mu} \frac{\partial G(p-q)}{\partial q^\nu} \frac{\partial G^{-1}(p)}{\partial p^\rho} \right].
$$
(2.42)

Using the relation

$$
\frac{\partial G(p)}{\partial p} = G(p) \frac{\partial G^{-1}(p)}{\partial p} G(p),
$$
(2.43)

we obtain

$$
C = \frac{i\varepsilon^{\mu\nu\rho}}{24\pi^2} \int d^3 p\, \mathrm{Tr}\left[G(p) \frac{\partial G^{-1}(p)}{\partial p^\mu} G(p) \frac{\partial G^{-1}(p)}{\partial p^\nu} G(p) \frac{\partial G^{-1}(p)}{\partial p^\rho} \right],
$$
(2.44)

which is the Chern number expressed in terms of the Green function [40]. This is called the Wess-Zumino-Witten (WZW) term. It is to be remarked that various interactions and also impurity effects may be included into the Green function. Furthermore we may derive the WZW action for any Hamiltonian systems provided that $C \neq 0$.

2.3.3 WZW Term and Chern Number

We proceed to derive the TKNN formula [26] of the Chern number from the WZW
term (2.44), following [41]. The gapped Hamiltonian is written in the form

$$H(\mathbf{k}) = \sum_{\alpha=1}^{N} \varepsilon_\alpha |\alpha, \mathbf{k}\rangle \langle \alpha, \mathbf{k}|, \tag{2.45}$$

where ε_α are the energy eigenvalues arranged in the following order,

$$\varepsilon_1(\mathbf{k}) \leq \varepsilon_2(\mathbf{k}) \leq \cdots \leq \varepsilon_M(\mathbf{k}) < 0 < \varepsilon_{M+1}(\mathbf{k}) \leq \cdots \leq \varepsilon_N(\mathbf{k}). \tag{2.46}$$

The Hamiltonian is diagonalized by the unitary transformation

$$H(\mathbf{k}) = U(\mathbf{k})D(\mathbf{k},0)U^\dagger(\mathbf{k}), \tag{2.47}$$

with the diagonal matrix

$$D(\mathbf{k},0) = \text{diag.}(\varepsilon_1(\mathbf{k}), \ldots, \varepsilon_M(\mathbf{k}), \varepsilon_{M+1}(\mathbf{k}), \ldots, \varepsilon_N(\mathbf{k})). \tag{2.48}$$

We define an adiabatic pass for $t \in [0, 1]$, defining

$$E_\alpha(\mathbf{k}, t) = \begin{cases} (1-t)\varepsilon_\alpha(\mathbf{k}) + t\varepsilon_G, & 1 \leq \alpha \leq M \\ (1-t)\varepsilon_\alpha(\mathbf{k}) + t\varepsilon_E, & M < \alpha \leq N \end{cases}. \tag{2.49}$$

In this adiabatic pass, the gap never closes, and hence the topological properties
do not change. It is enough to investigate the topological properties of the deformed
Hamiltonian instead of analyzing the original Hamiltonian.

The diagonal matrix $D(\mathbf{k}, 0)$ is transformed into

$$D(\mathbf{k},1) = \text{diag.}(\varepsilon_G(\mathbf{k}), \ldots, \varepsilon_G(\mathbf{k}), \varepsilon_E(\mathbf{k}), \ldots, \varepsilon_E(\mathbf{k})). \tag{2.50}$$

The deformed Hamiltonian is given by

$$H(\mathbf{k},1) = U(\mathbf{k})D(\mathbf{k},1)U^\dagger(\mathbf{k}). \tag{2.51}$$

In summary, we have proven that the gapped Hamiltonian can be adiabatically
connected to the two-level Hamiltonian of the form of

$$H(\mathbf{k},1) = \varepsilon_G P_G(\mathbf{k}) + \varepsilon_E P_E(\mathbf{k}), \tag{2.52}$$

where

$$P_G(\mathbf{k}) = \sum_{\alpha=1}^{M} |\alpha, \mathbf{k}\rangle \langle \alpha, \mathbf{k}|, \quad P_E(\mathbf{k}) = \sum_{\beta=M+1}^{N} |\beta, \mathbf{k}\rangle \langle \beta, \mathbf{k}|, \tag{2.53}$$

and $P_{\mathrm{G}}(\mathbf{k})$ is the projection operator into the occupied band and $P_{\mathrm{E}}(\mathbf{k})$ is the projection operator into the unoccupied band,

$$P_{\mathrm{G}}(\mathbf{k}) + P_{\mathrm{E}}(\mathbf{k}) = 1, \quad P_{\mathrm{E}}(\mathbf{k})P_{\mathrm{G}}(\mathbf{k}) = 0,$$
$$P_{\mathrm{G}}^2(\mathbf{k}) = P_{\mathrm{G}}(\mathbf{k}), \quad P_{\mathrm{E}}^2(\mathbf{k}) = P_{\mathrm{E}}(\mathbf{k}). \tag{2.54}$$

The Hamiltonian is decomposed into the occupied and unoccupied bands. Using them, the Green function is represented in the Lehmann form

$$G(\mathbf{k}, \omega) = \frac{1}{\omega + i\delta - \varepsilon_{\mathrm{G}}P_{\mathrm{G}}(\mathbf{k}) - \varepsilon_{\mathrm{E}}P_{\mathrm{E}}(\mathbf{k})} = \frac{P_{\mathrm{G}}(\mathbf{k})}{\omega + i\delta - \varepsilon_{\mathrm{G}}} + \frac{P_{\mathrm{E}}(\mathbf{k})}{\omega + i\delta - \varepsilon_{\mathrm{E}}}. \tag{2.55}$$

On the other hand, we have

$$\frac{\partial G^{-1}(\mathbf{k}, \omega)}{\partial \omega} = 1, \quad \frac{\partial G^{-1}(\mathbf{k}, \omega)}{\partial k_i} = (\varepsilon_{\mathrm{E}} - \varepsilon_{\mathrm{G}})\frac{\partial P_{\mathrm{G}}(\mathbf{k})}{\partial k_i}, \tag{2.56}$$

Substituting (2.56) into (2.44), we find

$$C = \frac{i\varepsilon^{ij}}{8\pi^2} \int d^3p \sum_{st} \mathrm{Tr} \frac{(\varepsilon_{\mathrm{E}} - \varepsilon_{\mathrm{G}})^2}{(\omega + i\delta - \varepsilon_s)^2(\omega + i\delta - \varepsilon_t)} P_s \frac{\partial P_{\mathrm{G}}(\mathbf{k})}{\partial k_i} P_t \frac{\partial P_{\mathrm{G}}(\mathbf{k})}{\partial k_j}. \tag{2.57}$$

By using

$$P_{\mathrm{E}} \frac{\partial P_{\mathrm{G}}(\mathbf{k})}{\partial k_i} = -\frac{\partial P_{\mathrm{E}}(\mathbf{k})}{\partial k_i} P_{\mathrm{G}} = \frac{\partial P_{\mathrm{G}}(\mathbf{k})}{\partial k_i} P_{\mathrm{G}},$$
$$P_{\mathrm{G}} \frac{\partial P_{\mathrm{G}}(\mathbf{k})}{\partial k_i} = -P_{\mathrm{G}} \frac{\partial P_{\mathrm{E}}(\mathbf{k})}{\partial k_i} = \frac{\partial P_{\mathrm{G}}(\mathbf{k})}{\partial k_i} P_{\mathrm{E}}, \tag{2.58}$$

and

$$P_{\mathrm{G}} \frac{\partial P_{\mathrm{G}}(\mathbf{k})}{\partial k_i} P_{\mathrm{G}}(\mathbf{k}) = P_{\mathrm{E}} \frac{\partial P_{\mathrm{G}}(\mathbf{k})}{\partial k_i} P_{\mathrm{E}}(\mathbf{k}) = 0, \tag{2.59}$$

it reads

$$C = \frac{i\varepsilon^{ij}}{8\pi^2} \int d^3p \sum_{st} \mathrm{Tr} \frac{(\varepsilon_{\mathrm{E}} - \varepsilon_{\mathrm{G}})^2}{(\omega + i\delta - \varepsilon_{\mathrm{E}})^2(\omega + i\delta - \varepsilon_{\mathrm{G}})} P_{\mathrm{E}} \frac{\partial P_{\mathrm{G}}(\mathbf{k})}{\partial k_i} P_{\mathrm{G}} \frac{\partial P_{\mathrm{G}}(\mathbf{k})}{\partial k_j}$$
$$+ \frac{(\varepsilon_{\mathrm{E}} - \varepsilon_{\mathrm{G}})^2}{(\omega + i\delta - \varepsilon_{\mathrm{E}})(\omega + i\delta - \varepsilon_{\mathrm{G}})^2} P_{\mathrm{G}} \frac{\partial P_{\mathrm{G}}(\mathbf{k})}{\partial k_i} P_{\mathrm{E}} \frac{\partial P_{\mathrm{G}}(\mathbf{k})}{\partial k_j}. \tag{2.60}$$

Using the relation

$$
\begin{aligned}
P_{\mathrm{E}}\partial P_{\mathrm{G}}P_{\mathrm{G}}\partial P_{\mathrm{G}} &= P_{\mathrm{E}}\partial P_{\mathrm{G}}\partial P_{\mathrm{G}}P_{\mathrm{E}} = P_{\mathrm{E}}\partial P_{\mathrm{G}}\partial P_{\mathrm{G}}, \\
P_{\mathrm{G}}\partial P_{\mathrm{G}}P_{\mathrm{E}}\partial P_{\mathrm{G}} &= \partial P_{\mathrm{G}}P_{\mathrm{E}}P_{\mathrm{E}}\partial P_{\mathrm{G}} = \partial P_{\mathrm{G}}P_{\mathrm{E}}\partial P_{\mathrm{G}} = P_{\mathrm{E}}\partial P_{\mathrm{G}}\partial P_{\mathrm{G}},
\end{aligned}
\tag{2.61}
$$

and carrying out the integral over ω, we obtain

$$
C = \frac{\varepsilon^{ij}}{2\pi}\int d^2k \mathrm{Tr}P_{\mathrm{E}}\frac{\partial P_{\mathrm{G}}(\mathbf{k})}{\partial k_i}\frac{\partial P_{\mathrm{G}}(\mathbf{k})}{\partial k_j}.
\tag{2.62}
$$

By using the relation

$$
\frac{\partial P_{\mathrm{G}}(\mathbf{k})}{\partial k_i} = \sum_{\alpha=1}^{M}\left[\frac{\partial|\alpha,\mathbf{k}\rangle}{\partial k_i}\langle\alpha,\mathbf{k}| + |\alpha,\mathbf{k}\rangle\frac{\partial\langle\alpha,\mathbf{k}|}{\partial k_i}\right],
\tag{2.63}
$$

we find

$$
\begin{aligned}
\mathrm{Tr}P_{\mathrm{E}}\partial P_{\mathrm{G}}\partial P_{\mathrm{G}} &= \mathrm{Tr}(1 - |\alpha,\mathbf{k}\rangle\langle\alpha,\mathbf{k}|)\left[\frac{\partial\langle\alpha,\mathbf{k}|}{\partial k_i}\frac{\partial|\alpha,\mathbf{k}\rangle}{\partial k_j} + \frac{\partial|\alpha,\mathbf{k}\rangle}{\partial k_i}\frac{\partial\langle\alpha,\mathbf{k}|}{\partial k_j}\right] \\
&= \mathrm{Tr}P_{\mathrm{E}}(\mathbf{k})\left[\frac{\partial\langle\alpha,\mathbf{k}|}{\partial k_i}\frac{\partial|\alpha,\mathbf{k}\rangle}{\partial k_j} + \frac{\partial|\alpha,\mathbf{k}\rangle}{\partial k_i}\frac{\partial\langle\alpha,\mathbf{k}|}{\partial k_j}\right].
\end{aligned}
\tag{2.64}
$$

On the other hand, we define the non-Abelian Berry connection by

$$
a_i^{\alpha\beta}(\mathbf{k}) = -i\langle\alpha,\mathbf{k}|\frac{\partial}{\partial k_i}|\beta,\mathbf{k}\rangle,
\tag{2.65}
$$

in which $\alpha,\beta\,(=1,2,\ldots,M)$ represent the occupied band index. Then the $U(M)$ Berry curvature is given by

$$
\begin{aligned}
f_{ij}^{\alpha\beta} &= \partial_i a_j^{\alpha\beta} - \partial_j a_i^{\alpha\beta} + i[a_i . a_j]^{\alpha\beta} \\
&= -i\left(\frac{\partial\langle\alpha,\mathbf{k}|}{\partial k_i}\frac{\partial|\beta,\mathbf{k}\rangle}{\partial k_j} - \frac{\partial\langle\alpha,\mathbf{k}|}{\partial k_j}\frac{\partial|\beta,\mathbf{k}\rangle}{\partial k_i}\right) \\
&\quad + i\left(\frac{\partial\langle\alpha,\mathbf{k}|}{\partial k_i}\sum_{\gamma=1}^{M}|\gamma,\mathbf{k}\rangle\langle\gamma,\mathbf{k}|\frac{\partial|\beta,\mathbf{k}\rangle}{\partial k_j} - \frac{\partial\langle\alpha,\mathbf{k}|}{\partial k_j}\sum_{\gamma=1}^{M}|\gamma,\mathbf{k}\rangle\langle\gamma,\mathbf{k}|\frac{\partial|\beta,\mathbf{k}\rangle}{\partial k_i}\right) \\
&= -i\left(\frac{\partial\langle\alpha,\mathbf{k}|}{\partial k_i}\frac{\partial|\beta,\mathbf{k}\rangle}{\partial k_j} - \frac{\partial\langle\alpha,\mathbf{k}|}{\partial k_j}\frac{\partial|\beta,\mathbf{k}\rangle}{\partial k_i}\right)P_{\mathrm{E}}(\mathbf{k}).
\end{aligned}
\tag{2.66}
$$

In operator form, we have

$$
\sum_{\alpha,\beta=1}^{M}|\alpha,\mathbf{k}\rangle f_{ij}^{\alpha\beta}\langle\beta,\mathbf{k}| = -i\left(\frac{\partial P_{\mathrm{G}}}{\partial k_i}P_{\mathrm{E}}\frac{\partial P_{\mathrm{G}}}{\partial k_j} - \frac{\partial P_{\mathrm{G}}}{\partial k_j}P_{\mathrm{E}}\frac{\partial P_{\mathrm{G}}}{\partial k_i}\right)
\tag{2.67}
$$

Comparing (2.64) and (2.66), finally we obtain the TKNN formula [26],

$$C_n = \frac{1}{2\pi} \int d^2k \mathrm{Tr}\, \varepsilon^{ij} f_{ij}^{\alpha\alpha} = \frac{1}{2\pi} \int d^2k\, F_n(\mathbf{k}). \tag{2.68}$$

where we have defined $F_n(\mathbf{k}) = f_{xy}^{nn} - f_{yx}^{nn}$.

2.4 Hall Conductance

We have derived the TKNN formula of the Hall conductance,

$$\sigma_{xy} = \frac{e^2}{2\pi h} \sum_n \int d^2k \Theta(\mu - E_n(\mathbf{k})) F_n(\mathbf{k}), \tag{2.69}$$

where $F_n(\mathbf{k})$ is the Berry curvature defined by

$$F_n(\mathbf{k}) = i\left[\langle \partial_{k_x}\psi_n(\mathbf{k})|\partial_{k_y}\psi_n(\mathbf{k})\rangle - \langle \partial_{k_y}\psi_n(\mathbf{k})|\partial_{k_x}\psi_n(\mathbf{k})\rangle\right], \tag{2.70}$$

and $\Theta(x)$ is the step function with $\Theta(x) = 0$ for $x < 0$ and $\Theta(x) = 1$ for $x > 0$. The integration is taken in the region for $k > k_c$ with

$$k = \frac{1}{\hbar v_F}\sqrt{\mu^2 - \left(\Delta_s^\eta\right)^2}, \tag{2.71}$$

which is the solution of $E(\mathbf{k}) = \mu$.

For the massive Dirac Hamiltonian (2.10), it is calculated as

$$F_s^\eta(\mathbf{k}) = -\eta \frac{\Delta_s^\eta}{2\left((\hbar v_F k)^2 + \left(\Delta_s^\eta\right)^2\right)^{3/2}}. \tag{2.72}$$

The Hall conductance as a function of the chemical potential is given by

$$\sigma_{xy} = \begin{cases} -\frac{\eta}{2}\frac{e^2}{2\pi h}\mathrm{sgn}(\Delta_s^\eta) & \text{for } |\mu| < |\Delta_s^\eta| \\ -\frac{\eta}{2}\frac{|\Delta_s^\eta|}{|\mu|}\frac{e^2}{2\pi h}\mathrm{sgn}(\Delta_s^\eta) & \text{for } |\mu| > |\Delta_s^\eta| \end{cases}. \tag{2.73}$$

It is quantized between the bulk band gap $|\mu| < |\Delta_s^\eta|$, which is known as the quantum anomalous Hall effect. On the other hand, the Hall conductance is anti-proportional to the chemical potential outside the band gap $|\mu| > |\Delta_s^\eta|$.

The Hall conductance is given by the Chern number when the chemical potential is within the bulk band gap since the formula (2.75) is interpreted as the integral of the Berry curvature $F_n(\mathbf{k})$ over the first Brillouin zone, which is the total "magnetic flux",

$$\sigma_{xy} = \frac{e^2}{h}\sum_n \mathcal{C}_n, \tag{2.74}$$

where

$$\mathcal{C}_n = \frac{1}{2\pi}\int d^2k F_n(\mathbf{k}). \tag{2.75}$$

and n runs over all the occupied bands. In the Dirac fermion, there is only one occupied band. The Chern number is explicitly obtained as

$$\mathcal{C} = -\frac{\eta}{2}\mathrm{sgn}(\Delta_s^\eta), \tag{2.76}$$

which is determined by the valley index η and the sign of the Dirac mass. The massive Dirac Hamiltonian always has a half Chern number.

At finite temperature, the Hall conductance is replaced by

$$\sigma_{xy} = \frac{e^2}{2\pi h}\sum_n \int d^2k f(E_n(\mathbf{k}))F_n(\mathbf{k}), \tag{2.77}$$

where $f(E_n(\mathbf{k}))$ is the Fermi distribution function. We have numerically calculated the Hall conductance, which is shown in Fig. 2.5b. The Hall conductance is not quantized even for the bulk band gap but smoothly change as a function of the chemical potential. We show the Hall conductance at the Fermi energy $\mu = 0$ as a

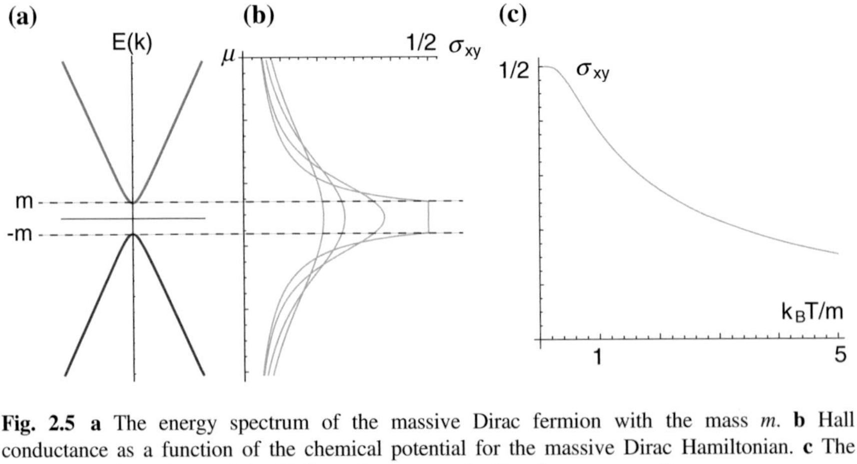

Fig. 2.5 **a** The energy spectrum of the massive Dirac fermion with the mass m. **b** Hall conductance as a function of the chemical potential for the massive Dirac Hamiltonian. **c** The temperature dependence of the Hall conductance at the Fermi energy $\mu = 0$

function of the temperature. It is half quantized for $k_B T < |\Delta_s^\eta|$ and smoothly decrease as the temperature increases.

2.4.1   Application to Honeycomb System

The honeycomb system consists of the four spin-valley dependent Dirac Hamiltonians. Introducing the spin-valley dependent gauge potential $a^{s\eta}$, we obtain the Chern-Simons action as

$$S_{CS} = -\frac{e^2}{8\pi\hbar} \sum_{s\eta} \eta \frac{\Delta_s^\eta}{|\Delta_s^\eta|} \varepsilon^{\mu\nu\lambda} a_\mu^{s\eta} \partial_\nu a_\lambda^{s\eta} = \frac{e^2}{8\pi\hbar} \sum_{IJ} \varepsilon^{\mu\nu\lambda} K_{I,J} a_\mu^I \partial_\nu a_\lambda^J, \qquad (2.78)$$

where I and J run over the spin and valley indices, and the K matrix is defined by $K = \mathrm{diag}(C_\uparrow^K, C_\downarrow^K, C_\uparrow^{K'}, C_\downarrow^{K'})$ with

$$C_s^\eta = -\frac{\eta}{2} \frac{\Delta_s^\eta}{|\Delta_s^\eta|}. \qquad (2.79)$$

We always have det $K = 1$, which implies there is no ground state degeneracy and there is no topological order and fractional excitations [42]. The spin-valley dependent current is given by

$$j_s^\eta = \frac{\partial S_{CS}}{\partial a_\mu^{s\eta}} = \frac{e^2}{h} \varepsilon^{\mu\nu\rho} C_s^\eta \partial_\nu a_\rho^J. \qquad (2.80)$$

We introduce the electromagnetic potential A_μ, the spin gauge potential A_μ^s, the valley gauge potential A_μ^v and spin-valley gauge potential A_μ^{sv} by

$$A_\mu = a_\mu^{\uparrow K} + a_\mu^{\downarrow K} + a_\mu^{\uparrow K'} + a_\mu^{\downarrow K'}, \qquad (2.81)$$

$$A_\mu^s = a_\mu^{\uparrow K} - a_\mu^{\downarrow K} + a_\mu^{\uparrow K'} - a_\mu^{\downarrow K'}, \qquad (2.82)$$

$$A_\mu^v = a_\mu^{\uparrow K} + a_\mu^{\downarrow K} - a_\mu^{\uparrow K'} - a_\mu^{\downarrow K'}, \qquad (2.83)$$

$$A_\mu^{sv} = a_\mu^{\uparrow K} - a_\mu^{\downarrow K} - a_\mu^{\uparrow K'} + a_\mu^{\downarrow K'}. \qquad (2.84)$$

The Chern-Simons action (2.78) may be rewritten in terms of these gauge potentials as

$$S_{CS} = \frac{e^2}{8\pi\hbar} \varepsilon^{\mu\nu\lambda} K_{I,J} A_\mu^I \partial_\nu A_\lambda^J, \qquad (2.85)$$

where K matrix is defined by

$$K_{I,J} = \begin{pmatrix} C & 2C_s & C_v & 2C_{sv} \\ 2C_s & C & 2C_{sv} & C_v \\ C_v & 2C_{sv} & C & 2C_s \\ 2C_{sv} & C_v & 2C_s & C \end{pmatrix}. \tag{2.86}$$

Then the current j, the spin current j_s, the valley current j_v and the spin-valley current j_{sv} are defined by

$$j_I = \frac{\partial S_{CS}}{\partial A_\mu^I}, \tag{2.87}$$

which result in

$$j = j_\uparrow^K + j_\downarrow^K + j_\uparrow^{K'} + j_\downarrow^{K'}, \tag{2.88}$$

$$j_s = j_\uparrow^K - j_\downarrow^K + j_\uparrow^{K'} - j_\downarrow^{K'}, \tag{2.89}$$

$$j_v = j_\uparrow^K + j_\downarrow^K - j_\uparrow^{K'} - j_\downarrow^{K'}, \tag{2.90}$$

$$j_{sv} = j_\uparrow^K - j_\downarrow^K - j_\uparrow^{K'} + j_\downarrow^{K'}. \tag{2.91}$$

A simple example is realized when there exists only the Kane-Mele SO interaction. In this case, the Chern-Simons action is given by

$$S_{CS} = \frac{e^2}{4\pi\hbar} \varepsilon^{\mu\nu\lambda} \left(A_\mu \partial_\nu A_\lambda^s + A_\mu^s \partial_\nu A_\lambda \right), \tag{2.92}$$

which is consistent with the previous result [43].

2.4.2 Classification of Topological Insulators

We have defined four independent Chern numbers C_s^η. Equivalently we may define the total Chern number C, the spin Chern number C_s [44–48], the valley Chern number [49–51] and the spin-valley Chern number [49, 52],

$$C = C_\uparrow^K + C_\uparrow^{K'} + C_\downarrow^K + C_\downarrow^{K'}, \tag{2.93}$$

$$C_s = \frac{1}{2}(C_\uparrow^K + C_\uparrow^{K'} - C_\downarrow^K - C_\downarrow^{K'}), \tag{2.94}$$

Table 2.2 Corresponding to the spin and valley degrees of freedom, there are 4 Chern numbers C_s^η, each of which takes $\pm\frac{1}{2}$, Equivalently they are given by the Chern, spin Chern, valley Chern and spin-valley Chern numbers C, C_s, C_v and C_{sv}

	$C_\uparrow^K$	$C_\uparrow^{K'}$	$C_\downarrow^K$	$C_\downarrow^{K'}$	C	$2C_s$	C_v	$2C_{sv}$
QAH	1/2	1/2	1/2	1/2	2	0	0	0
SQAH	1/2	1/2	1/2	-1/2	1	1	1	-1
SQAH	1/2	1/2	-1/2	1/2	1	-1	1	1
QVH	1/2	1/2	-1/2	-1/2	0	0	2	0
SQAH	1/2	-1/2	1/2	1/2	1	1	-1	1
QSH	1/2	-1/2	1/2	-1/2	0	2	0	0
QSVH	1/2	-1/2	-1/2	1/2	0	0	0	2
SQAH	1/2	-1/2	-1/2	-1/2	-1	1	1	1
SQAH	-1/2	1/2	1/2	1/2	1	-1	-1	-1
QSVH	-1/2	1/2	1/2	-1/2	0	0	0	-1
QSH	-1/2	1/2	-1/2	1/2	0	-2	0	0
SQAH	-1/2	1/2	-1/2	-1/2	-1	-1	1	-1
QVH	-1/2	-1/2	1/2	1/2	0	0	-2	0
SQAH	-1/2	-1/2	1/2	-1/2	-1	1	-1	-1
SQAH	-1/2	-1/2	-1/2	1/2	-1	-1	-1	1
QAH	-1/2	-1/2	-1/2	-1/2	-2	0	0	0

They are independently controlled by the four parameters λ_{SO}, λ_V, λ_Ω and λ_{SX}. Hence there are 16 states indexed by them. The genuine topological numbers are only C and C_s

$$C_v = C_\uparrow^K - C_\uparrow^{K'} + C_\downarrow^K - C_\downarrow^{K'}, \tag{2.95}$$

$$C_{sv} = \frac{1}{2}(C_\uparrow^K - C_\uparrow^{K'} - C_\downarrow^K + C_\downarrow^{K'}). \tag{2.96}$$

We note that the valley Chern number and the spin-valley Chern number are well defined only in the Dirac theory. Namely they are ill defined in the tight-binding model. Hence, we may call C and C_s the genuine Chern numbers.

We show all of the possible sets of topological numbers in Table 2.2. They includes QSH, QAH, spin-polarized quantum anomalous Hall (SQAH) insulators, the quantum valley Hall (QVH) insulator [53, 54] and the quantum spin-valley Hall (QSVH) insulator [13].

We comment on the relation between the $\mathbb{Z}_2$ index and the spin Chern number. The spin Chern number C_s is identical to the $\mathbb{Z}_2$ index by modulo 2 when there exists the time-reversal symmetry [44, 45]. The spin Chern number is well defined even when there is no time-reversal symmetry, while the $\mathbb{Z}_2$ index is well defined even when s_z is not a good quantum number.

2.5 Bulk-Edge Correspondence

We have revealed a rich variety of topological insulators which may arise in the
honeycomb Dirac system. Let us consider a system containing two different
topological insulators side by side. Such a system may be constructed by applying
an external field: See Fig. 2.6. Let us call the junction the inner edge. A prominent
feature is the emergence of a gapless edge mode along an inner edge, because the
inner edge separates two distinctive states. The topological number is defined only
for the gaped phase, and it cannot change its value as long as the gap opens since it
is quantized. Consequently, in order to change the value of the topological number,
the gap must close, which implies the emergence of gapless edge states at the phase
boundary.

We have mentioned in Sect. 4.2 that genuine Chern numbers are $\mathcal{C}$ and $\mathcal{C}_s$. These
numbers are well defined and zero in the vacuum. On the other hand, all the other
Chern numbers are ill defined in the vacuum. Consequently, when we consider a real
edge separating the topological insulator and the vacuum, a gapless edge state
appears only when the topological insulator carries genuine Chern numbers. For
instance, in Fig. 2.6, there are no edge states associated with the boundary between
the topological insulator (QVH insulator) and the vacuum. On the other hand, in
Fig. 2.8a, there are two gapless edge states corresponding to the upper and lower real
boundaries separating the vacuum from the topological insulator (QSH insulator).

2.5.1 Entanglement Theory

The entanglement theory provides us with other quantities demonstrating the
bulk-edge correspondence. We divide a nanoribbon into A and B subsystems as
shown in Fig. 2.7. We trace out the information of the B subsystem. In so doing, we

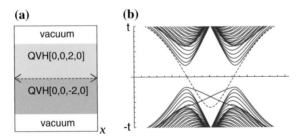

Fig. 2.6 a Silicene nanoribbon placed parallel to the x axis. When we apply the external field
$E_z(y)$ inhomogeneous along the y axis, $E_z(y) > E_{cr}$ for $y > 0$ and $E_z(y) < -E_{cr}$ for $y < 0$. The
region with $|E_z| > |E_{cr}|$ becomes a QVH insulator. An inner edge emerges between the two
different QVH parts. No gapless edge states appear along the real edge above and below the QVH
parts. **b** The band structure of a silicene nanoribbon. A *dotted curve* represents an inner edge state

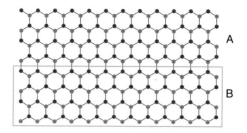

Fig. 2.7 We devide a nanoribbon into the A and B subsystems, where the B subsystem is traced out

may define the entanglement spectrum, the entanglement entropy and the quantum distance, about which we review in the subsequent subsections.

We show the entanglement spectrum and the entanglement entropy of a silicene nanoribbon in Fig. 2.8b, c. In the entanglement spectrum, the bulk bands are squashed to be 0 or 1. On the other hand, the edge states appear between 0 and 1. The edge states connect 0 and 1 for the QSH phase, while they do not connect 0 and 1 for the QVH phase. Recall that the QSH phase is indexed by the genuine Chern number but the QVH phase is not. We can determine the topological nature of the system by checking whether there are edge states connecting 0 and 1.

On the other hand, the entanglement entropy is not so good to determine the topological properties since the behavior is similar between the QSH and QVH phases although there are some peak structures at the topological phase transition point as shown in Fig. 2.8c.

Finally, the quantum distance shows similar behavior to the edge states of the entanglement spectrum as shown in Fig. 2.8d. The difference is that there are no bands at 0 and 1 corresponding to the bulk bands. We find that the winding number of the quantum distance gives us a topological number

$$N = \frac{1}{M} \int_0^{2\pi} \frac{d}{dk} d_{\Pi_A}^2(k) dk \tag{2.97}$$

which is 1 for the QSH phase and 0 for the QVH phase.

The entanglement spectrum and the quantum distance have clearer information about the bulk-edge correspondence than the edge spectrum of a nanoribbon, since the information of the bulk bands are deleted.

2.5.2 Definition of Entanglement Entropy

We divide the total system into subsystems A and B. The entanglement entropy S_A is defined by the von Neumann entropy,

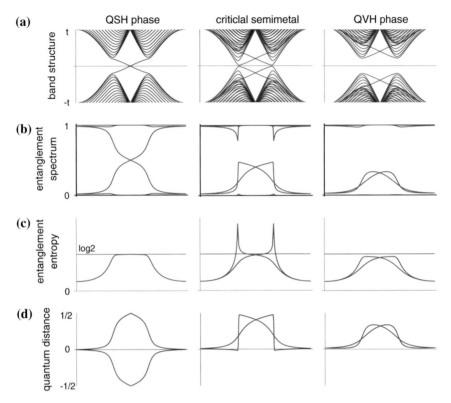

Fig. 2.8 a Band structures of a zigzag nanoribbon. The gapless edge states appear in the QSH phase but not in the QVH phase. **b** Entanglment spectrum of the correlation matrix ζ_ℓ. Two curves appear connecting 1 and 0 in the QSH phase but not in the QVH phase. **c** Entanglement entropy S_A. There are no clear difference between the QSH and QVH phases. **d** Quantum distance $\left(d_{\Pi_A}^2 - \frac{N_F}{2}\right)/M$. A clear similarity is observed for the edge states between the quantum distance and the entanglement spectrum, but the bulk spectrum is removed in the quantum distance. The *red* (*blue*) curves indicate the *up* (*down*) spin

$$S_A = -\mathrm{Tr}\,\rho_A \log \rho_A, \tag{2.98}$$

where ρ_A is the reduced density matrix defined by

$$\rho_A = \mathrm{Tr}_B |\Psi\rangle\langle\Psi|, \tag{2.99}$$

which is obtained by tracing over the states in subsystem B of the total density matrix,

$$\rho = |\Psi\rangle\langle\Psi|. \tag{2.100}$$

Namely, we trace out the information of the subsystem for the entanglement entropy and entanglement spectrum.

2.5.3   Entanglement Entropy and Entanglement Spectrum for Free-Fermion System

We first review the entanglement entropy for the free-fermion system [55–58]. We investigate a system of free fermions hopping between lattice sites with Hamiltonian

$$H = -t \sum_{\langle i,j \rangle s} c_{is}^{\dagger} c_{js}. \tag{2.101}$$

The canonical transformation

$$c_i = \sum_{j=1}^{N_{\mathrm{tot}}} U_{ij} \chi_j, \tag{2.102}$$

diagonalizes the Hamiltonian

$$H = \sum_{i=1}^{N_{\mathrm{tot}}} \varepsilon_i \chi_i^{\dagger} \chi_i. \tag{2.103}$$

The ground state is given by the Slater determinant describing the filled Fermi sea,

$$|\Psi_{\mathrm{FS}}\rangle = \prod_{i=1}^{N_{\mathrm{F}}} \chi_i^{\dagger} |0\rangle, \quad \chi_i|0\rangle = 0, \tag{2.104}$$

with $\varepsilon_I < \varepsilon_{\mathrm{F}}$. We consider the one-particle correlation function,

$$C_{ij} = \langle c_i^{\dagger} c_j \rangle = \langle \Psi_{\mathrm{FS}} | c_i^{\dagger} c_j | \Psi_{\mathrm{FS}} \rangle, \quad i,j = 1, 2, \ldots N_{\mathrm{tot}}. \tag{2.105}$$

By inserting (2.102) and (2.104) into (2.105), we find

$$C_{ij} = \sum_{i=1}^{N_{\mathrm{F}}} U_{ii'}^* U_{ji'}. \tag{2.106}$$

Then C_{ij} form a Hermitian matrix C and satisfies $C^2 = C$. All the higher correlation functions can be expressed by C. For example, a four-point correlation function can be expressed in terms of two-point correlation functions,

$$\langle c_i^\dagger c_j^\dagger c_k c_\ell \rangle = \langle c_i^\dagger c_\ell \rangle \langle c_j^\dagger c_k \rangle - \langle c_i^\dagger c_k \rangle \langle c_j^\dagger c_\ell \rangle. \tag{2.107}$$

Now consider a subsystem, for which we use the notation n, m. By definition, the reduced density matrix ρ_A reproduces all expectation values in the subsystem. Therefore, the one-particle function for the subsystem A is given by

$$(C_A)_{nm} = \mathrm{Tr}[\rho_A c_n^\dagger c_m]. \tag{2.108}$$

According to the Wick theorem, this property holds if ρ is the exponential of a free-fermion operator,

$$\rho_A = \frac{1}{Z} e^{-H_E}, \tag{2.109}$$

where

$$Z = \mathrm{Tr}\, e^{-H_E} \tag{2.110}$$

is the normalization constant assuring $\mathrm{Tr}\, \rho_A = 1$, and

$$H_E = \sum_{n,m} H_{n,m}^E c_n^\dagger c_m \tag{2.111}$$

is the entanglement Hamiltonian. It is shown that the eigenvalues of the correlation matrix

$$C_A |\phi_\ell\rangle = \zeta_\ell |\phi_\ell\rangle \tag{2.112}$$

and those of the entanglement Hamiltonian

$$H_E |\phi_\ell\rangle = \xi_\ell |\phi_\ell\rangle \tag{2.113}$$

are related by

$$\zeta_\ell = \frac{1}{e^{\xi_\ell} + 1}. \tag{2.114}$$

Hence we find $0 \leq \zeta_\ell \leq 1$. The entanglement entropy can be calculated by using the eigenvalues of the correlation matrix,

$$S_A = -\sum_\ell (1 - \zeta_\ell) \log(1 - \zeta_\ell) + \zeta_\ell \log \zeta_\ell, \tag{2.115}$$

which is useful for practical applications. We plot it for the case of silicene in Fig. 2.8c.

2.5.4 Quantum Distance

Next we show the entanglement spectrum is related to the quantum distance. The quantum distance between the states at momenta k_i and k_j (the Fubini-Study distance) is defined by [59, 60]

$$d(k_i, k_j) = \sqrt{N_F - \sum_{a,b=1}^{N_F} |\langle u_a(k_i)|u_b(k_j)\rangle|^2} \qquad (2.116)$$

for the lowest N_F bands occupied. We introduce a projector onto the occupied bands

$$\Pi(k) = \sum_{i=1}^{N_F} |u_i(k)\rangle\langle u_i(k)|. \qquad (2.117)$$

By using it, the quantum distance is rewritten as [61]

$$d^2(k_i, k_j) = N_F - \mathrm{Tr}\,\Pi(k_i)\Pi(k_j). \qquad (2.118)$$

We also introduce a projector onto the subsystem Π_A, which satisfies

$$C_A = \Pi_A C \Pi_A. \qquad (2.119)$$

We generalize the quantum distance to that of the projectors [61],

$$d^2(\Pi_i, \Pi_j) \equiv \frac{1}{2}\mathrm{Tr}[(\Pi_i - \Pi_j)^2] = \frac{1}{2}[\mathrm{Tr}\,\Pi_i + \mathrm{Tr}\,\Pi_j] - \mathrm{Tr}[\Pi_i\Pi_j]. \qquad (2.120)$$

When $\mathrm{Tr}\,\Pi_i = \mathrm{Tr}\,\Pi_j = N_F$, it is reduced to (2.116). Now we consider the quantum distance between the occupied bands Π and the subsystem Π_A

$$d_{\Pi_A}^2 \equiv d^2(\Pi, \Pi_A) = \frac{N_F + M}{2} - \mathrm{Tr}[\Pi_A \Pi], \qquad (2.121)$$

where we have used $\mathrm{Tr}\,\Pi(k) = N_F$ and assumed $\mathrm{Tr}\,\Pi_A = M$. By using the fact that $\Pi_A^2 = \Pi_A$ and the trace is invariant under cyclic permutation, we can rewrite it as

$$d_{\Pi_A}^2 = \frac{N_F + M}{2} - \mathrm{Tr}[\Pi_A \Pi(k)\Pi_A]. \qquad (2.122)$$

By using (2.119), we find

$$d_{\Pi_A}^2 = \frac{N_F + M}{2} - \mathrm{Tr}\,C_A = \frac{N_F + M}{2} - \sum_{\ell}^{M} \zeta_\ell, \qquad (2.123)$$

which relates the quantum distance with the entanglement spectrum of the corre-lation function. Namely, the quantum distance is given by the sum of the entan-glement spectrum. The quantum distance is bounded

$$\frac{N_F - M}{2} < d_{II_A}^2 < \frac{N_F + M}{2},$$ (2.124)

since $0 < \zeta_\ell < 1$. Then we find $-1/2 < \left(d_{II_A}^2 - \frac{N_F}{2} \right)/M < 1/2$. We plot it for the case of silicene in Fig. 2.8d.

2.6 Conclusion

We have explored the topological physics of massive Dirac fermions which are realized in silicene, germanene and stanene. The topological field theory of the honeycomb system is constructed by using the spin-valley dependent Chern-Simons action. We also show that the entanglement spectrum and the quantum distance provide information on the bulk-edge correspondence. These monolayer topological insulators will be useful for future topological electronics including topological field effect transistors [62].

References

1. M.I. Katsnelson, *Graphene: Carbon in Two Dimensions* (Cambridge University Press, Cambridge, 2012)
2. R. Saito, G. Dresselhaus, and M.S. Dresselhaus, in *Physical Properties of Carbon Nanotubes* (Imperial College Press, London, 1998)
3. K. Takeda, K. Shiraishi, Phys. Rev. B **50**, 075131 (1994)
4. G.G. Guzmán-Verri, L.C.L.Y. Voon, Phys. Rev. B **76**, 075131 (2007)
5. S. Cahangirov, M. Topsakal, E. Akturk, H. Sahin, S. Ciraci, Phys. Rev. Lett. **102**, 236804 (2009)
6. S. Lebegue, O. Eriksson, Phys. Rev. B **79**, 115409 (2009)
7. C.-C. Liu, W. Feng, Y. Yao, Phys. Rev. Lett. **107**, 076802 (2011)
8. M. Ezawa, New J. Phys. **14**, 033003 (2012)
9. N.D. Drummond, V. Zolyomi, V.I. Fal'ko, Phys. Rev. B **85**, 075423 (2012)
10. Z. Ni, Q. Liu, K. Tang, J. Zheng, J. Zhou, R. Qin, Z. Gao, D. Yu, J. Lu, Nano Lett. **12**, 113 (2012)
11. M. Ezawa, Phys. Rev. Lett. **109**, 055502 (2012)
12. M. Ezawa, Phys. Rev. Lett. **110**, 026603 (2013)
13. M. Ezawa, Phys. Rev. B **87**, 155415 (2013)
14. P. Vogt, P. De Padova, C. Quaresima, J. Avila, E. Frantzeskakis, M.C. Asensio, A. Resta, B. Ealet, G. Le Lay, Phys. Rev. Lett. **108**, 155501 (2012)
15. C.-L. Lin, R. Arafune, K. Kawahara, N. Tsukahara, E. Minamitani, Y. Kim, N. Takagi, M. Kawai, Appl. Phys. Express **5**, 045802 (2012)

16. B. Feng, Z. Ding, S. Meng, Y. Yao, X. He, P. Cheng, L. Chen, K. Wu, Nano Lett. **12**, 3507 (2012)
17. A. Fleurence, R. Friedlein, T. Ozaki, H. Kawai, Y. Wang, Y. Yamada-Takamura. Phys. Rev. Lett. **108**, 245501
18. L. Meng, Y. Wang, L. Zhang, S. Du, R. Wu, L. Li, Y. Zhang, G. Li, H. Zhou, W.A. Hofer, H.-J. Gao, Nano Lett. **13**, 685 (2013)
19. D. Chiappe, E. Scalise, E. Cinquanta, C. Grazianetti, B. van den Broek, M. Fanciulli, M. Houssa, A. Molle, Adv. Mat. **26**, 2096 (2014)
20. M.E. Davila, L. Xian, S. Cahangirov, A. Rubio, G. Le Lay, New J. Phys. **16**, 095002 (2014)
21. L. Li, S.-Z. Lu, J. Pan, Z. Qin, Y.-Q. Wang, Y. Wang, G.-Y. Cao, S. Du, H.-J. Gao, Adv. Mat. **26**, 4820 (2014)
22. M. Derivaz, D. DENTEL, R. Stephan, M.-C. Hxanf, A. Mehdaoui, P. Sonnet, C. Pirri, Nano Lett. **15**, 2510 (2015)
23. P. Bampoulis, L. Zhang, A. Safaei, R. van Gastel, B. Poelsema, H.J.W. Zandvliet, J. Phys. Condens. Matter **26**, 442001 (2014)
24. F. Zhu, W.-J. Chen, Y. Xu, C.-L. Gao, D.-D Guan, C. Liu, D. Qian, S.-C. Zhang, J.-F. Jia, Nat. Mat. **14**, 1020 (2015)
25. L. Tao et al., Nat. Nanotech. **10**, 227 (2015)
26. D.J. Thouless, M. Kohmoto, M.P. Nightingale, M. den Nijs, Phys. Rev. Lett. **49**, 405 (1982)
27. X.-G. Wen, Int. J. Mod. Phys. B6 1711 (1992). Adv. Phys. **44**, 405 (1995)
28. X.-G. Wen, Y.S. Wu, Y. Hatsugai, Nucl. Phys. B **422**, 476 (1994)
29. H.B. Nielsen, M. Ninomiya, Nucl. Phys. B **185**, 20 (1981)
30. C.-C. Liu, H. Jiang, Y. Yao, Phys. Rev. B **84**, 195430 (2011)
31. C.L. Kane, E.J. Mele, Phys. Rev. Lett. **95**, 226801 (2005); ibid **95**, 146802 (2005)
32. Y. Xu, B. Yan, H.-J. Zhang, J. Wang, G. Xu, P. Tang, W. Duan, S.-C. Zhang, Phys. Rev. Lett. **111**, 136804 (2013)
33. X. Li, T. Cao, Q. Niu, J. Shi, J. Feng, PNAS **110**, 3738 (2013)
34. Q.-F. Liang, L.-H. Wu, X. Hu, New J. Phys. **15**, 063031 (2013)
35. G.V. Dunne, Aspects Of Chern-Simons theory "in "Aspects topologiques de la physique en basse dimension. Topological aspects of low dimensional systems. Les Houches - Ecole d \primeEte de Physique Theorique **69**, 177–263 (1999)
36. Y. Zhong, Y.-F. Wang, H.-T. Lu, H.-G. Luo, Phys. Rev. B **88**, 235111 (2013)
37. K. Ishikawa, T. Matsuyama, Nucl. Phys. B **280**, 523 (1987)
38. M.F.L. Golterman, K. Jansen, D.B. Kaplan, Phys. Lett. B **391**, 219 (1993)
39. S. Coleman, B. Hill, Phys. Lett. B **159**, 184 (1985)
40. G.E. Volovik, *The Universe in a Helium Droplet* (Oxford University Press, New York, 2003)
41. X.-L. Qi, T.L. Hughes, S.-C. Zhang, Phys. Rev. B **78**, 195424 (2008)
42. Y.-M. Lu, A. Vishwanath, Phys. Rev. B **86**, 125119 (2012)
43. T. Grover, T. Senthil, Phys. Rev. Lett. **100**, 156804 (2008)
44. E. Prodan, Phys. Rev. B **80**, 125327 (2009)
45. E. Prodan, New J. Phys. **12**, 065003 (2010)
46. D.N. Sheng, Z.Y. Weng, L. Sheng, F.D.M. Haldane, Phys. Rev. Lett. **97**, 036808 (2006)
47. Y. Yang, Z. Xu, L. Sheng, B. Wang, D.Y. Xing, D.N. Sheng, Phys. Rev. Lett. **107**, 066602 (2011)
48. L. Sheng, D.N. Sheng, C.S. Ting, F.D.M. Haldane, Phys. Rev. Lett. **95**, 136602 (2005)
49. F. Zhang, J. Jung, G.A. Fiete, Q. Niu, A.H. MacDonald, Phys. Rev. Lett. **106**, 156801 (2011)
50. F. Zhang, A.H. MacDonald, E.J. Mele, Proc. Natl. Acad. Sci. U.S.A. **110**, 10546 (2013)
51. J. Li, A.F. Morpurgo, M. Bütiker, I. Martin, Phys. Rev. B **82**, 245404 (2010)
52. M. Ezawa, Phys. Rev. B **88**, 161406 (R) (2013)
53. W.-K. Tse, Z. Qiao, Y. Yao, A.H. MacDonald, Q. Niu, Phys. Rev. B **83**, 155447 (2011)
54. Z. Qiao, H. Jiang, X. Li, Y. Yao, Q. Niu, Phys. Rev. B **85**, 115439 (2012)
55. I. Peschel, J. Phys. A Math. Gen. **36** L205 (2003)
56. I. Peschel, Braz. J. Phys. **42**, 267 (2012)
57. P.-Y. Chang, C. Mudry and S. Ryu, J. Stat. Mech. **09014** (2014)

58. A. Alexandradinata, T.L. Hughes, B.A. Bernevig, Phys. Rev. B **84**, 195103 (2011)
59. J. Provost, G. Vallee, Commun. Math. Phys. **76**, 289 (1980)
60. A.K. Pati, Phys. Lett. A **159**, 105 (1991)
61. M. Legner, T. Neupert, Phys. Rev. B **88**, 115114 (2013)
62. M. Ezawa, Appl. Phys. Lett. **102**, 172103 (2013)

Chapter 3
Free-Standing Multilayer Silicene: Molecular-Dynamics and Density Functional Theory Studies

Tetsuya Morishita and Michelle J.S. Spencer

Abstract In this chapter, we will focus on theoretical attempts to characterize multilayer silicene. Particular focus will be given to the structural and electronic properties of multilayer silicene in vacuum, which are quite different to those of single-layer silicene, especially at finite temperature. We will begin with a molecular dynamics study that predicts the possible formation process of bilayer silicene (BLS) in a slit nanopore upon quenching. The electronic properties of BLS will be discussed within the framework of density-functional theory. The structural and electronic properties of multilayer silicene having more than two layers will then be presented. It has been found that surface reconstruction plays an important role in characterizing multilayer silicene at finite temperature, which will also be detailed in this chapter.

3.1 Introduction

Since silicene consists of an atomic layer of silicon in the honeycomb lattice, theoretical calculations that can directly deal with atomic scale materials are suitable for examining its properties. Density functional theory (DFT) calculations have been carried out to determine the most stable atomic geometry and the electronic properties of silicene. Cahangirov et al. [1] have estimated the structural parameters and electronic structures of free-standing single-layer silicene by DFT calculations, following

T. Morishita (✉)
Research Center for Computational Design of Advanced Functional Materials,
National Institute of Advanced Industrial Science and Technology (AIST),
1-1-1 Umezono, Tsukuba, Ibaraki 305-8568, Japan
e-mail: t-morishita@aist.go.jp

M.J.S. Spencer (✉)
School of Science, RMIT University, GPO Box 2476, Melbourne, VIC 3001, Australia
e-mail: michelle.spencer@rmit.edu.au

© Springer International Publishing Switzerland 2016
M.J.S. Spencer and T. Morishita (eds.), *Silicene*, Springer Series
in Materials Science 235, DOI 10.1007/978-3-319-28344-9_3

63

the previous theoretical work by Takeda and Shiraishi [2], Guzmàn-Verri and Voon [3], and Lebegue and Eriksson [4]. Molecular dynamics (MD) calculations, especially in combination with DFT, have also been widely performed to examine the structural stability of silicene at finite temperature and to calculate its dynamical properties such as its vibrational density of state [5–7]. MD calculations incorporating DFT, and often called first-principles MD (FPMD), however, can be performed for up to a few hundred ps only. Meanwhile, MD calculations using model interatomic potentials allow us to trace the formation processes of nanoscale structures that need more than nanoseconds to complete [8–11]. Thus, with well-developed interatomic potentials, formation of silicene-related materials can be simulated and even predicted on an atomistic level using MD simulations.

Recently, the effect of confinement in nanopores upon freezing has been extensively examined in experiments and in computer simulations [12]. It has been suggested that formation of exotic nanoscale structures could be realised via freezing processes inside a nanopore (such as zeolite, porous silica and carbon nanotube) that acts as a template for nanocrystals, nanowires and nanosheets. For example, novel one- and two-dimensional nanoscale ice structures have been found to form inside nanopores upon freezing [8, 9, 13], which is likely to be effective also for exploring two-dimensional (2D) Si nanostructures [10, 11, 14].

In this chapter, we review recent theoretical studies of silicene having more than one honeycomb layer using DFT and MD calculations. While multilayer silicene (MLS) has been experimentally synthesized on metal substrates [15, 16], pristine MLS or free-standing MLS has not been well characterized. We start with an MD result that shows a possible formation process of MLS inside a slit pore [10, 11]. DFT results on the electronic properties of the MLS thus obtained then follow [17–19]. Possible surface reconstructions in MLS are also discussed [5, 20], and a short summary is given at the end of this chapter.

3.2 Structural and Electronic Properties of Bilayer Silicene

3.2.1 Formation of Bilayer Silicene from Molten Si

While single-layer silicene (SLS) has been synthesized and is well characterized on metal substrates [21–24], experimental attempts to synthesize and characterize MLS have only commenced recently [15, 16]. In the meantime, molecular-dynamics simulation studies have demonstrated that MLS structures can be formed by exploiting slit pores [10, 11, 14], wherein the spontaneous formation of two-dimensional crystalline structures can be induced by quenching molten Si. In this section, the formation process of bilayer silicene (BLS), which has an AA stacking of the honeycomb layer, in molecular-dynamics calculations are presented.

In 2008 Morishita et al. [10] performed MD simulations of molten Si sand-wiched by two parallel plane walls that act as a chemically inert confinement for the Si atoms. The interaction between Si atoms was described by the Tersoff model [25], and the interaction between the Si atom and the wall was described by the 9-3 Lennard-Jones potential [12, 26].

Molten Si consisting of 512 atoms was prepared in the slit nanopore of width $h_z = 9.3$ Å with free boundary conditions in the directions parallel to the walls (i.e. zero lateral pressure). This confined molten Si was equilibrated for ~ 1 ns at 2400 K, and was then quenched to 0 K at a rate of 2×10^{10} K/s (i.e. 10 K per 500 ps) in the MD simulation. A layered structure was finally formed upon freezing in the slit nanopore.

Figure 3.1a shows the temperature dependence of the potential energy, U, in the quenching process for the confined molten Si. A sharp drop of U is clearly seen at ~ 1700 K, indicating a first-order transition from a liquid to a solid state. The temperature dependence of the atomic mobility was also examined as a function of the temperature. The root mean square displacement (RMSD) parallel to the wall over a 10 ps time period was calculated for each Si atom, which was then averaged over all the atoms for 50 time origins at each temperature. The temperature dependence of the 2D RMSD is given in the inset of Fig. 3.1a. A sharp drop is again seen at ~ 1700 K, at which the RMSD decreases down to less than ~ 1 Å, indicating solidification at ~ 1700 K.

Figure 3.1b shows the 2D radial distribution function $g(r_{xy})$ for the confined Si. The first peak (except for the peak at the origin, which is attributed to the contribution from other atoms above or underneath the central atom) is seen at ~ 2.35 Å at 2400 K, but the second and the third peaks at 3.5–6 Å are much broader than the first peak. In contrast, $g(r_{xy})$ at temperatures $< \sim 400$ K exhibits clear peaks in the whole r_{xy} range, indicating that a well-ordered 2D crystalline structure is formed in the slit pore.

The number density profile perpendicular to the wall also reflects the liquid-crystal transition (Fig. 3.1c). We see a broad distribution with two peaks visible (at ~ 3 and ~ 6 Å) at 2400 K. These two peaks grow in intensity and split into two sub-peaks upon quenching to 400 K. The broad distribution between 3.8 and 5.5 Å disappears at 400 K, reflecting the formation of a bilayer structure.

The atomic configuration obtained (after quenching to 0 K) for the bilayer structure is displayed in Fig. 3.1d1 (top view) and d2 (side view). It is clearly seen that two honeycomb layers are formed parallel to the walls, which results in a BLS structure with an AA stacking configuration. The MD result by Morishita et al. [10] thus demonstrated two-dimensional nanostructures of Si can be synthesized by exploiting slit pores. In fact, a recent experimental study [27] shows that intercalated Si atoms form two- or three-layer structures in a layered inorganic compound crystal under certain conditions, strongly supporting the theoretical predictions of the formation of the BLS (or MLS) structure inside a slit pore.

Bai et al. [11] also performed MD simulations of quenching molten Si confined in a slit pore. They employed the Stillinger-Weber model [28], as well as the Tersoff model, to describe the interaction between the Si atoms. Two types of slit

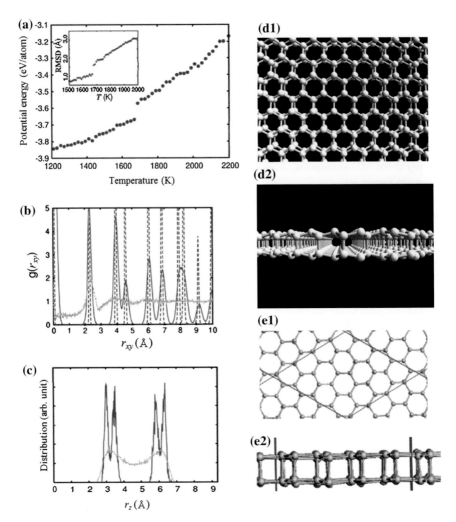

Fig. 3.1 **a** Potential energy of the confined Si as a function of temperature (*Inset* 2D RMSD of Si atoms as a function of temperature). **b** 2D $g(r_{xy})$ for the confined Si at 2400 K (*dotted line*), 400 K (*solid lines*), and 20 K (*dashed line*). **c** Number density profile of the confined Si perpendicular to the slit walls at 2400 K (*dotted line*) and 400 K (*solid lines*). **d1** and **d2** *Top* and *side views*, respectively, of the BLS structure obtained in the simulation by Morishita et al. [10]. **e1** and **e2** *Top* and *side views*, respectively, of the BLS structure obtained in the simulation by Bai et al. [11]. Panels **a–d** are reprinted with permission from Morishita et al. [10]. Copyright (2008) by the American Physical Society. Panels **e** are reproduced with kind permission from Springer Science +Business Media: Bai et al. [11], Fig. 3

walls were employed to examine the effect of the details of the wall surface; one was a plane wall, as in the MD simulations by Morishita et al. [10], while the other was a structured wall that consisted of a single-layer graphene sheet. The Tersoff model was used to describe both the Si–Si and Si–C interactions in the latter case.

They, however, found the detailed structure of the surface of the wall had little effect on the resultant structure of BLS formed in the slit pore during the quenching process. A slit width of 7–8 Å was employed and the lateral pressure was maintained at 50 MPa throughout their MD simulations [11].

Figure 3.1e1 and e2 show the top and side views, respectively, of the BLS obtained in the simulation by Bai et al. [11]. Interestingly, while two hexagonal layers were formed in the same way as in the simulation by Morishita et al. [10], the buckling of the honeycomb layer was totally reduced, becoming almost identical to the single-layer of graphene. The resultant BLS still possessed an AA stacking configuration, so we called this structure "plane AA" (p-AA). The reason that a slightly different BLS structure was formed in their simulation may be attributed to the narrower slit width and the imposed lateral pressure of 50 MPa. We will see later that p-AA actually has a lower energy than the BLS with AA or AB stacking without surface reconstruction.

The slit width and lateral pressure are found to have a significant impact on the resultant structures formed in a pore. Morishita et al. [10] found that one- or three-layer silicene can also be formed in narrower or wider pores, respectively. Jonston et al. [14] simulated formation processes of a variety of two-dimensional Si structures, including quasicrystals, under various lateral pressures. Their simulations indicated that silicene with the buckled honeycomb layer is not a unique two-dimensional structure of Si, which will again be discussed in the light of surface reconstruction in Sect. 3.2.3.

3.2.2 Bilayer Silicene with Ideal Surfaces

We have seen that two types of BLS, having an AA stacking, are found to be formed by quenching molten Si confined in a slit pore [10, 11]. The structural and electronic properties of these BLS structures at 0 K are discussed in this section. In addition to an AA and p-AA stacking, we also consider a configuration with an AB stacking, which is consistent with the cubic diamond structure of bulk crystalline Si (c-Si). These three types of the BLS structure are displayed in Fig. 3.2 (referred to as AA, AB and p-AA). DFT-based electronic state calculations for these three structures have been performed by several theoretical groups [17–19] to estimate the in-plane and inter-plane Si–Si bond lengths in the geometrically optimized structures for them. All the calculations based on DFT presented in this chapter employ the PBE functional [29] for the exchange-correlation energy. Some structural properties were also estimated using a van der Waals type functional (optB86b-vdW [30]), showing good agreement with the PBE results [19].

The in-plane Si–Si bond in both AA and AB BLS were estimated to be 2.32 Å [17–19]. The bond length in the bulk c-Si is 2.352 Å, thus the in-plane Si–Si bond in AA and AB are slightly shorter than that in the bulk c-Si. On the other hand, the inter-plane bond is longer than the Si–Si bond in c-Si; 2.46–2.47 Å for AA and 2.51–2.54 Å for AB [17–19]. Since the outermost Si atoms are three-fold

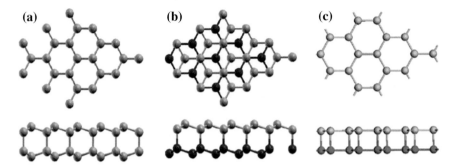

Fig. 3.2 *Top* and *side views* of **a** AA, **b** AB, and **c** p-AA BLS (see text)

coordinated in BLS, it is not surprising that the in-plane and inter-plane bond
lengths are not the same. In other words, there exists dangling bonds on these
under-coordinated (surface) Si atoms, which results in degradation of the sp^3 hybridization and thus of the tetrahedral bonding. Note, however, that two silicene
layers are still covalently bonded judging from the inter-plane bond length and the
charge density distribution [18], in contrast to stacked two-layer graphene.

p-AA BLS was calculated to have a longer in-plane bond (2.39 Å) and a shorter
inter-plane bond (2.41 Å) than for the AA and AB BLS structures [19]. This
structure, however, had no buckling. Thus, the nature of the sp^3 hybridization, i.e.
covalency, would be further reduced, resulting in the longer in-plane bond length.

Kamal et al. [18] and Fu et al. [19] estimated the cohesive energies of these BLS
structures, which were calculated using DFT, showing the relative stability of these
structures from an energetic viewpoint. p-AA BLS was found to have the highest
cohesive energy among the three structures, while AB and AA BLS had lower
cohesive energies by 0.06 eV/Si atom and 0.1 eV/atom, respectively, than that of
p-AA. Thus, p-AA BLS is energetically more stable than AA and AB BLS (note in
passing that the cohesive energy of p-AA is lower than that of the bulk c-Si
by ∼0.45 eV/Si atom). It should, however, be remarked that p-AA is actually not
the most stable structure among all allotropes of BLS (as will be discussed in
Sect. 3.2.3).

Kamal et al. [18] obtained the electronic band dispersion relations for AA and
AB BLS (presented in Fig. 3.3). We see that linear dispersions, which are characteristic of the band dispersion for SLS remain in AA BLS, while parabolic
dispersion is instead observed in AB BLS. Notably, this trend is also seen in AA
and AB graphene [31]. The linear dispersion in AA BLS is observed not at the high
symmetry K point (Fig. 3.3a), in contrast to that in SLS, but at the points along the
Γ-K and Γ-M lines. And the intersection points do not lie exactly on the Fermi
level. Kamal et al. [18] also estimated the band dispersion relations for the AA and
AB stacking arrangements with artificially elongated inter-plane distances. It was
found from their calculations, that the substantial change in the band dispersion
from that of SLS is due to the formation of covalent inter-plane bonds of ∼2.5 Å

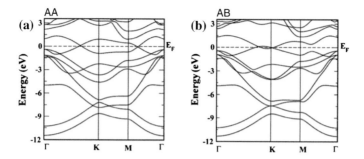

Fig. 3.3 Band dispersion relations for **a** AA BLS and **b** AB BLS. Adapted and reproduced from [18]

long. With an inter-plane distance of 3–4 Å, the dispersion relations for the AA and AB silicene became similar to those for AA and AB graphene, respectively, where a weak van der Waals interaction comes into play, confirming that there exists a strong covalent interaction between the two silicene layers in AA and AB BLS, with an original inter-plane distance of ∼2.5 Å. For an interlayer separation distance longer than 7 Å, the dispersion becomes identical to that for SLS, where any inter-layer interactions are negligible.

Morishita et al. [17] reported that the bands crossing the Fermi level basically come from the dangling bond on the undercoordinated (surface) Si atoms. They found that these bands can be removed by saturating the dangling bonds, e.g. by hydrogenation or substitutional doping.

Parabolic dispersion was also found in the band dispersion relation for p-AA BLS [11, 19] (as seen in Fig. 3.4 that compares the band dispersion relations for p-AA and AB BLS). Interestingly, no band crosses the Fermi level, so p-AA BLS can be seen as a zero-gap semiconductor. In fact, a DFT calculation with the HSE06 functional predicts that there exists a tiny gap of 0.11 eV at the Fermi level in p-AA BLS [19].

3.2.3 Bilayer Silicene with Surface Reconstruction

Although we saw in the previous section that p-AA BLS was energetically more stable than the other two BLS structures, we have not discussed their stability at finite temperature. In fact, a more stable BLS structure was discovered by Morishita et al. [5] in their first-principles MD (FPMD) simulations at 300 K. In this section, the structural and electronic properties of this new BLS will be detailed.

It is well known that the cleaved Si(111) surface reconstructs at finite temperature in vacuum, creating a 7 × 7 or 2 × 1 surface structure [32]. Such surface reconstruction is triggered by the driving force to reduce the number of dangling bonds on the surface. Since the BLS structures discussed so far contain the Si(111)-1 × 1 surface, it is not surprising that surface reconstruction would take place in

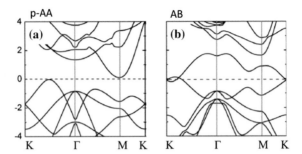

Fig. 3.4 Band dispersion relations for **a** p-AA BLS and **b** AB BLS. Adapted and reproduced from [19]

BLS. The questions to be considered here are what surface structures can be formed in comparison with the bulk Si(111) surface and whether a reconstructed BLS structure is more stable than the non-reconstructed honeycomb lattice seen in the AA, AB and p-AA BLS structures.

Morishita et al. [5] carried out a series of FPMD calculations to address these questions. Their FPMD runs, most lasting for ∼15 ps, were started with an AA or AB stacking consisting of 64 Si atoms at 300 K. They found that either the AA or the AB structure is transformed to a new BLS structure exhibiting surface reconstruction in all FPMD runs. Some metastable reconstructed structures, as well as the most stable structure, were formed depending on the initial conditions, however, these metastable structures could be converted to the lowest energy structure by heating up to ∼1200 K. We thus focus only on the reconstructed BLS (re-BLS) with the lowest energy among those obtained in their calculations.

Figure 3.5a and b show the top and side views, respectively, of the re-BLS structure, which was formed in the 300 K MD run followed by subsequent relaxation at 0 K. The periodicity in the x- and y-directions was found to be doubled upon reconstruction. Hence, the resultant surface arrangement can be labeled as Si (111)-2 × 2, which has importantly not been observed on bulk surfaces. We see that there are two types of 3-fold coordinated (undercoordinated) atoms on both sides of re-BLS; "atop" and "sp2-like" atoms, as labeled in Fig. 3.5b. In fact, all of the outermost atoms in the AA or AB configuration are 3-fold coordinated before reconstruction, but the number of these undercoordinated atoms is halved after reconstruction. It is worth noting that an "sp2-like" configuration, formed by a central atom ("sp2-like" atom) and its three nearest neighbors, is formed on re-BLS in the same way as on the bulk Si(111)-2 × 1 surface.

It was shown that re-BLS has a lower energy, by ∼0.06 eV/atom, than p-AA BLS, thus indicating that re-BLS is the most stable BLS form in vacuum. In fact, p-AA BLS is still stable at ∼600 K, thus, the transition barrier from p-AA to re-BLS is relatively high (600 K or higher) compared to that from AA or AB BLS (note that AA or AB are found to be immediately transformed to re-BLS at 300 K [5]). It appeared that AA or AB BLS could exist only with supporting walls to

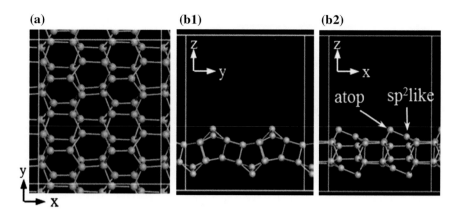

Fig. 3.5 a Top and **b1, b2** *side views* of the re-BLS structure. Reprinted from Morishita et al. [5], Copyright (2011) with permission from Elsevier

sandwich them. Lian and Ni [33] found in their first-principles calculations that re-BLS could be transformed to p-AA BLS by homogenous tensile strain. Their calculations showed that p-AA BLS became more stable than re-BLS under tensile condition (~ 7 %), while it was only metastable without tension.

The bond-length and bond-angle distributions for re-BLS are shown in Fig. 3.6a and b, respectively. In contrast to c-Si, both properties have a wide distribution ranging from 2.30 to 2.45 Å and 80–140°. The bonds linking 3-fold coordinated atoms are relatively short (2.30 and 2.33 Å), especially those in the "sp2-like" triangle configuration, which are comparable to the bond lengths in the "sp2-like" configuration on the Si(111)-2 × 1 surface (2.31 and 2.34 Å). Moreover, most of the bond-angles associated with 3-fold coordinated atoms are far from the tetrahedral angle of 109.5° (red circles in Fig. 3.6b). In particular, the tetrahedral angles associated with "atop" atoms (Fig. 3.5b) are smaller than 91°. Although "atop" atoms were also formed on the Si(111)-2 × 1 surface, the corresponding tetrahedral angles were 103–112°, being much closer to the regular tetrahedral angle of 109.5°. This is in stark contrast to what we found in re-BLS, indicating that the tetrahedral order in BLS is highly distorted upon reconstruction.

Figure 3.7 compares the band dispersion relation for re-BLS and the Si(111)-2 × 1 bulk surface. It was shown that re-BLS has an indirect band gap (~ 0.33 eV) in contrast to the AA, AB or p-AA BLS structures. Therefore reconstruction induces a metal-nonmetal transition in BLS. The overall shape of the total electronic density of states (DOS) for re-BLS (Fig. 3.8a) was similar to that for bulk amorphous Si (*a*-Si) [34, 35] rather than for c-Si [36]. This can be understood by the fact that both *a*-Si and re-BLS exhibit wide bond-length and bond-angle distributions (Fig. 3.6). In the partial DOS for 4-fold coordinated atoms (Fig. 3.8b), almost equal contributions from p_x, p_y, and p_z electrons to the valence bands could be seen. Figure 3.8c and d, however, show that the p_z contribution to the partial DOS for 3-fold coordinated atoms was distinct from the p_x and p_y contributions.

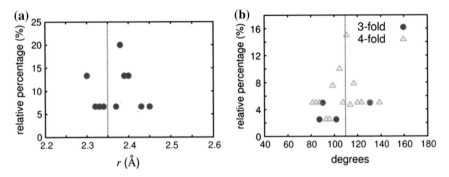

Fig. 3.6 Distributions of **a** bond-lengths and **b** bond-angles in re-BLS [5]. *Vertical dashed lines* indicate the bond-length (2.35 Å) and bond-angle (109.5°) in bulk c-Si. Reprinted from Morishita et al. [5], Copyright (2011) with permission from Elsevier

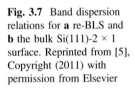

Fig. 3.7 Band dispersion relations for **a** re-BLS and **b** the bulk Si(111)-2 × 1 surface. Reprinted from [5], Copyright (2011) with permission from Elsevier

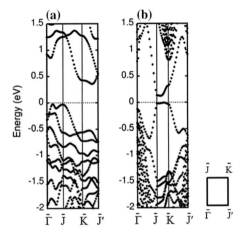

Surprisingly, the bands associated with p_z electrons in the partial DOS for "atop" atoms were highly localized around the top of the valence band, while those in the partial DOS for "sp2-like" atoms were dominant in the conduction band. In fact, this trend has also been found in the DOS for Si(111)-2 × 1, a-Si [34, 35] and MLS (see Sect. 3.3.2). We thus consider that Fig. 3.8c and d reflect a general feature of Si that p_z bands associated with undercoordinated tetrahedral atoms are localized in the valence band, while those associated with "sp2-like" configurations play a significant role in the conduction band.

The discovery of re-BLS indicates that BLS could offer various types of Si surfaces depending on the conditions under which the BLS is formed. In fact, a variety of metastable BLS structures have been reported quite recently [37, 38]. This means a variety of functionalization or chemical reactions could be induced on the BLS surfaces. In the next sections, we will see another surface reconstruction can also take place in MLS.

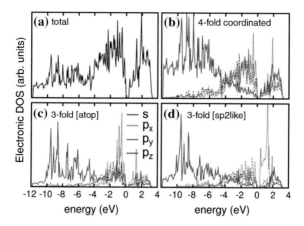

Fig. 3.8 Electronic DOS for re-BLS. **a** Total DOS. **b** Decomposed DOS for 4-fold coordinated Si.
c Decomposed DOS for 3-fold coordinated "atop" Si. **d** Decomposed DOS for 3-fold coordinated
"sp2-like" Si. Reprinted from [5], Copyright (2011) with permission from Elsevier

3.3 Structural and Electronic Properties of Multilayer Silicene

3.3.1 Multilayer Silicene with Ideal Surfaces

Firstly, in this section, MLS with the honeycomb lattice (no reconstruction) will be
discussed.

Similar to BLS, two types of stacking of buckled honeycomb layers were
considered; one was an ABC stacking that is consistent with the stacking in the
cubic diamond structure of c-Si and another was an AAA stacking that is consistent
with the stacking in the hexagonal diamond structure of, e.g., ice. The cohesive
energy, lattice constant and overall thickness of stacked layers were calculated by
Kaml et al. [18] for optimized ABC and AAA configurations, respectively, using
DFT calculations up to 10 silicene layers.

The cohesive energy was shown to increase as the number of layer increased,
both in the ABC and AAA stacked structures. It was found that, up to 10 layers, the
ABC stacking always had a higher cohesive energy than that of AAA, though both
cohesive energies were closer to the bulk value as the layer number increased. For
10 layers, the cohesive energy of the ABC MLS was higher by 15 meV/atom than
that of the AAA MLS. The lattice constant of ABC MLS was found to be always
slightly larger than that of AAA MLS by ~ 0.01 Å with 4 layers or more. In
contrast, the thickness of ABC was always smaller than that of AAA; with the
difference in the thickness increasing with the layer number from 0.1 Å (4 layers) to
0.35 Å (10 layers). The cohesive energy difference between 10-layer ABC stacked
MLS and bulk c-Si was ~ 0.14 eV/atom, which is much lower than that between
two-layer AB BLS and bulk c-Si (~ 0.5 eV/atom).

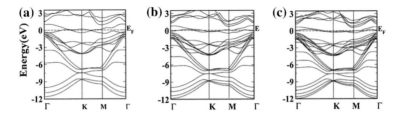

Fig. 3.9 Band dispersion relations for **a** 3-layer, **b** 4-layer, and **c** 5-layer ABC MLS. Adapted and reproduced from [18]

The band dispersion relations were calculated for ABC and AAA MLS with 3–10 layers [18]. Since they are very similar, only the band dispersion relations for 3-, 4- and 5-layer ABC MLS are shown in Fig. 3.9. There always exist two bands that cross the Fermi level, which were attributed to the existence of dangling bonds on the MLS surface. As the number of layers increased, the width of the bands from the dangling bonds decreased. This can be understood by the fact that the two surfaces of each MLS get further apart as the number of layer increases, resulting in less interaction between them.

3.3.2 Multilayer Silicene with Surface Reconstruction

We have already seen that BLS exhibits surface reconstruction at finite temperatures, which leads to the formation of the novel surface structure, Si(111)-2 × 2, which has unique electronic properties [5]. We show, in this subsection, surface reconstruction also takes place in MLS with more than two layers.

The MLS structures (AAA and ABC) discussed in the previous section possess the bulk Si(111) surface as in the honeycomb BLS. Spencer et al. [20] carried out a series of FPMD calculations for 3- to 5-layer MLS, and found that the Si(111)-2 × 1 surface was formed at finite temperature, which is different from the novel Si surface structure labelled as Si(111)-2 × 2 in re-BLS [5], but is the same as the Si (111)-2 × 1 reconstruction formed on the bulk c-Si. They found that the Si(111)-2 × 1 surface reconstruction took place on both sides of the MLS, and also found the reconstructed surfaces could be arranged in two ways; one with the Si(111)-2 × 1 surface formed along the same surface direction (Figs. 3.10a and 3.11a), and the other having the reconstruction aligned along different directions to each other (Figs. 3.10b and 3.11b), with one being rotated 60° relative to the other (referenced to the direction of the π-bonded chain [39]). While the latter structure is less stable by 0.01 eV/atom for 3-layer MLS, it is more stable by 0.01 eV/atom for 4-layer MLS [20]. The structural properties of this alternative arrangement are, however, essentially the same, indicating that the rotation of one of the MLS surface layers has little effect on the overall structural details [20].

The π-bonded configuration consisting of atoms labeled 1 and 2 (in Figs. 3.10 and 3.11), which is also found on the bulk Si(111)-2 × 1 surface, was seen on both

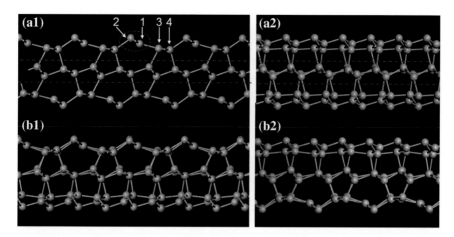

Fig. 3.10 Structures of 3-layer MLS with **a** π-bonded chain oriented along the same direction, and **b** π-bonded chain oriented along different directions. Views obtained by rotating the MLS in **a1** and **b1** 60° around the c axis (perpendicular to the silicene layers) are shown in **a2** and **b2**, respectively. All structures were found to form in the FPMD simulations by Spencer et al. [20]. Reproduced from [20] with permission from the Royal Society of Chemistry

surfaces of the MLS, comprising alternating 5- and 7-membered rings. It was shown that both positive and negative buckling could form in this reconstruction, where the negatively buckled structure has "atom 2" topmost on the surface, while the positive structure has "atom 1" topmost on the surface (see Fig. 3.11b1). The calculations by Spencer et al. [20] indicated that the negatively buckled geometry was more stable (as shown in Fig. 3.10a), which is consistent with the previous calculation by Zitzlsperger [39] that showed this topology is 2.7 meV/(surface atom) lower in energy (i.e. more stable) than the positively buckled topology.

On the outer surface layers (top and bottom) there were 3 unique atom types which can be classified according to their coordination and geometry. The topmost atoms on the surface (labelled as 2 in Fig. 3.10a in the case of the 3-layer MLS) are 3-fold coordinated having 3 bonds of almost equal length ($\sim$2.25 Å). These atoms have a similar bonding arrangement to the "atop" atoms in re-BLS and hence are referred to as "atop". The atom labelled 1 in Fig. 3.10a is also 3-fold coordinated having equal bond lengths of 2.32 Å, with the bonded atoms lying almost in a plane, and hence are identified as an "sp2-like" atom, as in re-BLS and the bulk Si (111)-2 × 1 surface. The other two topmost surface atoms are tetrahedrally coordinated, with the bond lengths being 2.37–2.42 Å (atom labelled 3 in Fig. 3.10a) and 2.37–2.39 Å (atom labelled 4 in Fig. 3.10a). In order to form the 2 × 1 surface reconstruction on MLS with 3-layers, the atoms in the middle layer (lying within the 2 red dotted lines in Fig. 3.10a) need to form part of highly distorted tetrahedra (with angles of up $\sim$136°), greater than occurs on the bulk Si(111)-2 × 1 surface

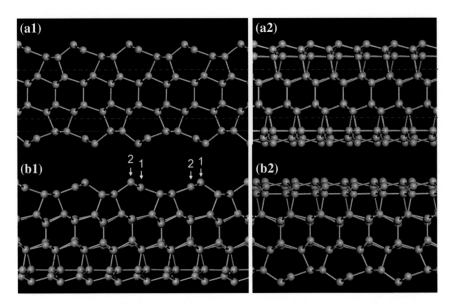

Fig. 3.11 Same as Fig. 3.10, but for 4-layer MLS. Reproduced from [20] with permission from the Royal Society of Chemistry

($\sim 131°$). This larger distortion allows the 2 × 1 surface reconstruction to form on both sides of the 3-layer MLS.

For the 4-layer MLS, both positive and negative buckling was observed on the topmost layer of the MLS, consistent with the small energy barrier between the two geometries and indicating that both can co-exist on the surface at finite temperature. Figure 3.11a shows the MLS with both positive and negative buckling on the bottom surface and negative buckling only on the top surface. The "atop" and "sp2-like" atoms were again 3-fold coordinated having bond lengths of ~ 2.29 Å. The tetrahedrally coordinated atoms in the top layer all have similar bond lengths as in the 3-layer MLS and the Si(111)-2 × 1 bulk surface. Unlike the 3-layer MLS, however, the two middle layers retained their tetrahedral order (see Fig. 3.11a) as in bulk Si(111)-2 × 1, forming bond angles up to $\sim 130°$. Hence, it seems that only 2 middle layers are needed to retain the bulk configuration having the Si(111)-2 × 1 surface geometry, though only 1 middle layer is needed to create the 2 × 1 surface reconstruction itself.

For the 5-layer MLS, the Si(111)-2 × 1 surface structure again stably formed on both sides of the MLS. As mentioned before, there are two possible arrangements of the reconstructed surface; one where the π-bonded chains are oriented along the same direction and the other where the π-bonded chains are rotated 60° to each other. Interestingly, the energy difference between these structural arrangements was found to be very small (the structure with the π-bonded chain oriented along the same direction had a lower energy only by 1.3×10^{-4} eV/atom) for the 5-layer

Fig. 3.12 Electronic DOS for **a** 2-layer, **b** 3-layer, **c** 4-layer reconstructed MLS, and **d** the Si(111)-2 × 1 surface. Reproduced from [20] with permission from the Royal Society of Chemistry

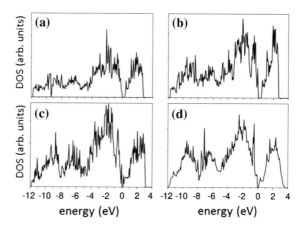

MLS. Hence, both structures are likely to exist, indicating that the effect of the π-chain orientation on the structural stability becomes negligible once the MLS reaches 5 layers.

The electronic density of states (DOS) for the reconstructed MLS (re-MLS) with 3 and 4 layers [20] are presented in Figs. 3.12, 3.13, 3.14 and 3.15. Only the DOS for the 3- and 4-layer structures having the π-bonded chains oriented along the same direction are shown as there are no significant differences compared to the alternate structures, except for the band gap which will be discussed later. While the DOS for the 5-layer MLS is not shown, the DOS of re-BLS and the bulk Si(111)-2 × 1 surface are shown for comparison. We show the total DOS as well as the atom resolved DOS for the "atop", "sp2-like", and the 4-fold coordinated atoms.

From the total DOS (Fig. 3.12) it can be seen that the overall shape becomes similar to that of the bulk Si(111)-2 × 1 surface as the number of layers in MLS increases, indicating that the overall features become similar to the bulk surface once a thickness of 4-layers is reached. The calculated band gap was 0.40, 0.22 and 0.19 eV for the 3-, 4- and 5-layered MLS, respectively, as compared to the gap of 0.13 eV for the bulk Si(111)-2 × 1 surface. Though the band gap is not yet converged to that of the bulk surface value with 5-layers, the trend is that the gap decreases with increasing layer thickness. Interestingly, when the π-bonded chain is oriented 60° to each other, the band gap was 0.13 eV for all thicknesses examined (>2 layers), and was the same as the gap for the bulk Si(111)-2 × 1 surface. This indicates that the relative orientation of the π-chain affects the size of the band gap. That is, if the sheet is substantially thin, the band gap is dependent on the π-bonded chain orientation which may alter the electronic properties (other than the DOS) of MLS.

From the DOS of the outermost atoms ("atop" atoms), it could be seen that the p_z orbital primarily dominated the valence band just below the highest occupied state in all systems, including the bulk Si(111)-2 × 1 surface (Fig. 3.13). This trend has already been pointed out in the discussion of the DOS for re-BLS, again confirming

Fig. 3.13 Decomposed DOS for "atop" Si atoms in **a** 2-layer, **b** 3-layer, **c** 4-layer reconstructed MLS, and **d** the Si(111)-2 × 1 surface. Reproduced from [20] with permission from the Royal Society of Chemistry

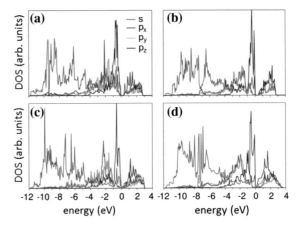

Fig. 3.14 Same as Fig. 3.13, but for "sp2-like" Si atoms. Reproduced from [20] with permission from the Royal Society of Chemistry

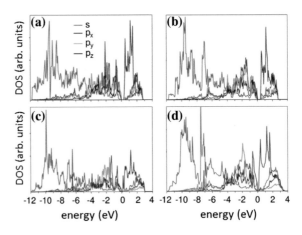

it is a general trend in Si. As the number of layers increased, this state became more localized. On the other hand, for the "sp2-like" atoms (Fig. 3.14), the conduction band was primarily dominated by the p_z orbitals for all systems. Hence, the 3-fold coordinated atoms on re-MLS of all thickness as well as the bulk Si surface are characterized by p_z orbitals, the details of which depends on the surface morphology. For the 4-fold coordinated atoms (Fig. 3.15) the s-band is split at ∼ −8 eV, with the split becoming more visible as the number of layers increased. This indicates that the s orbitals are sensitive to the distortion of the tetrahedral order and thus the thickness of the sheet. It may be possible to measure and confirm the split in the s-orbital peak experimentally using a spectroscopic technique, such as X-ray photoelectron spectroscopy; this may be a plausible way to experimentally estimate the thickness of MLS.

Fig. 3.15 Same as Fig. 3.13, but for 4-fold coordinated Si atoms. Reproduced from [20] with permission from the Royal Society of Chemistry

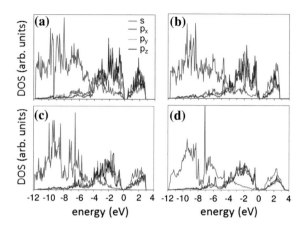

3.4 Summary

Theoretical attempts to characterize free-standing silicene consisting of more than one silicene layer have been reviewed in this chapter. For BLS structures without surface reconstruction, three types of stacking configurations (AA, AB and p-AA) were considered, which all exhibit band dispersion relations substantially different from those for SLS. The interaction between the layers is found to play a significant role in characterizing the electronic properties of BLS compared to two-layer graphene.

One of the particularly notable points is that MLS exhibits surface reconstructions at finite temperatures, which does not apply to SLS. It was shown that BLS has a unique reconstructed surface, which has not been seen on the bulk Si(111) surface, and is labeled as Si(111)-2 × 2, while MLS (having more than 2 layers) shows the Si(111)-2 × 1 surface reconstruction, which is a metastable surface structure on bulk c-Si. It is interesting that only three layers is required for the 2 × 1 reconstruction to form on MLS.

We have seen that the electronic properties of MLS strongly depend on the surface structure and the number of silicene layers. This indicates that MLS could be used as building blocks for a variety of electronic or optical nanodevices without introducing impurities. The reconstructed surfaces of MLS could also raise the possibility that novel surface reactions or adsorption processes would be induced. We thus believe that MLS has an extremely wide applicability to various nanodevices, which now attracts research from many theoretical and experimental groups.

Acknowledgements Some of the work presented in this chapter was undertaken with the assistance of the following computational facilities: the National Computational Infrastructure (NCI) through the National Computational Merit Allocation Scheme supported by the Australian Government, the V3 Alliance (formerly VPAC—the Victorian Partnership for Advanced

Computing), the Multi-modal Australian ScienceS Imaging and Visualisation Environment (MASSIVE), the Pawsey Supercomputing Centre (formerly iVEC), and facilities at the Research Center for Computational Science, National Institute of National Sciences, and the Research Institute for Information Technology, Kyushu University, Japan.

References

1. S. Cahangirov, M. Topsakal, E. Akturk, H. Sahin, S. Ciraci, Phys. Rev. Lett. **102**, 236804 (2009)
2. K. Takeda, K. Shiraishi, Phys. Rev. B **50**, 14916 (1994)
3. G.G. Guzmàn-Verri, L.C.L.Y. Voon, Phys. Rev. B **76**, 075131 (2007)
4. S. Lebegue, O. Eriksson, Phys. Rev. B **79**, 115409 (2009)
5. T. Morishita, M.J.S. Spencer, S.P. Russo, I.K. Snook, M. Mikami, Chem. Phys. Lett. **506**, 221 (2011)
6. M.J.S. Spencer, T. Morishita, M. Mikami, I.K. Snook, Y. Sugiyama, H. Nakano, Phys. Chem. Chem. Phys. **13**, 15418 (2011)
7. T. Morishita, M.J.S. Spencer, S. Kawamoto, I.K. Snook, J. Phys. Chem. C **117**, 22142 (2013)
8. K. Koga, H. Tanaka, X.C. Zeng, Nature **408**, 564 (2000)
9. K. Koga, G.T. Gao, H. Tanaka, X.C. Zeng, Nature **412**, 802 (2001)
10. T. Morishita, K. Nishio, M. Mikami, Phys. Rev. B **77**, 081401(R) (2008)
11. J. Bai, H. Tanaka, X.C. Zeng, Nano Res. **3**, 694 (2010)
12. C. Alba-Simionesco, B. Coasne, G. Dosseh, G. Dudziak, K. E. Gubbins, R. Radhkrishnan, M. Sliwinska-Bartkowiak. J. Phys. Condens. Matter **18**, R15 (2006)
13. Y. Maniwa, H. Kataura, M. Abe, A. Udaka, S. Suzuki, Y. Achiba, H. Kira, K. Matsuda, H. Kadowaki, Y. Okabe, Chem. Phys. Lett. **401**, 534 (2005)
14. J.C. Johnston, S. Phippen, V. Molinero, J. Phys. Chem. Lett. **2**, 384 (2011)
15. P. De Padova, P. Vogt, A. Resta, J. Avila, I. Razado-Colambo, C. Quaresima, C. Ottaviani, B. Olivieri, T.T. Bruhn, T. Hirahara, T. Shirai, S. Hasegawa, M.C. Asensio, G. Le Lay, Appl. Phys. Lett. **102**, 163106 (2013)
16. P. Vogt, P. Capiod, M. Berthe, A. Resta, P. De Padova, T. Bruhn, G. Le Lay, B. Grandidier, Appl. Phys. Lett. **104**, 021602 (2014)
17. T. Morishita, S.P. Russo, I.K. Snook, M.J.S. Spencer, K. Nishio, M. Mikami, Phys. Rev. B **82**, 045419 (2010)
18. C. Kamal, A. Chakrabarti, A. Banerjee, S. K. Deb. J. Phys. Condens. Matter **25**, 085508 (2013)
19. H. Fu, J. Zhang, Z. Ding, H. Li, S. Meng, Appl. Phys. Lett. **104**, 131904 (2014)
20. M.J.S. Spencer, T. Morishita, I.K. Snook, Nanoscale **4**, 2906 (2012)
21. B. Aufray, A. Kara, S. Vissini, H. Oughaddou, C. Leandri, B. Ealet, G. Le Lay, Appl. Phys. Lett. **96**, 183102 (2010)
22. P. Vogt, P. De Padova, C. Quaresima, J. Avila, E. Frantzeskakis, M.C. Asensio, A. Resta, B. Ealet, G. Le Lay, Phys. Rev. Lett. **108**, 155501 (2012)
23. A. Fleurence, R. Friedlein, T. Ozaki, H. Kawai, Y. Wang, Y. Yamada Takamura, Phys. Rev. Lett. **108**, 245501 (2012)
24. L. Meng, Y. Wang, L. Zhang, S. Du, R. Wu, L. Li, Y. Zhang, G. Li, H. Zhou, W.A. Hofer, H.-J. Gao, Nano Lett. **13**, 685 (2013)
25. J. Tersoff, Phys. Rev. B **39**, R5566 (1989)
26. W.A. Steele, Surf. Sci. **36**, 317 (1973)
27. R. Yaokawa, T. Ohsuna, T. Morishita, Y. Hayasaka, M.J.S. Spencer, H. Nakano, Nature Com. (in press)
28. F.H. Stillinger, T.A. Weber, Phys. Rev. B **31**, 5262 (1985)
29. J.P. Perdew, K. Burke, M. Ernzerhof, Phys. Rev. Lett. **77**, 3865 (1996)

30. J. Klimes, D.R. Bowler, A. Michaelides, Phys. Rev. B **83**, 195131 (2011)
31. C.J. Tabert, E.J. Nicol. Phys. Rev. B **86**, 075439 (2012)
32. D. Hanemann, Rep. Prog. Phys. **50**, 1045 (1987)
33. C. Lian, J. Ni, AIP Adv. **3**, 52102 (2013)
34. P.A. Fedders, D.A. Drabold, S. Nakhmanson, Phys. Rev. B **58**, 15624 (1998)
35. T. Morishita, J. Chem. Phys. **130**, 194709 (2009)
36. M.T. Yin, M.L. Cohen, Phys. Rev. B **26**, 5668 (1982)
37. J.E. Padilha, R.B. Pontes, J. Phys. Chem. C **119**, 3818 (2015)
38. Y. Sakai, A. Oshiyama, Phys. Rev. B **91**, 201405(R) (2015)
39. M. Zitzlsperger, R. Honke, P. Pavone, U. Schroder, Surf. Sci. **377**, 108 (1997)

Part II
Modified Silicene

Chapter 4
Soft Chemical Synthesis of Functionalized Silicene

Hideyuki Nakano and Masataka Ohashi

Abstract In this chapter, soft chemical syntheses of functionalized silicene are reviewed. Free standing silicene with sub-nanometer thicknesses has been prepared by exfoliating layered silicon compounds, and they are found to be composed of crystalline single-atom-thick silicon layers. Organic modified silicenes can be prepared by modifying a layered polysilane (Si_6H_6), which has an analogous structure to that of graphite, and this allows the properties of the silicene to be controlled in order to make it suitable for particular applications. The potential applications of these silicenes and organic modified silicenes are also reviewed.

4.1 Introduction

Silicon nanomaterials have attracted increasing attention, not only because of the many technological applications they currently have and that are anticipated, but also in terms of research into the fundamental principles relating to them. For example, highly porous silicon produced by electrochemical dissolution exhibited visible photoluminescence at room temperature [1, 2]. Low-dimensional nanomaterials are of particular interest because they may exhibit anisotropic and/or dimension-tunable properties, both of which are important attributes in nanodevice applications [3]. Recently, zero-dimensional crystal Si nanomaterials, i.e. Si nanoparticles [4, 5], and one-dimensional nanomaterials, i.e. Si nanowires [6, 7], and Si nanotubes [8, 9], have been successfully synthesized. Although controlling the dimensions of the materials produced is difficult because the melting point of

H. Nakano (✉) · M. Ohashi
Toyota Central R&D Labs., Inc., 41-1 Yokomichi, Nagakute
Aichi 480-1192, Japan
e-mail: hnakano@mosk.tytlabs.co.jp

© Springer International Publishing Switzerland 2016
M.J.S. Spencer and T. Morishita (eds.), *Silicene*, Springer Series
in Materials Science 235, DOI 10.1007/978-3-319-28344-9_4

silicon is higher than 1400 °C, the fabrication techniques have been studied intensively with the aim of improving the size and chemical stabilities of the materials produced and to achieve high yields. However, two-dimensional (2D) crystal silicon nanomaterials, i.e. silicenes, have not been extensively developed.

In 1996, nanosheets were defined in a detailed study on the exfoliation of layered titanium oxide [10]. Nanosheets are generally characterized as having a thickness of the order of nanometers and lateral dimensions ranging from the submicrometer to the micrometer scale. The most important feature of nanosheets is their extraordinarily high specific surface areas, which makes them promising candidates for various applications, depending on their specific physical properties. With respect to materials science, nanosheets can be used as building blocks for novel supramolecular assemblies with applications in nanoscience and nanotechnology [11, 12]. Since 1996, many kinds of nanosheets have been prepared and their unique properties have been demonstrated. A process involving cation exchange has been used to create inorganic nanosheets of layered compounds, clay minerals [13], metal chalcogenides [13, 14], phosphonates [15, 16], and layered transition-metal oxides [17–19]. Extensive research has also been performed on exfoliating anion-exchangeable layered double hydroxides [20–22]. Moreover, few layered van der Waals materials, such as MoS_2 and $NbSe_2$, have been prepared using a mechanical exfoliation process and have had their properties studied since the 1960s [23, 24]. In 2004, graphene was produced by the micromechanical cleavage of graphite [25]. Graphene is a two-dimensional material with unique electronic properties that are qualitatively different from those of standard semiconductors such as silicon [26]. On the other hand, the fabrication of silicene by the ultrahigh vacuum deposition of silicon atoms onto a metallic substrate has recently been reported [27, 28]. Silicene, as discussed in the earlier chapters, is a relatively new allotrope of silicon and can be viewed as the silicon version of graphene. Therefore, silicene shares many of the fascinating properties of graphene, such as the so-called Dirac electronic dispersion. However, the structure of silicene is slightly different from that of graphene, leading to a few major differences in properties, such as its ability to open a bandgap in the presence of an electric field or on a substrate, which is a key property for digital electronic applications.

As for previously reported syntheses, the top-down approach is a reliable method of synthesizing silicenes, if suitable 2D Si structures are available. Recently, an attempt to produce silicenes from Zintl-phase $CaSi_2$, layered polysilane (Si_6H_6), and siloxene $(Si_6H_3(OH)_3)$, which are crystalline layered silicon compounds that each have a backbone of puckered 2D Si layers similar to the (111) planes in 3D crystalline Si has been tried. The single silicon atom layers in these layered silicon compounds only weakly interact with each other, but they are not easily exfoliated and do not dissolve in either organic or inorganic solvents. In this chapter, the soft chemical synthesizing silicenes, organic modified silicenes, and their potential applications are mainly described.

4.2   Layered Silicon Compounds

4.2.1   Growth and Electronic Properties of Calcium Intercalated Silicene, CaSi$_2$

Calcium disilicide (CaSi$_2$) is a so-called Zintl phase, i.e. an intermetallic compound formed between a strongly electropositive metal, such as alkali metals, alkaline earth metals and lanthanides, and a somewhat less electropositive metal. In a Zintl silicide, silicon is the anionic part of the structure and fulfills the simple role of typical valence compounds, resulting in a variety of sub-network structures with Si–Si covalent bonds; each Si atom achieves an octet shell with covalent bonds and lone-pair electrons according to the 8-N rule. CaSi$_2$ has a 2D silicon sub-network resembling buckled Si(111) planes, in which the Si$_6$ rings are interconnected with sp^3 bonds; the puckered (Si)$_n$ polyanion layers are separated by planar monolayers of Ca^{2+} as shown in Fig. 4.1. CaSi$_2$ has a trigonal structure (space group R-3 m) at ambient pressure, which can only vary in its stacking sequence: tr6 (a = 3.855 Å, c = 30.62 Å) [29] and tr3 (a = 3.829 Å, c = 15.90 Å) [30]. In the tr6 structure, the stacking of trigonal Ca layers follows an AABBCC sequence with a six-layer repeat distance. The tr3 structure has a three-layer repeat distance, with Ca layers stacked in an ABC sequence.

Figure 4.2 shows the top surfaces of Ca$_{1.05}$Si$_2$, Ca$_{1.00}$Si$_2$ and Ca$_{0.90}$Si$_2$ ingots cooled at a rate of 1, 10 and 100 °C min^{-1}, respectively. Plate-like large grains grew on the surface of the Ca$_{1.00}$Si$_2$ ingot cooled at 10 °C min^{-1} (Fig. 4.2e). XRD analysis revealed that these plate-like grains consisted of single-phase CaSi$_2$ whose crystal orientation coincided with the {001} facet planes. For all cooling rates, Ca$_{1.00}$Si$_2$ ingots were composed entirely of the CaSi$_2$ (tr6) phase, while Ca$_{1.05}$Si$_2$ ingots consisted of the phases CaSi$_2$, CaSi and Si. It should be mentioned that the CaSi$_2$ single-phase ingot can be obtained from the stoichiometric composition, since CaSi$_2$ is a line compound that has no solid solution range [31].

CaSi$_2$ provides a good opportunity to investigate the intrinsic electronic structure of genuine silicene, as the Si sub-lattice in CaSi$_2$ form the hexagonal 2D layer while silicene synthesized on a silver substrate has been reported to have the dumbbell structure different from the hexagonal structure [31]. In CaSi$_2$, Ca atoms are

Fig. 4.1 a Schematic illustration and **b** STEM image of tr6-CaSi$_2$

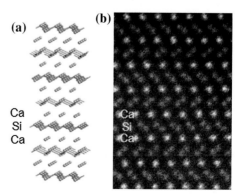

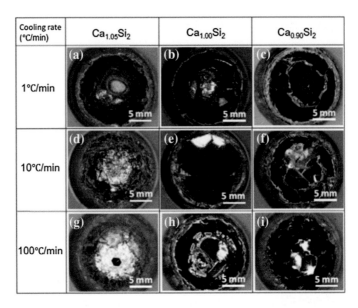

Fig. 4.2 a–i Top surface of ingots fabricated from $Ca_{1.05}Si_2$, $Ca_{1.00}Si_2$ and $Ca_{0.90}Si_2$ composition melts at a cooling rate of 1, 10 and 100 °C min^{-1}, respectively

considered to play a role in stabilizing the buckled honeycomb network structure of silicon atoms by donating electrons to the silicene layers. Figure 4.3a and b display the XRD patterns from the powders and the single crystal, showing the high quality of the grown single crystal. The low-energy-electron diffraction (LEED) measurement of the cleaved surface shows a clear 1 × 1 pattern, indicating that no reconstruction of the crystal structure takes place at the surface (Fig. 4.3c). Using high-resolution angle resolved photoemission spectroscopy (ARPES) on tr3-$CaSi_2$, the massless Dirac-cone π-electron dispersions at the K(H) point in the Brillouin zone was observed, together with the σ-band dispersions at the Γ point (Fig. 4.3d). The Dirac point is located about 2 eV from the Fermi level (E_F), showing a substantial charge transfer from Ca atoms to silicene layers (Fig. 4.3e and f). The present ARPES results indicate that the sp^2 bonding-framework essentially holds in $CaSi_2$ to produce the massless Dirac-cone state at the K-point despite the strongly buckled structure in the silicene layer. This indicates that the graphene-like electronic structure with a massless Dirac cone is stably formed in the Ca intercalated multilayer silicene, $CaSi_2$ [32].

4.2.2 Deintercalation of Ca from $CaSi_2$

Treating $CaSi_2$ (Fig. 4.4) with concentrated aqueous HCl causes it to be topotactically transformed into a green–yellow solid, which is a layered siloxene ($Si_6H_3(OH)_3$), with the release of hydrogen gas, as described in reaction (4.1) [33].

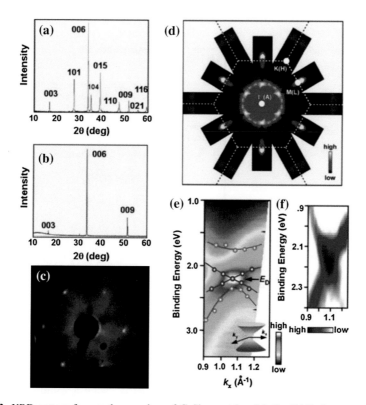

Fig. 4.3 XRD pattern from **a** the powders of CaSi$_2$ crystal and **b** the (111) cleavage plane of CaSi$_2$ single crystal. **c** LEED pattern from the (111) cleavage plane of CaSi$_2$ crystal measured with the primary electron energy of 80 eV. **d** Fermi surface of tr3-CaSi$_2$ obtained by plotting the ARPES spectral intensity integrated within ±50 meV with respect to E$_F$ and folded by the crystal symmetry. **e** ARPES-intensity plot around K(H) point as a function of wave vector and binding energy. Peak position in the ARPES spectra obtained by fitting with Lorentzians is shown by *blue* and *red circles*. **f** Second-derivative ARPES intensity plot around E$_D$

$$3CaSi_2 + 6HCl + 3H_2O \rightarrow Si_6H_3(OH)_3 + 3CaCl_2 + 3H_2 \qquad (4.1)$$

The crystalline sheets have puckered 2D Si layers that are similar to crystalline Si(111) layers, and the Si atoms in the layers are stabilized by terminal hydrogen or hydroxide groups that point out of the layer plane. At temperatures below −30 °C, CaSi$_2$ was found to change into the layered polysilane (Si$_6$H$_6$) without hydrogen being evolved, as shown in reaction (4.2) [34, 35]. This means that the additional reactions between the Si layer and water do not occur.

$$3CaSi_2 + 6HCl \rightarrow Si_6H_6 + 3CaCl_2 \qquad (4.2)$$

The structures of Si$_6$H$_6$ and Si$_6$H$_3$(OH)$_3$ are similar, but in Si$_6$H$_6$ all of the Si atoms are terminated with a hydrogen atom only, and there are no oxygen atoms.

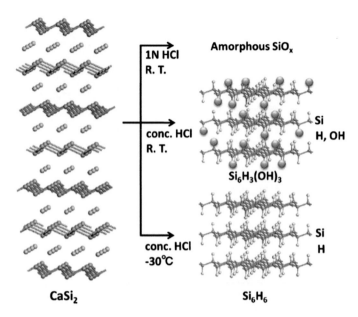

Fig. 4.4 Schematic illustration of layered silicon compounds

Interlayer bonding between adjacent polymer sheets is weaker in both $Si_6H_3(OH)_3$ and Si_6H_6 than in $CaSi_2$, because the bonding consists of only weak van der Waals forces in Si_6H_6 and hydrogen bonds in $Si_6H_3(OH)_3$. Using a solution of HCl in methanol, ethanol, butanol, $C_{12}H_{25}$, benzyl alcohol, or CH_2COOMe, corresponding alkoxide-terminated organosiloxenes (Si_6H_5OR, where R = methanol, ethanol, butanol, $C_{12}H_{25}$, benzyl alcohol, or CH_2COOMe) were also obtained [36, 37].

4.3 Exfoliation of Layered Silicon Compounds

Layered siloxene $Si_6H_3(OH)_3$ was successfully exfoliated into individual sheets using a surfactant, sodium dodecylsulfate (SDS: $C_{12}H_{25}OSO_3Na$) [38]. This suspension remained stable for two months, showing no precipitation. The XRD pattern for the wet colloidal aggregate centrifuged from the suspension reveals that the parent siloxene was exfoliated into individual sheets, a small portion of which was stacked to give the small angle scattering (Fig. 4.5b). On drying, almost all of the siloxene nanosheets were restacked as a 5–20 sheet unit with a basal spacing of 3.8 nm, which represents a large expansion from the original value of 0.63 nm for siloxene and can be identified as belonging to a dodecylsulfate intercalated compound (Fig. 4.5c). The transmission electron micrograph (TEM) images showed that the siloxene nanosheets had a lateral dimension of less than 200 nm (Fig. 4.5d). Although almost transparent ultrathin sheets are observed, relatively thick fragments were also observed in small quantities. It could be seen that cleavage

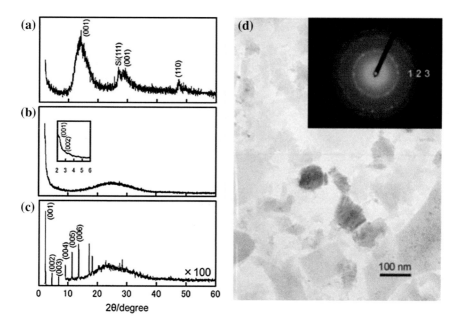

Fig. 4.5 X-ray diffraction patterns for **a** $Si_6H_3(OH)_3$, **b** the wet colloidal aggregate separated from the exfoliated $Si_6H_3(OH)_3$ suspension by centrifugation, and **c** the dried colloidal aggregate. **d** TEM image of $Si_6H_3(OH)_3$ nanosheets and electron diffraction pattern (*inset*)

occurred in many layers throughout the particle, rather than on a layer-by-layer basis starting from the surface. Evidence for preservation of the two-dimensional silicon network during exfoliation was also provided by electron diffraction (Fig. 4.5d inset). Atomic force microscopy (AFM) revealed that the average nanosheet thickness was 0.7 nm, which is close to the crystallographic thickness calculated to be 0.65 nm. The close match between this value and that obtained by AFM indicated that the nanosheets consisted of single siloxene layers. Room-temperature absorption spectra for the nanosheet suspension were characterized by Lambert-Beer behavior and showed the absorption for the colloidal nanosheets was significantly blue shifted (4.6 eV) compared with those for bulk parent siloxene (2.4 eV). The large energy shift suggests that the spectral changes upon delamination are likely to be associated with size quantization. The colloidal suspension method with SDS can synthesize $Si_6H_3(OH)_3$ nanosheets, but the silicon skeleton was partially oxidized.

The $CaSi_2$ formula with the formal charges included is $Ca^{2+}(Si^-)_2$; thus, the electrostatic interactions between the Ca^{2+} and Si^- layers are strong. It is, therefore, very important to decrease the amount of negative charge on the silicon layers to allow exfoliation to occur. To achieve this, Mg-doped $CaSi_2$ was used as a starting material to decrease the strength of the interactions between the Ca and Si layers. The immersion of bulk $CaSi_{1.85}Mg_{0.15}$ in a solution of propylamine hydrochloride led to the deintercalation of the calcium ions, which was accompanied by the

Fig. 4.6 **a** AFM image of the
silicene. **b** Line profile taken
along the *white line* in (**a**).
c Atomically resolved AFM
image of the silicene (Color
figure online)

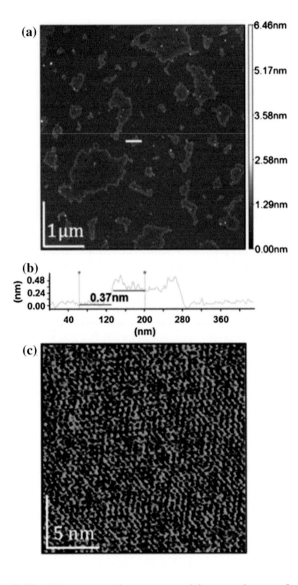

evolution of hydrogen. The CaSi$_{1.85}$Mg$_{0.15}$ was then converted into a mixture of
Mg doped silicene and an insoluble black metallic solid [39]. A light-brown suspension containing silicene was obtained after the sediment had been removed from
the bottom of the flask. The composition of the silicene obtained was determined to
be Si:Mg: O = 7.0:1.3:7.5, indicating that the silicenes were preferentially exfoliated from sections of the silicon layers where magnesium atoms were present.

The dimensions of the silicenes were determined by direct observations using
AFM, and the sheets were found to be 0.37 nm thick with lateral dimensions of
200–500 nm (Fig. 4.6a and b). The crystallographic thickness of the silicene was

calculated, from its atomic architecture, to be 0.16 nm, and the difference between this value and that obtained using AFM indicates that the surfaces of the silicenes were stabilized by capping with oxygen atoms. High resolution AFM images revealed that the closest distance between the atoms (dot-like marks in the AFM image) was 0.41 ± 0.02 nm (Fig. 4.6c), which is slightly larger than the distance between Si atoms in the Si(111) plane in bulk crystalline silicon (0.38 nm). The sheets are considered to be the most similar to free standing silicene.

4.4 Organo-Modifications of Silicene

Capping of layered silicon compounds with organic groups is supposed to promote their exfoliation, and assist in the control of the interlayer spacing. The introduction of organic groups on Si_6H_6 through $\equiv Si-C$ or $\equiv Si-N$ bonds is a plausible method to obtain silicenes capped with organic moieties, because of their high dispersion in organic solvents and their stability with respect to oxidation and/or hydrolysis, due to the formation of a hydrophobic surface on silicon. In contrast to carbon, silicon is less electronegative than hydrogen; hence, $\equiv Si-H$ bonds are polarized in the opposite sense, as $\equiv Si^{\delta+}-H^{\delta-}$. This implies that nucleophilic attack on silanes usually occurs at the silicon centers. Silanes are also much more reactive than the corresponding hydrocarbon analogues due to larger polarization. The high reactivity of silanes and hydride-terminated silicon crystalline surfaces are also due to weak $\equiv Si-H$ bonds, which are significantly weaker than C–H bonds. Based on this characteristic, layered polysilane, Si_6H_6, is a suitable starting material for the preparation of silicenes by organic modification (Fig. 4.7).

4.4.1 Hydrosilylation of Layered Polysilane (Si_6H_6)

Surface modification of silicon particles or substrates with organic groups, can be achieved through hydrosilylation of hydrogen-terminated silicon surfaces with unsaturated organic groups, such as alkenes and alkynes [40]. The $\equiv Si-H$ bonds as reactive sites are not readily accessible because they are located in the interlayer spacing of the layered starting material. Therefore, for the reaction to happen, it is imperative that the organic moiety be intercalated into the interlayer. Using a Pt-catalyzed hydrosilylation reaction, Si_6H_6 was functionalized with 1-hexene to produce a stable colloidal silicene suspension [41]. The obtained product was soluble in typical organic solvents, such as hexane, chloroform, acetone, and ether, and insoluble in water and ethanol.

FTIR and Si K-edge X-ray absorption near-edge structure (XANES) analyses revealed that the 1-hexene does not merely attach to Si_6H_6, but reacts with it to form $\equiv Si-C$ linkages. The IR spectrum (Fig. 4.8a) shows an alkyl stretching and bending absorption bands between 2856 and 2954 cm^{-1} and between 1259 and

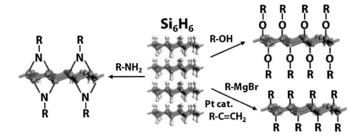

Fig. 4.7 Schematic illustration of the layered silicon compound (Si_6H_6) and its modification into organomodified silicenes

Fig. 4.8 **a** FTIR spectra of [$Si_6H_3(C_6H_{13})_3$]. **b** Si K-edge XANES spectra of [$Si_6H_3(C_6H_{13})_3$] (*brown line*), Si(111) wafer and SiO_x (*black line*). **c** XPS Si 2p spectra of [$Si_6H_3(C_6H_{13})_3$] (*brown line*), Si(111) wafer and SiO_x (*black line*) (Color figure online)

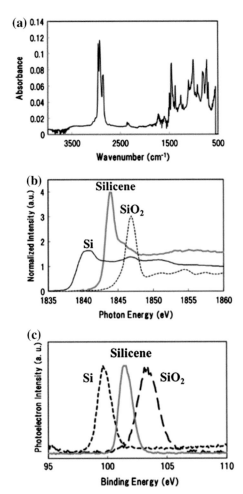

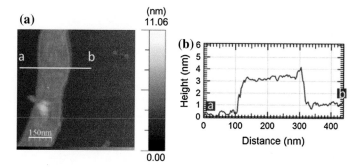

Fig. 4.9 **a** AFM image of [Si$_6$H$_3$(C$_6$H$_{13}$)$_3$]. **b** Line profile taken along the white line in (**a**)

1459 cm^{-1}, respectively. In addition, a band attributable to $\equiv$Si–CH$_2$ vibrational scissoring was observed at 729 cm^{-1}. The presence of characteristic vibrations for organic molecules in the sheets after the reaction, along with a reduction in the intensity of $\equiv$Si–H at 2110 cm^{-1} and the absence of bands characteristic of a terminal double bond (C=C; 1600 cm^{-1}), indicate that the organic molecules were covalently attached to the silicene surface. XANES spectra of the prepared silicene (Fig. 4.8b) exhibited features at an energy of 1844 eV, which is intermediate between those measured for Si and SiO$_2$ samples. This peak was assigned to the $\equiv$Si–C bond, which is the same bond energy as that for a tetra-coordinated silicon atom-bonded organic group. The XPS spectrum displayed a Si 2p peak with a binding energy of 102.0 eV (Fig. 4.8c), which is different from that of bulk Si (99.0 eV) and SiO$_2$ (104 eV). Compared to XPS data for decyl-capped silicon nanocrystals reported recently [42], this binding energy (102.0 eV) was attributed to a Si-organo group.

As shown in Fig. 4.9, the AFM image of the sheet had a thickness of 3.1 nm, which was compared to a thickness of 2.3 nm calculated from the model structure optimized using force field calculations. The model structure was constructed on the basis of the ideal formula [Si$_6$H$_3$(C$_6$H$_{13}$)$_3$]. The small thickness indicated that the sample was composed of monolayer sheets. From the XRD data, the silicene possesses molecular-scale periodicity of 0.71 nm, consisting of silicon-hexane composites. The same value (0.71 nm) was calculated from the proposed structural model for the periodicity of hexyl groups on the sheet surface. The silicene was stabilized though termination of hexane groups pointing out of the (111) plane structure of bulk silicon since it was prepared from layered polysilane Si$_6$H$_6$.

4.4.2 Grignard Reagent

The reactivity of Si$_6$H$_6$ must be increased by halogenation, but it would be undesirable because the use of highly toxic PCl$_5$, CCl$_4$, or Cl$_2$ gas is required. In this

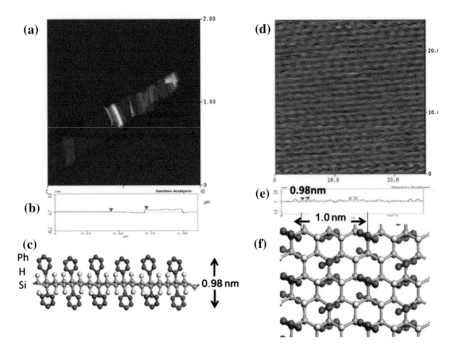

Fig. 4.10 **a** Atomic force microscopy (AFM) image of [Si$_6$H$_4$Ph$_2$] in contact mode. **b** Line profile along the *black line* in image (**a**). **c** *Side view* of the model structure of [Si$_6$H$_4$Ph$_2$]. **d** Atomically resolved AFM image of the surface of [Si$_6$H$_4$Ph$_2$]. **e** Line profile along the *black line* in image (**d**). **f** *Top view* of the model structure for [Si$_6$H$_4$Ph$_2$] (Color figure online)

section, phenyl terminated silicenes (Si$_6$H$_4$Ph$_2$) are introduced, which were successfully synthesized by the reaction of Si$_6$H$_6$ with phenyl magnesium bromide (PhMgBr) [43]. Normally, an intermediate halogenation step is needed for the introduction of organic groups using the Grignard reagent on bulk Si(111) because of their low reactivity [44], but the synthesis of the silicenes succeeded without this step. The obtained material is a colorless paste and is soluble in typical organic solvents such as hexane, chloroform, acetone, or ether. FTIR and 1H NMR revealed that Si$_6$H$_4$Ph$_2$ includes not only ≡Si–Ph bonds but also ≡Si–H bonds. 1H NMR also showed that there are approximately double the number of ≡Si–H bonds as ≡Si–Ph bonds. XANES exhibited two peaks derived from the ≡Si–Si bond and the ≡Si–Ph bond, but no features related to a ≡Si–O bond were indicated even in measurements performed after storing the samples under air for 1 day. Si$_6$H$_4$Ph$_2$ was found to be stable toward oxidation and the underlying surface, and thus the material could be handled easily under air. Such increased stability of the material by the introduction of ≡Si–C bonds is supported by previous reports on porous silicon [45] and SiNWs [42]. The PL emission of Si$_6$H$_4$Ph$_2$ in 1,4-dioxane using excitation at 350 nm showed a peak at 415 nm, which is almost the same as other silicenes. The AFM image of Si$_6$H$_4$Ph$_2$ revealed a completely flat plane surface and the thickness of the sheet was measured as 1.11 nm (Fig. 4.10a, b). The thickness and flatness of the surface of $Si_6H_4Ph_2$ indicated that the sample was composed of monolayer sheets. Interestingly, the sample had the shape of a band and showed a folded moiety. This indicates that the sample is very thin and flexible, similar to silk. Atomically resolved AFM images showed a periodic structure of the phenyl groups as atom-like dots on the silicon surface (Fig. 4.10d). The closest distance between the dots was measured to be 0.96 ± 0.02 nm (Fig. 4.10e). The periodicity of phenyl groups on the $Si_6H_4Ph_2$ structural model was estimated to be 1.0 nm (Fig. 4.10f), which is in good agreement with the closest distance between the dots in the AFM image. A periodicity of ~ 1.0 nm was also shown by XRD.

4.4.3 Reaction with Amine Groups

The reaction of an amine with $\equiv$Si–H corresponds to that of n-hexylamine and hydrogen terminated SiNPs [46]. It is thought that Si atoms can easily accept the nucleophilic attack of a lone pair from N atoms in the amine because the $\equiv$Si–H bonds in Si_6H_6 look like silyl-hydride, i.e. the bonds are polarized as $\equiv Si^{\delta+}-H^{\delta-}$. Thermo gravimetric analysis (TGA) and FTIR for the heated sample above 400 °C revealed that the moieties of alkyl chains connected to Si layers remain. TGA also suggests that the silicon layers in C10-silicene are densely covered with n-decylamine residues with few defects. A small amount of C10-silicene is dispersible in chloroform and exfoliates to form individual sheets with lateral dimension in the range of 1–2 μm, as observed by AFM. As shown in Fig. 4.11, the C10-silicenes have very flat and smooth surfaces due to the dense coverage of n-decylamine with a thickness of 7.5 nm. The thickness of this silicene is 10–20 times larger than that of the other organo modified silicene or the Mg-doped silicene because of the long alkyl chain on the surfaces as well as the incomplete exfoliation (two or three layer stacking). It seems to be difficult to capture the monolayer structure because C10-silicene and other amine-modified silicenes have a strong tendency to stack when they are dried from solution.

Spectrophotometric measurements of dilute C10-silicene solutions reveal characteristics of the nanosheet structure. UV–vis absorption spectra for the solutions were characterized by Lambert-Beer behavior and showed a peak at 276 nm with a shoulder around 283 nm, which is very close to that for the Mg-doped silicene (268 nm) (Fig. 4.12a). It indicates that the C10-silicenes are dispersed into individual sheets in the solution. Moreover, the absorption edge of the C10-silicene solution at approximately 380 nm corresponds to that of the nanoclusters (360–390 nm). A broad main peak appears at 438 nm (2.8 eV) in the photoluminescence (PL) spectrum obtained using an excitation wavelength of 350 nm (Fig. 4.12b). The peak positions are mostly the same as that for Mg-doped silicon nanosheet or a silicon network polymer with hexyl side chains. The band gap energy and the broadness of the PL spectra reflect the structure of two-dimensional silicon nanomaterials.

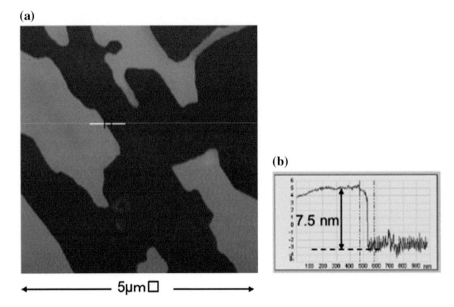

Fig. 4.11 **a** AFM image of a C10-silicene on a HOPG substrate. **b** Line profile measured along the *white line* denoted in (**a**) (Color figure online)

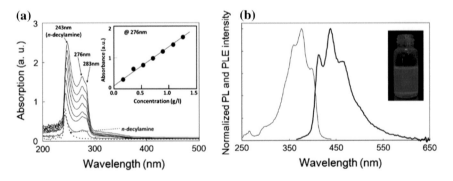

Fig. 4.12 **a** UV–vis spectra of chloroform solutions of C10-silicenes at various concentrations. The *dotted line* is the spectrum of n-decylamine in chloroform. The *inset* shows absorbance at 276 nm plotted against the concentration of the nanosheets. **b** PL (*thick line*) and PLE (*thin line*) spectra of C10-silicenes in chloroform. Excitation and fluorescence wavelengths are 350 and 450 nm, respectively. The *inset* photograph shows the fluorescent emission from a chloroform solution of C10-silicenes exposed to UV with a wavelength of 365 nm

4.4.4   Regular Stacking Structure in Amino-Modified Silicene

From the viewpoint of their applications, useful silicenes must have high affinities for planar integrated circuits and be easily fabricated using microprocess techniques. Silicenes made of Si are, therefore, suitable for use in conventional Si-based integrated circuits [47]. Therefore, stacking structures are suitable for fabricating electrochemical devices using organic modified silicenes.

Amino modification of silicene was conducted by stirring Si_6H_6 with primary n-alkylamines such as n-butyl-(C4), n-hexyl-(C6), n-decyl-(C10), n-dodecyl-(C12), and n-hexadecyl-(C16) amine in chloroform for 12 h at 60 °C in a nitrogen atmosphere [48]. The yellow-colored Si_6H_6 precipitate powder rose to the solvent surface after the reaction and became reddish in color. Figure 4.13a presents X-ray diffraction (XRD) patterns for the various Cm-silicene samples, in which the intensity is shown on a logarithmic scale. Samples were prepared by dropping a chloroform solution of Cm-silicene onto a glass plate and then evaporating off the solvent. Strong and narrow XRD peaks are evident in the lower-angle region, and many peaks are aligned at almost equal intervals in all of the patterns. The peaks at the lowest angles in each pattern correspond to the (001) planes, and the peaks at almost equal intervals are considered to be higher-order peaks, namely, (00l) peaks, where l is a positive integer, for the (001) planes in each sample. The interlayer distances for the m = 4, 6, 10, 12, and 16 samples were calculated from the (001) reflection as d = 1.30, 1.72, 2.48, 2.81, and 3.35 nm, respectively. The very sharp (001) peaks and the appearance of many (00l) peaks indicate a large number of silicenes that are regularly stacked parallel to the glass plate surface by self-assembly. The thicknesses of the stacked structures were estimated to be around 100 nm based on the full widths at half-maximum of the (00l) peaks and the Scherrer equation. Figure 4.13b shows the proportional relationship between the d spacing of the stacked Cm-silicene structures and m, where the slope and the intercept of the line are 0.172 and 0.674 nm, respectively. This relationship suggests that alkyl chains included in the stacked Cm-silicene materials are regularly arrayed and take the same conformation, probably an all-trans conformation. Then the slope of 0.172 nm indicates a bilayered alkyl-chain structure at a tilt angle of ca. 47° with respect to the stacking layers (Fig. 4.13c).

After the reaction of Si_6H_6 with the α,ω-diaminoalkanes, the following silicene were obtained: 1,2-diaminoethane (DiC2), 1,3-diaminopropane (DiC3), 1,6-diaminohexane (DiC6), and 1,12-diaminododecane (DiC12). The (001) reflection peak appeared at 2θ = 6.49° in the XRD pattern for DiC12-silicene (Fig. 4.13d). Higher-order (00l) peaks were also observed, which suggests that DiC12-silicene consists of a regularly stacked layer structure. The (001) peaks for DiC6-silicene and DiC2-silicene appeared at 2θ = 8.95° and 10.45°, respectively. The relationship between the d spacing for DiCm-silicene and m are shown in Fig. 4.13e. Although there are only three data points, they are on a straight line, and

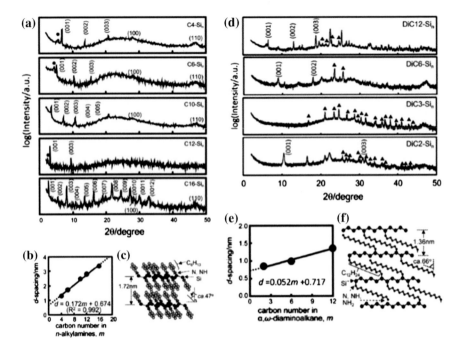

Fig. 4.13 **a** XRD patterns of Cm-silicene (m = 4, 6, 10, 12, and 16). *Peaks* marked with *circles* are derived from the (001) plane of Cm-HCl salts. **b** Relationship between the d spacing and m for Cm-silicene stacked structures. **c** Structural model of regularly stacked C6-silicene. **d** XRD patterns of DiCm-silicene (m = 2, 3, 6, and 12). *Peaks* marked with *triangles* are coincident with those for the DiCm-HCl salts. **e** Relationship between the d spacing and m for DiCm-silicene. **f** Structural model of DiC12-silicene

the slope and intercept of the line are 0.052 and 0.717 nm, respectively. This proportional relationship suggests that the alkyl chains included in the DiCm-silicene materials take the same conformation, probably an all-trans conformation, as in the case of Cm-silicene, although the alkyl chains form a single layer between the silicon layers. The slope of 0.052 nm indicates a highly tilted alkyl-chain structure at an angle of ca. 66° with respect to the silicon layers (Fig. 4.13f). The intercept of 0.717 nm is slightly larger than that for Cm-silicene. The smaller intercept of Cm-silicene might therefore be due to the dense packing at the interface of the two organic layers between the silicon layers. It is considered that, in many cases, both ends of the diamines bind to the silicene surfaces.

The regular stacking structure mentioned above will have a high affinity for planar integrated circuits and can be easily fabricated using microprocess techniques. Therefore, the electron transport properties of silicenes were investigated using AFM with a conductive cantilever, which was used to determine the number of silicenes and to obtain the data required to construct current–voltage (I–V)

Fig. 4.14 a Atomic force microscopy images of C10-silicenes dispersed on a highly ordered pyrolytic graphite substrate. **b** Line profiles for the height of the silicenes at points A, B, and C marked on image **a**; *inset* schematic of a C_{10}-silicene. **c** I–V curves of the silicene at points A, B, and C marked on image **(a)**

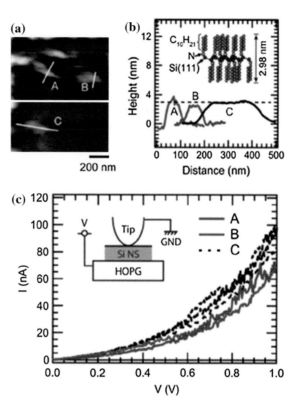

plots [49]. Figure 4.14a and c show the I–V relationships found for the C10-silicenes A, B, and C immediately after the topographs were obtained. All of the I–V curves had similar nonlinear shapes, but the currents in each curve were slightly different because of the variations in the contact area between the cantilever and the different C10-silicenes. The three I–V curves were almost of the same shape. The C10-silicenes were relatively stable, and showed similar I–V characteristics over a voltage of 0–1 V. However, the breakdown of the silicenes occasionally occurred at a voltage >1 V with a larger current being measured in the voltage sweep after the one in which the breakdown occurred, which indicated that the C10-silicenes had been destroyed. The Si(111) atomic layer comprises a periodic structure of Si atoms (see the inset in Fig. 4.14b), and it was the fundamental structure in the C10-silicenes. Organic molecules adsorbed on the Si(111) layers could create potential barriers against the band edges of the layers, resulting in C10-silicenes with double-barrier quantum-well structures. However, it is unlikely that electrons could tunnel through the confinement level in the quantum well (i.e. resonant tunneling) as the confinement level was too high, because of the presence of the ultrathin Si(111) layers. Therefore, the C10-silicenes functionalized with organic molecules could possibly act as single potential barriers.

4.4.5   Mechanochemical Lithiation of Silicene

The advantage of Si_6H_6 is the presence of highly reactive ≡Si–H groups on the surface and between the layers. Using this reactivity, densely lithiated Si_6H_6 (Si_6H_6/nLi) were prepared by a mechanochemical solid phase reaction as shown in Fig. 4.15a [50]. As shown in Fig. 4.15b, before milling, Si_6H_6 and Li were a light yellow powder and a silver metallic thin plate, respectively. After milling for 5 min, a greenish yellow powder and a ripped fragment of Li were observed. Over the next 15 min, the color changed from greenish yellow to faded green. Finally, after 30 min of milling, the Li fragments disappeared completely, and a dark green powder was obtained. The obtained composites were yellow to dark green, depending on the Li content (Fig. 4.15c). The XRD, FTIR, and XANES measurements revealed that the reaction mechanism is estimated to be the substitution of ≡Si–H in Si_6H_6 with Li, with the release of hydrogen gas, as shown.

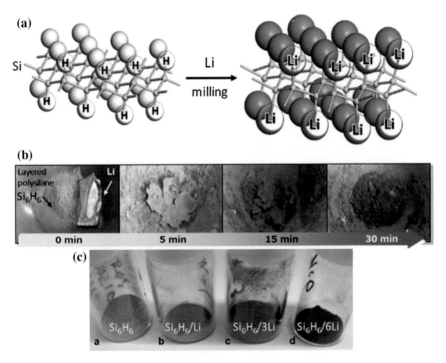

Fig. 4.15 a Schematic illustration of lithiation. **b** Photo images of lithiation and the mechanochemical reaction of Si_6H_6 with Li. **c** *Color* of the obtained composites: **a** Si_6H_6, **b** Si_6H_6/1Li, **c** Si_6H_6/3Li, and **d** Si_6H_6/6Li

4.5   Potential Applications

Silicenes are expected to be developed for use in various applications, such as in electronic and photonic devices and lithium ion battery electrodes. In this section, the potential applications of silicenes are reviewed. Silicenes functionalized with phenyl and larger aromatic molecules are expected to be useful in solar cells because such silicenes have UV–vis absorption maxima at around 270 nm (Fig. 4.16b). In particular, as shown in Fig. 4.16a, Ph-capped silicene ($Si_6H_4Ph_2$) produced a light-induced current when it was directly irradiated using a xenon lamp. The photocurrent disappeared when the irradiation was blocked using a shutter, and no photocurrent was observed when a filter was used to block wavelengths of less than 420 nm. The photocurrent was, therefore, wavelength dependent, which indicated that the origin of the photocurrent was a bandgap transition.

Silicon nanomaterials are also expected to find applications in lithium ion battery anodes in order to increase the energy density. Silicon is known to form lithium alloys, and when Li–Si alloys are formed electrochemically, the maximum theoretical capacity is approximately 4200 mAh g^{-1}, which is much higher than the capacity achieved using conventional carbonaceous materials (372 mAh g^{-1}). However, the volume of a silicon anode changes greatly during the charge-discharge cycle, leading to cracking and pulverization of the anode and the loss of silicon particles over numerous cycles, resulting in significant capacity fade. A composite anode containing silicon nanomaterials is a plausible solution to this problem because such an anode could have minimal volume changes, avoiding the pulverization and failure of the Li alloy anode through charge-discharge cycling. Silicenes alone could also be candidate materials for use as anodes because of their small size. As the results of examining the electrical capacitance and structural changes, an electrode made of silicenes was found to have higher capacitance

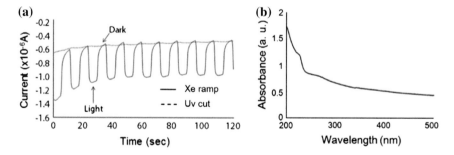

Fig. 4.16 a Light-induced photocurrent in the [$Si_6H_4Ph_2$] film. The *solid line* (Xe) indicates photocurrent in the sample with direct irradiation, and the *dotted line* (42 L) indicates photocurrent with irradiated light above 410 nm wavelength obtained using a filter. **b** UV–vis spectrum of the [$Si_6H_4Ph_2$] film

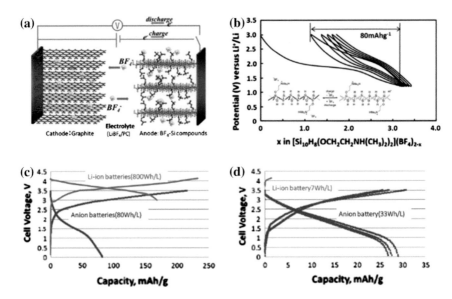

Fig. 4.17 **a** Schematic illustration of an anion battery. **b** Potential-composition profile. **c** and **d** Initial charge-discharge curves obtained by applying a constant current of 100 mA g^{-1} at 40 °C **b** and –30 °C. *Red lines* are an anion batteries. *Blue lines* are Li-ion batteries

(1620 mAh g^{-1}) and smaller volume change than a Si powder electrode during the lithium insertion and de-insertion process [51, 52].

The Li-ion batteries most commonly employed today use an electrode combination based on a graphite or silicon negative electrode and a LiCoO$_2$, LiMn$_2$O$_4$, or LiFePO$_4$ positive electrode. These Li-ion batteries are commercially available and dominate the consumer electronics market. However, their use in more demanding applications, such as sustainable transport, is prevented because of a number of drawbacks. The main drawback is safety during thermal runaway. Therefore, an anion battery has been proposed. It is based on the fabrication of tetrafluoroborate (BF$_4$) anion rocking-chair-type secondary batteries as an alternative approach to using Li cations (Fig. 4.17a). In this system, as the anode material, the Si-based compound [Si$_{10}$H$_8$(OCH$_2$CH$_2$NH(CH$_3$)$_2$)$_2$](BF$_4$)$_2$ was found to be capable of reacting with two BF$_4$ anions per formula unit at a potential of 1.8 V, giving a reversible capacity of 80 mAh g^{-1} (Fig. 4.17b). This anion battery also performed better than Li-ion batteries when they were thermally abused. Furthermore, anion batteries could operate at a temperature of −30 °C; however, at this temperature, Li-ion batteries cannot generally operate because Li-ions in such batteries become solvated by coordinating with basic organic molecules (Fig. 4.17c and d) [53].

4.6   Conclusions

A method for the soft synthesis of silicene by chemically modifying layered silicon compounds has been reviewed. The produced silicenes maintained their characteristic 2D crystalline structure and exhibited photoluminescent characteristics that were related to their structures. Phenyl-capped silicenes also exhibit light-induced photocurrent under direct irradiation with visible light. Although there is presently very little data on the chemical and physical properties of silicenes, the properties should be tunable by several different methods of surface or compositional modification. It is expected that this new synthetic approach will be developed by employing various techniques and theoretical calculations should be performed to predict the product properties. Silicenes have extraordinarily high specific surface areas and they can be used as building blocks for the fabrication of devices by solution processes. Silicenes are expected to be developed for applications in various fields that can take advantage of their specific characteristics.

References

1. L.T. Canham, Appl. Phys. Lett. **57**, 1046–1048 (1990)
2. V. Lehmann, U. Gçssele, Appl. Phys. Lett. **58**, 856–858 (1991)
3. B.K. Teo, X.H. Sun, Chem. Rev. **107**, 1454–1532 (2007)
4. R.A. Bley, S.M. Kauzlarich, J. Am. Chem. Soc. **118**, 12461–12462 (1996)
5. R.D. Tilley, J.H. Warner, K. Yamamoto, I. Matsui, H. Fujimori. Chem. Commun. 1833–1835 (255)
6. A.M. Morales, C.M. Lieber, Science **279**, 208–211 (1998)
7. W. Molnar, A. Lugstein, P. Pongratz, N. Auner, C. Bauch, E. Bertagnolli, Nano Lett. **10**, 3957–3961 (2010)
8. J. Sha, J. Niu, X. Ma, J. Xu, X. Zhang, Q. Yang, D. Yang, Adv. Mater. **14**, 1219–1221 (2002)
9. M. De Crescenzi, P. Castrucci, M. Scarselli, M. Diociaiuti, P.S. Chaudhari, C. Balasubramanian, T.M. Bhave, S.V. Bhoraskar, Appl. Phys. Lett. **86**, 231901 (2005)
10. T. Sasaki, M. Watanabe, H. Hashizume, H. Yamada, H. Nakazawa, J. Am. Chem. Soc. **118**, 8329–8335 (1996)
11. L. Wang, T. Sasaki, Chem. Rev. **114**, 9455–9486 (2014)
12. R. Ma, T. Sasaki, Adv. Mater. **22**, 5082–5104 (2010)
13. D.M.C. MacEwan, M.J. Wilson, in Crystal Structures of Clay Minerals and Their X-ray Identification, ed. G.W. Brindley, G. Brown, Mineralogical Society: London, 1980
14. L.F. Nazar, A.J. Jacobson, Chem. Commun. 570–571 (1986)
15. G. Alberti, M. Casciola, U. Costantino, J. Colloid Interface Sci. **107**, 256–263 (1985)
16. N. Yamamoto, T. Okuhara, T. Nakato, J. Mater. Chem. **11**, 1858–1863 (2001)
17. M.M.J. Treacy, S.B. Rice, A.J. Jacobson, J.T. Lewandowski, Chem. Mater. **2**, 279–286 (1990)
18. T. Sasaki, M. Watanabe, J. Am. Chem. Soc. **120**, 4682–4689 (1998)
19. R.E. Schaak, T.E. Mallouk, Chem. Mater. **14**, 1455–1471 (2002)
20. T. Hibino, W. Jones, J. Mater. Chem. **11**, 1321–1323 (2001)
21. L. Li, R. Ma, Y. Ebina, N. Iyi, T. Sasaki, Chem. Mater. **17**, 4386–4391 (2005)
22. R. Ma, K. Takada, K. Fukuda, N. Iyi, Y. Bando, T. Sasaki, Angew. Chem. Int. Ed. **47**, 86–89 (2008)
23. R.F. Frindt, J. Appl. Phys. **37**, 1928–1929 (1966)

24. R.F. Frindt, Phys. Rev. Lett. **28**, 299–301 (1972)
25. K.S. Novoselov, A.K. Geim, S.V. Morozov, D. Jiang, Y. Zhang, S.V. Dubonos, I.V. Grigorieva, A.A. Firsov, Science **306**, 666–669 (2004)
26. H. Yang, J. Heo, S. Park, H.J. Song, D.H. Seo, K.E. Byun, P. Kim, I. Yoo, H.J. Chung, K. Kim, Science **336**, 1140–1143 (2012)
27. G.G. Guzmán-Verri, L.C. Lew Yan Voon. Phys. Rev. B **76**, 075131 (2007)
28. A. Fleurence, R. Friedlein, T. Ozaki, H. Kawai, Y. Wang, Y. Yamada-Takamura, Phys. Rev. Lett. **108**, 245501 (2012)
29. J. Evers, J. Solid State Chem. **28**, 369–377 (1979)
30. S. Dick S, G. Öhlinger, Z. Krist. New Cryst Struct **213**, 232 (1998)
31. R. Yaokawa, H. Nakano, M. Ohashi, Acta Mater. **81**, 41–49 (2014)
32. E. Noguchi, K. Sugawara, R. Yaokawa, T. Hitosugi, H. Nakano, T. Takahashi, Adv. Mater. **27**, 856–860 (2015)
33. M.S. Brandt, G. Vogg, M. Stutzmann, in Silicon Chemistry, eds: P. Jutzi, U. Schubert, Wiley-VCH, Weinheim 194–213 (2003)
34. J.R. Dahn, B.M. Way, F. Fuller, J.S. Tse, Phys. Rev. B **48**, 17872–17877 (1993)
35. S. Yamanaka, H. Matsu-ura, M. Ishikawa, Mater. Res. Bull. **31**, 307–316 (1996)
36. Y. Sugiyama, H. Okamoto, H. Nakano, Chem. Lett. **39**, 938–939 (2010)
37. A. Weiss, G. Beil, H. Meyer. Z. Naturforsch. **34b**, 25–30 (1979)
38. H. Nakano, M. Ishii, H. Nakamura. Chem. Commun. (Camb.) 2945–2947 (2005)
39. H. Nakano, T. Mitsuoka, M. Harada, K. Horibuchi, H. Nozaki, N. Takahashi, T. Nonaka, Y. Seno, H. Nakamura, Angew. Chem. Int. Ed. **45**, 6303–6306 (2006)
40. J.M. Buriak. Chem. Commun. 1051–1060 (1999)
41. H. Nakano, M. Nakano, K. Nakanishi, D. Tanaka, Y. Sugiyama, T. Ikuno, H. Okamoto, T. Ohta, J. Am. Chem. Soc. **134**, 5452–5455 (2012)
42. H. Haick, P.T. Hurley, A.I. Hochbaum, P. Yang, N.S. Lewis, J. Am. Chem. Soc. **128**, 8990–8991 (2006)
43. Y. Sugiyama, H. Okamoto, T. Mitsuoka, T. Morikawa, K. Nakanishi, T. Ohta, H. Nakano, J. Am. Chem. Soc. **132**, 5946–5947 (2010)
44. A. Bansal, X. Li, S.I. Yi, W.H. Weinberg, N.S. Lewis, J. Phys. Chem. B **105**, 10266–10277 (2001)
45. J.M. Buriak, Chem. Rev. **102**, 1271–1308 (2002)
46. Y.-C. Liao, J.T. Roberts, J. Am. Chem. Soc. **128**, 9061–9065 (2006)
47. S.M. Sze, Semiconductor Devices, Physics and Technology, Wiley, London (2002)
48. H. Okamoto, Y. Sugiyama, K. Nakanishi, T. Ohta, T. Mitsuoka, H. Nakano, Chem. Mater. **27**, 1292–1298 (2015)
49. T. Ikuno. H. Okamoto, Y. Sugiyama, H. Nakano, F. Yamada, I. Kamiya, Appl. Phys. Lett. **99**, 023107 (2011)
50. M. Ohashi, H. Nakano, T. Morishita, M.J.S. Spencer, Y. Ikemoto, C. Yogi, T. Ohta, Chem. Commun. **50**, 9761–9764 (2014)
51. Y. Kumai, H. Kadoura, E. Sudo, M. Iwaki, H. Okamoto, Y. Sugiyama, H. Nakano, J. Mater. Chem. **21**, 11941–11946 (2011)
52. Y. Kumai, S. Shirai, E. Sudo, J. Seki, H. Okamoto, Y. Sugiyama, H. Nakano, J. Power Sources **196**, 1503–1507 (2011)
53. H. Nakano, Y. Sugiyama, T. Morishita, M.J.S. Spencer, I.K. Snook, Y. Kumai, H. Okamoto, J. Mater. Chem. A **2**(20), 7588–7592 (2014)

Chapter 5
Theoretical Studies of Functionalised Silicene

Michelle J.S. Spencer and Tetsuya Morishita

Abstract It was discussed in Chap. 4 how surface modification of silicene with different chemical groups has been shown to alter its properties. In this Chapter we focus on density functional theory (DFT) and first principles calculations of functionalised silicene and its structural, electronic and dynamic properties. We review the work on silicene that has been functionalized with different molecular groups and atoms and also discuss how a modeling approach can be used to examine how these layers stack together and interact with each other.

5.1 Introduction

It is well known that modification of the surface of a material can be used to alter it properties. The electronics industry has been using chemical functionalization of silicon (Si) for years to stabilize the Si surface and to provide properties conducive for electronic devices [1]. By altering the chemical groups attached to the nano-material surface, the stability, structure, adhesion, electronic properties can be tuned.

For silicon to be used as components in semiconductors or electronic type devices, its surface has to be modified, or capped, in order to prevent oxidation. An oxidised surface is usually not suitable for electronic applications because it provides an insulating barrier.

M.J.S. Spencer (✉)
School of Science, RMIT University, GPO Box 2476,
Melbourne, VIC 3001, Australia
e-mail: michelle.spencer@rmit.edu.au

T. Morishita (✉)
Research Center for Computational Design of Advanced Functional Materials,
National Institute of Advanced Industrial Science and Technology (AIST),
1-1-1 Umezono, Tsukuba, Ibaraki 305-8568, Japan
e-mail: t-morishita@aist.go.jp

© Springer International Publishing Switzerland 2016
M.J.S. Spencer and T. Morishita (eds.), *Silicene*, Springer Series
in Materials Science 235, DOI 10.1007/978-3-319-28344-9_5

When the bulk silicon surface is cleaved to reveal a surface, the topmost Si atoms lose their 4-fold coordination and are instead under coordinated. As a result, they can reconstruct so as to increase their coordination, with the 2×1 or 7×7 being common reconstructions occurring on the (111) surface of Si. The bulk cleaved surfaces are also highly reactive and therefore can bond to other chemical species as they attempt to rebalance the coordination lost due to formation of the surface. Adsorption of foreign atoms or molecules to saturate the dangling bonds can be used to stabilize the surface.

The same principle applies to silicene, which as discussed in previous chapters can be considered as resembling a single layer of a Si(111) surface, where all the atoms in the layer are under-coordinated. This material in the isolated or free form (as opposed to adsorbed on a surface) is yet to be characterized experimentally. However, as discussed in Chap. 4, the single layer material can be stabilized by termination of its surface by different chemical groups. Experimentally, functionalization of silicene with a mixed termination of H atoms and organic groups has been shown to produce a material that is stable in air (see Chap. 4 for details).

Using a theoretical approach a number of studies have examined functionalization of silicene and thicker nanosheets with a variety of elements and chemical groups. Attachment of hydrogen atoms to silicene or thicker nanosheets is the simplest termination that has been considered, with other elements and small organic groups also having been modeled. Some or all of the under-coordinated Si atoms have been bonded to atoms composed of the same element or molecular group, with the chemical modification being made to either both or only one side of the nanosheet.

In this chapter we review the studies that have examined functionalized silicene using a computational approach. There are also a number of studies that have examined the adsorption of small molecules on silicene (see for example adsorption of NH_3, NO and NO_2 [2], $MnCl_3$ [3], NO, NO_2, O_2, NH_3, SO_2, CO_2, and CO [4], CH_2O [5], organic molecules (acetone, acetonitrile, ammonia, benzene, methane, methanol, ethanol, and toluene [6]), and Li [7] to mention a few. Some of these are discussed in Chap. 11. The reader is also referred to a review on modified silicene, which also covers silicene doped with different atoms [8]. The focus here, however, will be on groups that are chemically attached to the surface at high, or specifically, full coverage. We discuss the structure, and associated properties of these materials.

5.2   Computational Methods Used to Examine Functionalized Silicene

Density functional theory is the most commonly employed method to examine functionalized silicene and is the focus of the majority of the studies reviewed here. Primarily, the generalized gradient approximation (GGA) is preferred over the local density approximation (LDA), with a variety of functionals having been employed.

For LDA, the Perdew-Zunger [9] functional has been used, while for the GGA, the exchange-correlation functionals have included those of Perdew-Burke-Ernzerhhof (PBE) [10] and Perdew-Wang (PW91) [11]. In order to include the effect of van der Waals forces, the vdW-DF [12–14] and vdW-DF2 [15] methods as well as the DFT-D2 method of Grimme [16] have been employed. The primary basis sets have included plane waves (PW) and double zeta basis set (DZ-BS) with polarization functions [17]. The ionic cores and valence electrons have been described using Trouiller-Martins (TM) norm-conserving pseudopotentials (NC-PP) [18], ultrasoft pseudopotentials (USPP) [19] and projector augmented wave functions (PAW) [20]. More recently a number of hybrid functionals have been used, in part to try and calculate more accurate band gaps, as DFT is well known to underestimate the size of the band gap. Some of these functionals have included M06-L [21, 22] and HSEH1PBE [23, 24] which were used with Pople's 6-31G* basis set [25]. The main codes used in these studies have included VASP [26–28], QuantumEspresso [29], SIESTA [17] and CASTEP [30].

5.3 Atom-Functionalised Silicene

We start by reviewing the studies that have used a computational approach to examine silicene terminated by different atoms, with the surface coverage consisting of one element or a mixture of elements. The most common termination for silicene that has been modelled is hydrogen, for example [31–34], with other elements that have been examined including silicene terminated with halogens [31–33, 35] and alkali earth and light metals [36–39], such as Li and Mg.

5.3.1 Hydrogen

Functionalisation of silicene with hydrogen can be thought of as the simplest modification of the surface. There is a large number of studies that have modeled hydrogen terminated silicene using DFT. Table 5.1 summarises these structures, together with their reported band gaps. The fully terminated structure is called silicane, in the same vain as graphane, which is hydrogen terminated graphene having a formula unit $(CH)_n$.

Hydrogenation of silicene is shown to open up a gap in the band structure, and may be a way to tune the electronic properties of silicene. Most studies are in agreement that increasing the coverage of H on the silicene surface opens up the band gap, with the band gap at full coverage being ~ 2.3 eV.

The group of Voon et al. [46] examined hydrogenated silicene in two different configurations; (a) the chair-like configuration, where neighboring H atoms are alternating on both sides of the sheet; and (b) the boat-like configuration, where H atoms are alternating in pairs. They found the chair-like configuration to be more

Table 5.1 Calculated band gap of silicene functionalised with hydrogen at full coverage, unless otherwise stated

Year	Ref.	Method	Band gap (eV)	Direct/indirect	Comments
2015	[31]	Vdw-DF, NC-PP, DZ-BS MO6-L, 6-31G* HSEH1PBE, 6-31G*	2.4 2.9 3.2	Direct Direct Direct	vdw-DF value M06-L HSE
2015	[34]	GGA (PBE/HSE06), PAW	2.94	Indirect	Chair-like conformation
2015	[40]	GGA (PBE), PAW	2.2	Indirect	Chair-like
2015	[41]	GGA (PBE), USPP	2.3	NA	
2012	[42]	GGA, PW	0.95	Direct	H atoms attached to 1 side
2012	[43]	GGA (PW91), PAW HSE06	2.36/1.6 3.51/2.41	Indirect/direct Indirect/direct	Chair-like/Boat-like (full) Chair-like/Boat-like
2012	[35]	GGA (sX-LDA), NC-PP	2.764	Indirect	
2011	[44]	Tight binding (TB)	2.2	Indirect	
2011	[45]	LDA (DZ-BS) NC-PP, HSE (Heyd-Scuseria-Ernzerhof) GW	2.0/1.6 4.0/3.3 3.8/2.9	Indirect/direct Indirect/direct Indirect/direct	Chair-like/boat-like Chair-like/boat-like Chair-like/boat-like
2010	[46]	LDA&GGA, TM-PP	∼2.0	Indirect	Chair-like

stable by 25 meV/atom. They then examined the band structure of silicane in more detail using tight binding (TB) calculations [44]. They showed that silicane is a semiconductor with an indirect gap of 2.2 eV, with the lowest point of the conduction band (CB) occurring at the M-point and the highest point of the valence band (VB) occurring at the Γ-point. The indirect nature of the band gap is mainly due to the band filling effect, rather than a gap opening of silicene, with the degeneracy at the Dirac point not being lifted by hydrogenation. Further investigations of partially hydrogenated silicane [47], with hydrogenation ratios between 3.1 and 100 %, showed that the more energetically favored structures were those that minimised unpaired electrons and lattice tension. The band structure could be tuned so as to undergo transformation from as zero gap semiconductor to an insulator by varying the hydrogenation ratio.

Houssa et al. [45] also examined the chair- and boat- like configurations of silicane as well as a third configuration (referred to as the top configuration) where H atoms are adsorbed on 1 side of the silicene sheet (see Fig. 5.1).

The chair-like and boat-like configurations were found to be almost degenerate (within ∼ 10 meV/atom), with both being ∼ 0.35 eV/atom more stable than the top configuration. Both configurations were determined using phonon calculations to be local energy minima as they had no imaginary frequencies. The band gap was found to open up upon hydrogenation due to the transition from sp^2 to sp^3 hybridization of

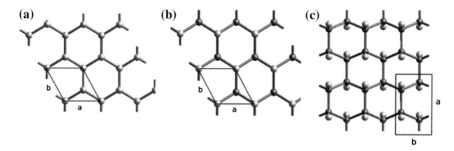

Fig. 5.1 Three different atomic configurations of silicane and germanane: **a** top configuration, where the H atoms are on the same side of the silicene/germanene plane, **b** chair-like configuration, where H atoms are alternating on both sides of the plane, and **c** boat-like configuration, where the H atoms are alternating in pairs. Si and H atoms are represented by *gray* and *white spheres*, respectively. Reprinted with permission from [45]. Copyright [2011], AIP Publishing LLC

the Si atoms. Interestingly, the gap was indirect for the chair-like configuration (2.0 eV), in agreement with Voon et al. [46] and direct for the boat-like configuration (1.6 eV). Using the hybrid Heyd-Scuseria-Ernzerhof (HSE) functional, these band gaps were calculated to be 4.0 and 3.3 eV, respectively.

Zhang and Yan [42] showed using DFT calculations, that it is possible to introduce ferromagnetism into silicene by surface modification with hydrogen at partial coverages. Half-hydrogenation of silicene (where only 1 side of the sheet was functionalized with H atoms) was shown to break the extended π-bonding network of silicene, leaving the electrons in the unsaturated Si atoms unpaired and localized. This structure had a band gap of 0.95 eV and exhibited ferromagnetic semiconducting behavior.

Zhang et al. [43] also showed the chair-like conformer to be more stable, having an indirect gap of 2.36 eV, whereas the boat-like conformer has a direct gap of 1.6 eV. Three configurations of silicane were studied with half-hydrogenation. For these 3 structures the H atoms were removed from 1 side of the silicene sheet, while arranging the H atoms on the other side in the following arrangements: chair-like (H atoms alternating), boat-like (H atoms distributed in pairs) and zigzag (H atoms adsorbed in zigzag chains along the silicene sheet). The band gaps for these 3 structures were 0.84 eV (indirect), 0.53 eV (direct) and no gap (metal), respectively. Using the hybrid HSE06 functional, the band gaps were all larger, as expected, however, the gap was calculated to be direct for the half-hydrogenation chair-like configuration. Overall, only the chair-like configuration with half-hydrogenation showed ferromagnetism.

Nguyen et al. [41] showed that the hydrogen atoms are more strongly bound to silicene than to graphene. The Si atoms in silicane are positively charge, in contrast to the C atoms in graphane, which are negatively charged.

Using a hybrid functional, Rupp et al. [34] calculated an indirect gap of 2.94 eV for silicane in the chair-like conformation. The structure has a buckling distance of 0.72 Å and binding energy of -3.39 eV (defined as the difference between the total energy of the stable structure and the total energies of the respective isolated atoms in their neutral charge state).

Morishita et al. [48] showed that functionalisation of the double layer (DL) silicene nanosheet with hydrogen results in an indirect band gap of 1.2 eV. The presence of hydrogen atoms removes the bands associated with the dangling bonds of Si which then contribute to the formation of sigma bonds between the Si and H atoms. As the hydrogenation ratio of the structure increases, the band gap is seen to open up with gap states appearing at concentrations over ~ 50 %. Over a concentration of 80 %, the gap states become highly localized and hence, the partially hydrogenated DL silicene can be considered a p-type semiconductor. While not necessarily considered a functionalized nanosheet, doping of the DL silicene with P also gives a p-type semiconductor material, which is in contrast to P doping of bulk Si. Interstitial doping of the DL-SiH nanosheet with Na or S results in an n-type semiconductor as they contribute donor states near the conduction band minimum. Hence, it is possible to obtain n- or p-type Si nanosheet based semiconductors depending on the functionalization and doping.

Huang et al. [48] also examined the bilayer structure and showed that hydrogenation converts the band gap from indirect to direct. By varying the hydrogen concentration and whether one or both sides are hydrogenated, they showed that the band gap was tunable, showing the material could be promising for optoelectronic applications.

Denis [31] calculated the interaction energy (per Si atom) of 2 stacked layers of silicane was 0.0356 eV (using the M06-L functional), and had a direct band gap of 2.4 eV, the same as the single layer structure using the VDW-DF functional. Interestingly they determined the gap of the single layer silicane to be direct.

Further modifications to silicane, by doping, have also been examined theoretically. Some examples include doping with P and B [34, 49], N [34], Al [50], alkali and alkali earth metals (we note that while it is called doping the metals were adsorbed on the silicane) [51]. Adatom adsorption of Li, F, Sc, Ti and V on silicane have also been examined by van Der Broek et al. [52] (Table 5.2).

5.3.2 Halogens

DFT calculations have been used to model silicene that is functionalized with fluorine (F) [31, 32, 35], chlorine (Cl) [35], bromine (Br) [33, 35] and iodine (I) [35] (Table 5.2). Halogens, and in particular fluorine, have a greater electronegativity than hydrogen, and hence can be used as a way to tune the electronic and magnetic properties of nanomaterials.

Gao et al. [35] examined silicene terminated with F, Cl, Br and I (Fig. 5.2). All the Si atoms of the nanosheet were terminated on both sides with these atoms. These structures were calculated to have positive formation energies, so therefore are likely to be capable of synthesis experimentally.

Table 5.2 Calculated band gap of silicene functionalised with different halogens and Li

Atom	Year	Ref.	Method	Band gap (eV)	Direct/indirect	Comments
F	2015	[31]	Vdw-DF (DZ-BS), NC-PP MO6-L (6-31G*) HSEH1PBE (6-31G*)	1.1 1.2 1.6	Direct Direct Direct	vdw-DF M06-L HSE
	2012	[35]	GGA (X-LDA), NC-PP	1.469	Direct	
Cl	2012	[35]	GGA (sX-LDA), NC-PP	1.979	Direct	
Br	2012	[35]	GGA (sX-LDA), NC-PP	1.950	Direct	
	2012	[33]	GGA, PW	1.47	Direct	Both sides functionalised
I	2012	[35]	GGA (sX-LDA), NC-PP	1.194	Direct	
Li	2012	[36]	GGA (PBE) LDA (PZ), DZP, NN-PP	0.368 –	Direct	

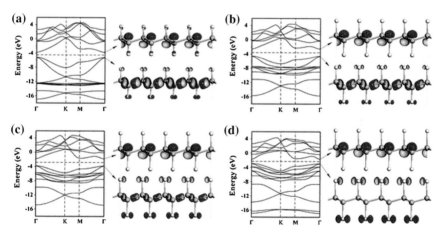

Fig. 5.2 Electronic band structures (*left*) and charge density distributions of the CBM (*right upper*) and the VBM (*right lower*) states at the Γ point for **a** F-silicene, **b** Cl-silicene, **c** Br-silicene, and **d** I-silicene. The Fermi level is labeled with a *dashed line*. The isosurface value is 0.025 au. Reproduced from [35] with permission of the PCCP Owner Societies

The presence of terminating halogens was shown to widen the band gap of silicene, while all of the halogen-terminated structures had a band gap smaller than that of silicane (see Table 5.2). As the size of the halogen increases, the band gap was found to increase at first, from 1.469 eV (F) to 1.979 eV (Cl), and then decrease to 1.950 eV (Br) and 1.194 eV (I). The widening of the band gap was due to the different electronegativity of the halogens. As the Si-X bonds are partly ionic in character, they weaken the Si–Si bonding, which decreases the bonding-antibonding splitting and hence the band gap. The decrease in the size of the band gap for Br and I was due to the competition between the energy of the Si–Si bonds and the Si-X bonds.

Zheng and Zhang [33] also modelled Br-terminated silicene. They showed that the fully saturated silicene exhibits nonmagnetic semiconducting behavior with a direct band gap of 1.47 eV, in general agreement with Gao et al. [35]; the larger band gap calculated by Gao et al. may be due to the fact they used a local exchange correlation functional and norm-conserving pseudopotentials. For saturation with Br atoms on one side of the silicene only, the band gap increased to 1.73 eV, and exhibited antiferromagnetic behavior.

Most recently, Denis [31] showed that for covalent functionalisation of silicene with F, compared to functionalisation with H, CH_3 or OH, the Si-F bond was strongest with a calculated binding energy value of 114.9 kcalmol^{-1}. For bilayer silicene, functionalisation with F was shown to alter the electronic properties of the material, reducing the band gap and inducing metallic properties for silicene. As the calculations were performed with hybrid functionals it is not possible to compare the band gaps directly with previous studies.

A number of other studies have examined silicene doped or functionalized with halogens (and OH) at different concentrations on one side of the sheet, including the work by Huang et al. [32].

5.3.3 Alkali Earth Metals

Osborn and Farajian [36] showed that functionlisation of silicene with lithium (Li) at full coverage (referred to as silicel) transforms silicene from a zero-gap semiconductor structure to a 0.368 eV bandgap semiconductor (Fig. 5.3). Molecular dynamics simulations at 300 and 900 K showed that the structure retained its stability at the elevated temperatures, but started to break apart at 1500 K unlike pristine silicene which still retains its structure.

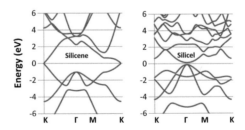

Fig. 5.3 Band structure for bare and fully lithiated silicene obtained by using GGA. The unit cells contain two Si atoms. Fermi energy is set to zero. Reprinted with permission from [36]. Copyright (2012) American Chemical Society

Huang et al. [53] showed that the adsorption energy of Li on silicene is significantly greater than on graphene, with the adsorption energy increasing with increasing concentration of Li. Regardless of Li concentration, the low diffusion energy barrier on silicene means that it is a suitable material for future Li ion battery anodes.

Hussain et al. [37] investigated functionalisation of silicene with a variety of metals including Li, Na, K, Be, Mg and Ca. They showed that the light alkali metals (Li and Na) bind more strongly (1.71 and 1.69 eV, respectively) than the alkaline earth metal adatoms (K, Be, Mg, Ca) that have binding energy values per adatom of 1.05, 1.56, 1.60 and 1.24 eV, respectively. Molecular dynamics simulations showed that these structures remain intact and are stable at 400 K for up to 4 ps indicating that they may be useful for practical applications at reasonably high temperatures. For Li and Na, both adatoms transfer charge to the silicene nanosheet, producing a cationic (Li, Na)/anionic(Si) structure. For both of these structures they were shown to be capable of adsorbing up to five H_2 molecules with average adsorption energy values lying within the range for practical hydrogen storage systems under ambient conditions. A structure functionalized with a mixture of Mg and H atoms [37] was also shown to adsorb up to 6 H_2 molecules per Mg adatom. Interestingly, Song et al. [38] determined that Ca prefers to adsorb in a hollow site on silicene, however, they showed that this material can also adsorb 9 H_2 molecules per Ca atom (6.4 wt%). Application of an external electric field could effectively enhance the hydrogen storage-release on the Ca-silicene system. Wang et al. [39] also showed that silicene decorated with Na, K, Mg and Ca can have high storage capacity.

5.4 Molecular-Functionalised Silicene

There are a variety of organic molecular fragments that have been used to modify the surface of silicene, including hydroxyl (OH) [31], methyl (CH_3) [31, 54], phenyl, (C_5H_5) [40, 55, 56], phenol (C_5H_5OH) [57], 2-(dimethylamino)ethanol or deanol ($OCH_2CH_2NH(CH_3)_2$) [58], acetylene (C_2H_2), ethylene (C_2H_4), styrene ($C_6H_5CH=CH_2$) [59], hexyl (C_6H_{13}) [54], propyl (CH_2–CH_2–CH_3), amine ($CH_3CH_2NH_2$) and ethoxy (CH_3CH_2O) [40] groups. These are discussed in detail in this section, with the calculated band gaps values (where available) shown in Table 5.3.

5.4.1 Hydroxyl and Methyl

Functionalization of silicene with hydroxyl (OH) and methyl (CH_3) groups has been considered by Denis [31]. For OH-terminated silicene, the groups were shown to cause larger structural changes due to the presence of intralayer hydrogen bonding (see Fig. 5.4), which does not occur for the CH_3-terminated nanomaterial. For CH_3 groups, a small gap semiconductor is achieved, while termination with OH

Table 5.3 Calculated band gap of silicene functionalized with different molecular groups

Chemical group	Year	Ref.	Method	Band gap (eV)	Direct/indirect	Comments
OH	2015	[31]	Vdw-DF (NC-PP)	0.7	Direct	vdw-DF
			MO6-L (6-31G*)	1.2	Direct	MO6-L
			HSEH1PBE (6-31G*)	1.5	Direct	HSE
Methyl	2015	[31]	Vdw-DF (DZ-BS) NC-PP	1.8	Direct	vdw-DF
			MO6-L (6-31G*)	2.2	Direct	MO6-L
			HSEH1PBE (6-31G*)	2.5	Direct	HSE
Phenyl	2015	[40]	GGA (PBE), PAW	2.0	Indirect	$Si_{18}C_{36}H_{42}$
Ethoxy	2015	[40]	GGA (PBE), PAW	1.7	Direct	$Si_{18}C_{12}O_6H_{42}$
Amine	2015	[40]	GGA (PBE), PAW	1.7	Direct	$Si_{18}C_{12}N_6H_{48}$
Propyl	2015	[40]	GGA (PBE), PAW	1.8	Indirect	$Si_{18}C_{18}H_{54}$
Nitrophenyl diazonium	2015	[60]	GGA (PBE) DFT-D2/HSE06, PAW	0.79	Indirect	NDP:Si ratio = 1:8
				0.30	NA	NDP:Si ratio = 1:18
				0.26	NA	NDP:Si ratio = 1:32
Deanol	2014	[58]	GGA (PBE), PAW	–	–	$Si_{10}H_8(OCH_2CH_2NH(CH_3)_2$
Methyl	2013	[54]	GGA (PBE)	1.67	Direct	$Si_8H_4(CH_3)_4$
Hexyl	2013	[54]	GGA (PBE)	1.66	Direct	$Si_8H_4(C_6H_{13})_4$
Phenol	2013	[57]	GGA (PBE) PAW	1.88	Direct	$Si_6H_4(C_6H_4OH)_2$, para-
Phenyl	2011	[55]	GGA (PBE) PAW	1.92	Direct	$Si_6H_4(C_6H_5)_2$

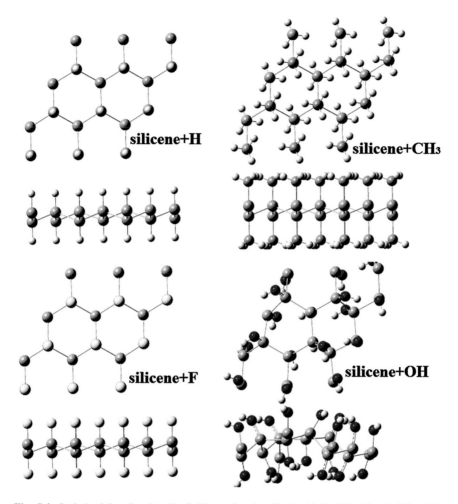

Fig. 5.4 Optimized 3 × 3 unit cell of silicene functionalized with H, CH₃, F and OH at full coverage, at the M06-L/6-31G* level (*top* and *side views* are presented for each structure). Reproduced from [31] with permission of the PCCP Owner Societies

groups results in silicene becoming metallic. Asymmetric, or 'Janus', functionalization of silicene, was also considered, where different molecules or atoms are attached to the two sides, or faces, of the silicene. Stacking of 2 layers of functionalised silicene was shown to reduce the band gap.

5.4.2   Phenyl-Modified Silicene

In Chap. 4, we saw that the group of Nakano et al. [61] synthesized
phenyl-modified silicene nanosheets, using an organic synthesis method and
characterized the structure and properties of the material using a variety of experimental techniques. They also performed forcefield (FF) calculations to determine
the possible structure and orientation of the attached phenyl groups. The structure
was determined to have two phenyl groups attached above and below the silicene
sheet with hydrogen atoms terminating the remaining under-coordinated Si atoms,
with a repeat unit of $[Si_6H_4Ph_2]$ (see Fig. 5.5). Subsequent DFT calculations of this
structure [55] showed that the phenyl groups are covalently bonded to the sheet and
prefer not to be arranged parallel to each other but on an angle or tilted. As the
phenyl groups are located more than 6.6 Å from each other on either side of the
sheet, the effect of long range forces on the orientation or interaction between
adjacent groups is likely to be negligible. For this reason van der Waals type forces
were not included in the calculations. Using ab initio molecular dynamics simulations, it was shown that the phenyl groups continually rotate and tilt on the sheet
at a simulation temperature of 300 K, due to the weak interaction between adjacent
groups, however, they remain covalently bonded, contributing to its overall
stability.

The vibrational density of states (VDOS) of this structure were in excellent
agreement with the measured IR spectrum, confirming the structure of the material
and also identifying modes associated with the rotation and bending motion of the
phenyl groups at ~ 30 and ~ 180 cm^{-1}, respectively, which had not been identified
experimentally.

The structure was calculated to have a direct band gap of 1.92 eV (Fig. 5.6),
which is larger than the corresponding DFT calculated value for bulk Si of 0.7 eV.
This gap is similar to the later study by Wang et al. [40] who also calculated an
indirect band gap of 2.0 eV. While the size of the gap is only ~ 0.08 eV smaller
than that for the hydrogen-terminated silicene structure, it is interesting to note that
the latter has an indirect gap. Hence, the introduction of phenyl groups produces a
material that would show interesting optical properties.

Fig. 5.5 Optimized structure
of phenyl-modified silicene

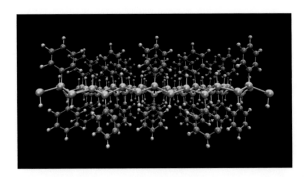

Fig. 5.6 Band structure of
the organosilicon nanosheet.
The band energy is measured
from the highest occupied
level at T = 0 K. Reproduced
from [55] with permission
from the PCCP Owner
Societies

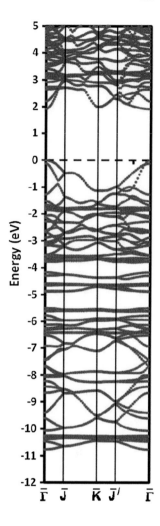

5.4.2.1   Stacking and Interaction Between Layers

Synthesis of phenyl-modified silicene results in a colorless paste composed of multiple layers within this product, with the likely separation distance between the layers being ~ 1.0 nm. Adhesion energy calculations examining the interaction between the nanosheets at different separation distances [56] were able to confirm that the optimum stacking separation is ~ 1 nm. As the interaction between the nanosheets is weak, a variety of methods that take into account van der Waals forces were used, including the DFT-D2 method of Grimme [16], and the vdW-DF [12–14] and vdW-DF2 [15] methods. The latter functionals included the original vdW-DF method (referred to as revPBE), the optimized vdW-DF methods (optPBE, optB88 and optB86b) and the vdW-DF2 method (rPW86). The PBE DFT

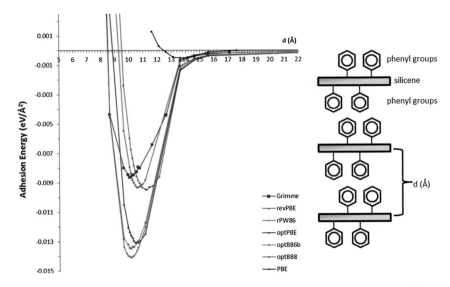

Fig. 5.7 *Left* Calculated adhesion energy curves for phenyl modified silicene at different separations, *d*, using the functional indicated (Reproduced with permission from [56]); *right* schematic picture of the stacking of the sheets, showing the interlayer separation (*d*)

functional was also used as a comparison. Adhesion energy curves using each of these functionals (see Fig. 5.7) clearly showed that there was little interaction between the sheets using the PBE functional with the minimum in the curve being at a separation distance of 13.64 Å.

For the other functionals, they showed a stronger interaction between the nanosheets. Overall, the interaction is still weak indicating that these sheets are held together by van der Waals type forces, similar to other weakly bonded inorganic compounds [62]. The DFT-DF2 method gave an adhesion energy of 8.6 meV $Å^{-2}$, the original vdw-DF method and the vdW-DF2 method gave similar adhesion energies ($\sim$9.4 meV $Å^{-2}$), and the three optimised DFT-D2 methods gave stronger adhesion energies ($\sim$13.5 meV $Å^{-2}$). Of the latter 3 methods, the optB88 gave the strongest adhesion energy at a separation of 10.14 Å which is in excellent agreement with the experimental XRD value. Using this method, it was shown that the gap increases slightly as the sheets are pushed closer together and compressed, and shifts from a direct gap to an indirect gap (see Table 5.1). At the same time some redistribution of the electron density results in a transfer of electrons back into the phenyl groups due to some π-π interaction between the phenyl groups, which was confirmed by the DOS which showed a greater contribution from the C $2p_x$ and $2p_y$ states at the bottom of the conduction band at the closer separation distances.

Hence, the DFT calculations were able to determine the adhesive force between the stacked layers, and how the separation distance between the sheets could modify the structural and electronic properties. It is also of importance to note that inclusion of van der Waals forces is required when modelling the interaction between these layered materials.

5.4.3 Phenol-Modified Silicene

In the previous section, the phenyl groups on the silicene can be considered as hydrophobic, which explains why the nanomaterial is soluble in organic solvents. In order to widen the applications of silicene it would be useful for it to be soluble in non-organic solvents and in particular water. One modification that can be made to the phenyl–modified silicene studied in the previous section is to substitute a hydroxyl group on the phenyl rings. While this nanomaterial is yet to be synthesized experimentally, Spencer et al. [57] performed DFT calculations of phenol modified silicene, where a hydroxyl group was substituted onto the rings of the phenyl modified-silicene structure, at one of three different substitution sites, otho-, meta- and para- (see Fig. 5.8). When van der Waals forces were not included in the calculations, the meta-substituted system was the most stable. However, after including VdWs forces by using the DFT-D2 functional by Grimme, the para-substituted structure was found to be the most stable, followed by the meta- and then ortho- systems. It was clear from the optimized structures that the phenol rings rotate and tilt on the silicene surface so as to form hydrogen bonds between adjacent phenol rings. While the groups remain covalently bonded to the silicene the presence of the OH groups enhances the binding between the groups. The calculated band gap was found to be 1.88 eV, which is similar to that of the phenyl-modified silicene. The lack of major contributions from the O atoms to the electronic states at the bottom or top of the conduction and valence bands, respectively, explains the similar electronic properties.

5.4.4 Other Organic Groups

As discussed in Chap. 4, hexyl-modified silicene has been synthesized by the group of Nakano et al. [63]. This material was subsequently modelled using DFT calculations and (first principles and force-field) molecular dynamics simulations by Li et al. [54]. This structure has hexyl groups attached to both sides of the silicene

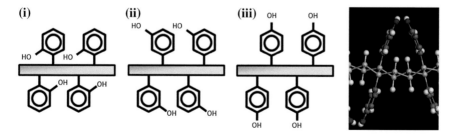

Fig. 5.8 Schematic diagram (*left*) of phenol modified silicene with the hydroxy (OH) groups attached to the phenyl ring in the (*i*) ortho-, (*ii*) meta-, and (*iii*) para-positions; structure of para-hydoxy phenyl-modified silicene (*right*)

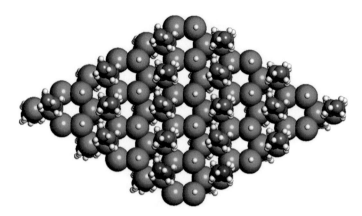

Fig. 5.9 Top down view of the hexyl-modified silicene. Reprinted with permission from [54]. Copyright (2013) American Chemical Society

sheet at a separation distance of 7.17 Å, which was determined from the experimental synthesis (see Fig. 5.9). Using FF-MD they showed that the hexyl-modified sheet maintained its structure up to a simulation temperature of 1000 K and time of 100 ps. The calculated band gap was determined to be direct with a size of 1.65 eV. The band gap could be changed from 1.95 to 1.13 eV by applying stress to the material during the simulation. The size of the band gap, however, was found to be insensitive to an increase in the length of the attached chains.

Rubio-Pereda and Takeuchi [64] examined silicene functionalised with the unsaturated hydrocarbons acetylene, ethylene, styrene (Fig. 5.10). They determined how alkyne and alkene molecules would react with hydrogenated silicene, showing that the rate determining step for this chain reaction is abstraction of a hydrogen atom. The intermediate state of the reaction consists of the unsaturated organic molecule attached to the surface forming an Si–C bond and hence a highly reactive carbon radical. The transition state corresponds to the point where the Si–H bond is broken and the final state occurs after the new C–H bond is formed and a new Si dangling bond is created.

They also showed that adsorption of a second molecule is achievable, and would allow a self-promoting process and hence formation of an organic monolayer. In order to do this, the first attached molecule has to rotate after the hydrogen abstraction to accommodate a second molecule in the newly created dangling bond. The energy loss due to steric interaction between two molecules is small, allowing the reaction to proceed.

Another organic functional group that has been used to functionalise silicene is deanol [58]. This modified silicene nanomaterial [$Si_{10}H_8(OCH_2CH_2NH(CH_3)_2)_2$] (see Fig. 5.11) was used in a rocking chair type secondary battery, as detailed in Chap. 4. The material was capable of reacting with two BF_4 anions per formula unit at a potential of 1.8 V, giving a reversible capacity of 80 mA h g^{-1}, in a battery set up with the modified silicene as the anode, and graphite as the cathode. Using DFT calculations, the structure and properties of the nanomaterial were modeled both in

Fig. 5.10 Potential energy surface along the minimum energy path for the reaction of: **a** acetylene, **b** ethylene, and **c** styrene with H-silicene. Silicon, carbon, and hydrogen atoms are represented by *black*, *red*, and *blue spheres*, respectively. Reprinted with permission from [64]. Copyright (2013), AIP Publishing LLC (Color figure online)

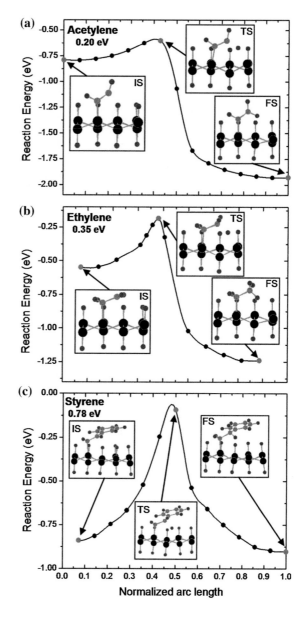

the presence and absence of the BF_4 anions in order to gain an insight into the attachment-detachment process. It was shown that dissociation of the HBF_4 molecule, which resulted in the protonated deanol group and BF_4 anion, was key to the variation in charge within the silicene nanomaterial.

Wang et al. [40] modelled propyl-, amine- and ethoxy- modified silicene with an approximate surface coverage of 33 %. The remaining under-coordinated Si atoms

(b)   **(c)**

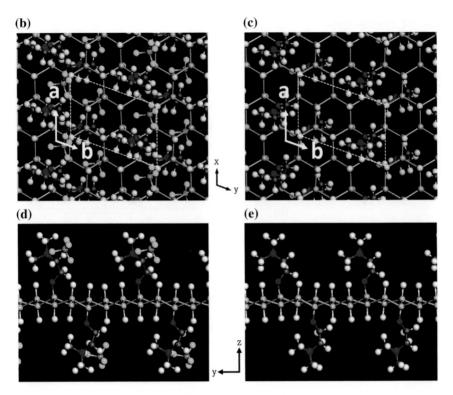

Fig. 5.11 b–e Top view and side view of the atomic arrangement of the deanol modified Si nanosheet with (**b, d**) and without (**c, e**) the presence of BF₄ anions; Si (*yellow spheres*), H (*white spheres*), O (*red spheres*), C (*gray spheres*), N (*blue spheres*), B (*pink spheres*), and F (*light blue spheres*). The *white dashed line* indicates the unit cell. Reproduced (in part) from [58] with permission from The Royal Society of Chemistry

were terminated by H atoms. All organic modifying groups led to a reduction in the size of the band gap compared to silicane. Propyl-modified silicene was an indirect band gap semiconductor, while the ethoxy- and amine- terminated silicene were direct band gap semiconductors. Changes in the optical adsorption induced by the organic modifications were correlated to the corresponding changes in the band structure. By increasing the coverage of the propyl groups, from 5 to 11 % and 33 to 44 %, the band gap varied in the range 1.6–1.9 eV, being either direct or indirect.

Dai and Zeng [60] modelled functionalization of silicene with nitrophenyl diazonium (NDP) for 3 different ratios of NDP to Si of 1:8, 1:18 and 1:32. They showed that the band gap decreased upon reducing the coverage of NDP. They also showed that the silicene loses about 0.65 e to each NPD functional group, and exhibits spin-polarized ferrimagnetism with a net magnetic moment of about 0.33 mB per supercell. The structure with a 1:8 ratio of NDP:Si was predicted to be a bipolar magnetic semiconductor, which could make it a potential candidate for spintronic applications.

5.5 Conclusion

This chapter has reviewed the different chemical groups or elements that have been used to functionalize silicene, using a computational modeling approach and specifically DFT calculations. Silicane (silicene terminated with H atoms) is the most widely studied chemical modification of silicene and is a wide band gap semiconductor, in contrast to the semi-metallic silicene. Functionalization of silicene and thicker nanosheets with hydrogen and other elements or chemical groups prevents reconstruction of the Si surface atoms, while allowing the electronic properties of the material and in particular the size of the band gap to be modified. In some cases functionalization with a combination of elements or groups can also be used to achieve different sized band gaps. In other cases, termination of the silicene on only one side can be used to tune the band gap. Overall, functionalisation is a highly useful approach for tuning the electronic properties of silicene, while generally retaining the underlying silicene structure. The flexibility provided by chemical modification of silicene, and the ability to modify the electronic properties suggest that this material is highly suitability for applications in optoelectronics, electronics, batteries, and potentially sensors.

Acknowledgements Some of the work in this chapter was undertaken with the assistance of the following computational facilities: the National Computational Infrastructure (NCI) through the National Computational Merit Allocation Scheme supported by the Australian Government, the V3 Alliance (formerly VPAC—the Victorian Partnership for Advanced Computing), the Multi-modal Australian ScienceS Imaging and Visualisation Environment (MASSIVE), the Pawsey Supercomputing Centre (formerly iVEC), and facilities at the Research Centre for Computational Science, National Institute of National Sciences, and the Research Institute for Information Technology, Kyushu University, Japan.

References

1. D.D.M. Wayner, R.A. Wolkow, J. Chem. Soc. Perkin Transac. **2**, 23–34 (2002). doi:10.1039/B100704L
2. W. Hu, N. Xia, X. Wu, Z. Li, J. Yang. Phys. Chem. Chem. Phys. **16**, 6957–6962 (2014). doi:10.1039/C3CP55250K
3. T. Zhao, S. Zhang, Q. Wang, Y. Kawazoe, P. Jena, Phys. Chem. Chem. Phys. **16**, 22979–22986 (2014). doi:10.1039/C4CP02758B
4. J.-w. Feng, Y.-j. Liu, H.-x. Wang, J.-x. Zhao, Q.-h. Cai, X.-z. Wang. Comp. Mater. Sci. **87**, 218–226 (2014). doi:10.1016/j.commatsci.2014.02.025
5. X. Wang, H. Liu, S.-T. Tu, RSC Adv. **5**, 65255–65263 (2015). doi:10.1039/C5RA12096A
6. T.P. Kaloni, G. Schreckenbach, M.S. Freund, J. Phys. Chem. C **118**, 23361–23367 (2014). doi:10.1021/jp505814v
7. J. Setiadi, M.D. Arnold, M.J. Ford, ACS Appl. Mat. Interfaces **5**, 10690–10695 (2013). doi:10.1021/am402828k
8. R. Wang, M.-S. Xu, X.-D. Pi, Chin. Phys. B **24**, 086807 (2015)
9. J.P. Perdew, A. Zunger, Phys. Rev. B **23**, 5048–5079 (1981)
10. J.P. Perdew, K. Burke, M. Ernzerhof, Phys. Rev. Lett. **77**, 3865–3868 (1996)

11. J.P. Perdew, Y. Wang, Phys. Rev. B **45**, 13244–13249 (1992)
12. K. Jiří, R.B. David, M. Angelos, J. Phys.: Condens. Matter **22**, 022201 (2010)
13. G. Román-Pérez, J.M. Soler, Phys. Rev. Lett. **103**, 096102 (2009)
14. M. Dion, H. Rydberg, E. Schröder, D.C. Langreth, B.I. Lundqvist, Phys. Rev. Lett. **92**, 246401 (2004)
15. K. Lee, É.D. Murray, L. Kong, B.I. Lundqvist, D.C. Langreth, Phys. Rev. B **82**, 081101 (2010)
16. S. Grimme, J. Comput. Chem. **27**, 1787–1799 (2006). doi:10.1002/jcc.20495
17. J.M. Soler, E. Artacho, J.D. Gale, A. Garcia, J. Junquera, P. Ordejon, D. Sanchez-Portal, J. Phys. Condens Matter **14**, 2745–2779 (2002)
18. N. Troullier, J.L. Martins, Phys. Rev. B **43**, 1993–2006 (1991)
19. D. Vanderbilt, Phys. Rev. B **41**, 7892–7895 (1990)
20. P.E. Blochl, Phys. Rev. B **50**, 17953–17979 (1994)
21. Y. Zhao, D. Truhlar, Theor. Chem. Acc. **120**, 215–241 (2008). doi:10.1007/s00214-007-0310-x
22. Y. Zhao, D.G. Truhlar. J. Chem. Phys. **125**, 194101 (2006). doi:10.1063/1.2370993
23. J. Heyd, G.E. Scuseria. J. Chem. Phys. **120**, 7274–7280 (2004). doi:10.1063/1.1668634
24. V. Barone, G.E. Scuseria. J. Chem. Phys. **121**, 10376–10379 (2004). doi:10.1063/1.1810132
25. R. Ditchfield, W.J. Hehre, J.A. Pople. J. Chem. Phys. **54**, 724–728 (1971). doi:10.1063/1.1674902
26. G. Kresse, J. Furthmuller. Phys. Rev. B **54**, 11169–11186 (1996)
27. G. Kresse, J. Furthmuller. Comp. Mater. Sci. **6**, 15–50 (1996)
28. G. Kresse, J. Hafner, Phys. Rev. B **47**, 558 (1993)
29. G. Paolo et al., J. Phys.: Condens. Matter **21**, 395502 (2009)
30. M.D. Segall, P.J.D. Lindan, M.J. Probert, C.J. Pickard, P.J. Hasnip, S.J. Clark, M.C. Payne, J. Phys. Condensed Matter **14**, 2717–2744 (2002)
31. P.A. Denis, Phys. Chem. Chem. Phys. **17**, 5393–5402 (2015). doi:10.1039/C4CP05331A
32. B. Huang, H.J. Xiang, S.-H. Wei, Phys. Rev. Lett. **111**, 145502 (2013)
33. F.B. Zheng, C.W. Zhang, Nanoscale Res. Lett. **7**, 422 (2012). doi:10.1186/1556-276x-7-422
34. C.J. Rupp, S. Chakraborty, R. Ahuja, R.J. Baierle, Phys. Chem. Chem. Phys. **17**, 22210 (2015). doi:10.1039/C5CP03489B
35. N. Gao, W.T. Zheng, Q. Jiang, Phys. Chem. Chem. Phys. **14**, 257–261 (2012). doi:10.1039/C1CP22719J
36. T.H. Osborn, A.A. Farajian, J. Phys. Chem. C **116**, 22916–22920 (2012). doi:10.1021/jp306889x
37. T. Hussain, S. Chakraborty, A. De Sarkar, B. Johansson, R. Ahuja. Appl. Phys. Lett. **105**, 123903 (2014). doi:10.1063/1.4896503
38. E.H. Song, S.H. Yoo, J.J. Kim, S.W. Lai, Q. Jiang, S.O. Cho, Phys. Chem. Chem. Phys. **16**, 23985–23992 (2014). doi:10.1039/C4CP02638A
39. Y. Wang, R. Zheng, H. Gao, J. Zhang, B. Xu, Q. Sun, Y. Jia. Int. J. Hydrogen Energy **39**, 14027–14032 (2014). doi:10.1016/j.ijhydene.2014.06.164
40. R. Wang, X. Pi, Z. Ni, Y. Liu, D. Yang, RSC Advances **5**, 33831–33837 (2015). doi:10.1039/C5RA05751E
41. M.-T. Nguyen, P.N. Phong, N.D. Tuyen. Chem. Phys. Chem, n/a–n/a (2015). doi:10.1002/cphc.201402902
42. C.-W. Zhang, S.-S. Yan, J. Phys. Chem. C **116**, 4163–4166 (2012). doi:10.1021/jp2104177
43. P. Zhang, X.D. Li, C.H. Hu, S.Q. Wu, Z.Z. Zhu. Phys. Lett. A **376**, 1230–1233 (2012). doi:10.1016/j.physleta.2012.02.030
44. G.G. Guzmán-Verri, L.C.L.Y. Voon, J. Phys.: Condens. Matter **23**, 145502 (2011)
45. M. Houssa, E. Scalise, K. Sankaran, G. Pourtois, V.V. Afanas'ev, A. Stesmans. Appl. Phys. Lett. **98** (2011). doi:10.1063/1.3595682
46. L.C.L.Y. Voon, E. Sandberg, R.S. Aga, A.A. Farajian. Appl. Phys. Lett. **97** (2010). doi:10.1063/1.3495786
47. T.H. Osborn, A.A. Farajian, O.V. Pupysheva, R.S. Aga, L.C. Lew Yan Voon. Chem. Phys. Lett. **511**, 101–105 (2011). doi:10.1016/j.cplett.2011.06.009

48. B. Huang, H.-X. Deng, H. Lee, M. Yoon, B.G. Sumpter, F. Liu, S.C. Smith, S.-H. Wei, Phys Rev X **4**, 021029 (2014)
49. X. Pi, Z. Ni, Y. Liu, Z. Ruan, M. Xu, D. Yang, Phys. Chem. Chem. Phys. **17**, 4146–4151 (2015). doi:10.1039/C4CP05196C
50. F. Sánchez-Ochoa, J. Guerrero-Sánchez, G. Canto, G. Cocoletzi, N. Takeuchi, J. Mol. Model. **19**, 2925–2934 (2013). doi:10.1007/s00894-013-1873-1
51. T. Hussain, T. Kaewmaraya, S. Chakraborty, R. Ahuja, Phys. Chem. Chem. Phys. **15**, 18900–18905 (2013). doi:10.1039/c3cp52830h
52. B. van den Broek, M. Houssa, E. Scalise, G. Pourtois, V.V. Afanas'ev, A. Stesmans, Appl. Surf. Sci. **291**, 104–108 (2014). doi:10.1016/j.apsusc.2013.09.032
53. J. Huang, H.-J. Chen, M.-S. Wu, G. Liu, C.-Y. Ouyang, B. Xu. Chin. Phys. Lett. **30** (2013). doi:10.1088/0256-307x/30/1/017103
54. F. Li, R. Lu, Q. Yao, E. Kan, Y. Liu, H. Wu, Y. Yuan, C. Xiao, K. Deng, J. Phys. Chem. C **117**, 13283–13288 (2013). doi:10.1021/jp402875t
55. M.J.S. Spencer, M. Morishita, M. Mikami, I.K. Snook, Y. Sugiyama, H. Nakano, Phys. Chem. Chem. Phys. **13**, 15418–15422 (2011)
56. M.J.S. Spencer, M.R. Bassett, T. Morishita, I.K. Snook, H. Nakano, New J. Phys. **15**, 125018 (2013)
57. M.J.S. Spencer, T. Morishita, M.R. Bassett, in *Density Functional Theory Calculations of Phenol-Modified Monolayer Silicon Nanosheets*, ed by J. Friend, H.H. Tan Micro/Nano Materials, Devices, and Systems, vol. 8923. Proceedings of SPIE (2013)
58. H. Nakano, Y. Sugiyama, T. Morishita, M.J.S. Spencer, I.K. Snook, Y. Kumai, H. Okamoto, J. Mater. Chem. A **2**, 7588–7592 (2014). doi:10.1039/C4TA00456F
59. P. Rubio-Pereda, N. Takeuchi. J. Chem. Phys. **138** (2013). doi:10.1063/1.4804545
60. J. Dai, X.C. Zeng, Phys. Chem. Chem. Phys. **17**, 17957–17961 (2015). doi:10.1039/C4CP04953E
61. Y. Sugiyama, H. Okamoto, T. Mitsuoka, T. Morikawa, K. Nakanishi, T. Ohta, H. Nakano, J. Am. Chem. Soc. **132**, 5946–5947 (2010). doi:10.1021/ja100919d
62. T. Björkman, A. Gulans, A.V. Krasheninnikov, R.M. Nieminen, Phys. Rev. Lett. **108**, 235502 (2012)
63. H. Nakano, M. Nakano, K. Nakanishi, D. Tanaka, Y. Sugiyama, T. Ikuno, H. Okamoto, T. Ohta, J. Am. Chem. Soc. **134**, 5452–5455 (2012). doi:10.1021/ja212086n
64. P. Rubio-Pereda, N. Takeuchi, J. Phys. Chem. C **117**, 18738–18745 (2013). doi:10.1021/jp406192c

Chapter 6
Interaction Between Silicene and Non-metallic Surfaces

Michel Houssa, André Stesmans and Valeri V. Afanas'ev

Abstract Silicene has so far been successfully grown on metallic substrates, like Ag(111), ZrB_2(0001) and Ir(111) surfaces. However, characterization of its electronic structure is hampered by the metallic substrate. In addition, potential applications of silicene in nanoelectronic devices will require its growth/integration with semiconducting or insulating substrates. In this chapter, we review recent theoretical works about the interaction of silicene with several non-metallic templates, distinguishing between the weak van der Waals like interaction of silicene with e.g. AlN or layered metal (di)chalcogenides, and the stronger covalent bonding between silicene and e.g. ZnS surfaces. Recent experimental results on the possible growth of silicene on MoS_2 are also highlighted and compared to the theoretical predictions.

6.1 Introduction

Very recently, the formation of silicene was reported on various metallic surfaces, like (111) Ag surfaces [1–6], (0001) ZrB_2 surfaces [7, 8] and (111) Ir surfaces [9]. The electronic properties of the silicene layer on Ag(111) was investigated using angle-resolved photoemission spectroscopy. These measurements revealed the presence of a linear dispersion in the band structure of silicene (so called Dirac cones) with a Fermi velocity of about 1.3×10^6 m/s, as theoretically predicted for free-standing silicene. A recent breakthrough paved the way to the possible realization of silicene-based field effect transistors operating at room temperature [10], presenting ambipolar current-voltage characteristics, as expected for a gapless semiconductor.

M. Houssa (✉) · A. Stesmans · V.V. Afanas'ev
Department of Physics and Astronomy, University of Leuven, Celestijnenlaan 200D,
3001 Leuven, Belgium
e-mail: michel.houssa@fys.kuleuven.be

© Springer International Publishing Switzerland 2016
M.J.S. Spencer and T. Morishita (eds.), *Silicene*, Springer Series
in Materials Science 235, DOI 10.1007/978-3-319-28344-9_6

The possible existence of silicene was so far reported on these metallic substrates. However, the characterization of the electronic and electrical properties of silicene on metallic substrates is very challenging, since these properties are then largely dominated by the metal. The growth of silicene on semiconducting or insulating substrates is required for their firm identification and complete characterization. In addition, potential applications of this novel 2D material in nano-electronic devices will also require its growth and integration on non-metallic substrates.

We review here recent theoretical results, based on density-functional theory (DFT) calculations, pertaining to the interaction of silicene with non-metallic surfaces. We first discuss the weak (van der Waals) interaction of silicene with e.g. AlN and layered dichalcogenide substrates. On these templates, silicene is predicted to be either metallic or semi-metallic (with preserved Dirac cones at the K-points), depending on its buckling. Recent experimental results on the possible growth of silicene on MoS_2 are discussed and compared to the theoretical predictions.

We next discuss the covalent bonding of silicene on e.g. ZnS(0001) surfaces. The charge transfer occurring at the silicene/ZnS(0001) interface leads to the opening of an indirect energy band gap in silicene. Very interestingly, it is found that the nature (indirect or direct) and magnitude of its energy band gap can be controlled by an external electric field, a results potentially very interesting for field-effect devices.

6.2 Silicene/Substrate Interaction: Weak van der Waals Bonding

Layered semiconducting materials, with strong intra-layer covalent bonding and weak inter-layer van der Waals bonding, are expected to interact weakly with silicene, potentially preserving its peculiar electronic properties [11–18]. We discuss here the weak interaction of silicene with two different type of layered materials, namely graphite-like AlN [11, 12] and semiconducting transition metal dichalcogenides [13–15].

AlN is an insulator (energy band gap of about 6.5 eV [19]) which crystallizes in the Wurtzite phase, with in-plane lattice parameters a = b = 3.11 Å and out-of-plane lattice parameter c = 4.98 Å [20]. Very interestingly, the polar AlN(0001) surface is predicted to evolve to a more stable graphite-like structure [11, 21], with the Al and N atoms adopting a sp^2-hybridization, as shown in Fig. 6.1.

The graphite-like form of AlN is insulating, with a computed energy band-gap of about 4.6 eV, and is more stable than the (0001) AlN polar surface by about 0.27 eV/atom [11]. This predicted structural and electronic "phase transition" in AlN is consistent with first-principles calculations on ultra-thin wurtzite films [21]. The driving force for the planarization and sp^2-hybridization of the AlN layers is the suppression of the strong dipole between the bottom and the top surface of the film, which are terminated either by anions or cations; this transition depends on the

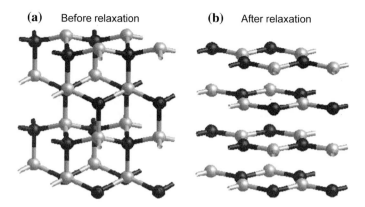

Fig. 6.1 Atomic configuration of a (0001) AlN slab model, **a** in wurtzite form (before structural relaxation) and **b** in graphite-like form (after relaxation). *Light gray* and *dark gray spheres* represent Al and N atoms, respectively

electronegativity difference between the anions and cations as well as the energy gap of the material, leading to a thickness dependence of this transition [21]. The graphite-like form of AlN is predicted to be more stable than the (0001) wurtzite structure up to 24 layers, corresponding to a layer thickness of about 2.6 nm. Very interestingly, the possible growth of graphite-like AlN on Ag(111) surfaces has been recently observed experimentally [22].

The possibility of inserting a silicene "flake" (i.e. a silicene ribbon terminated by H atoms) in-between a graphite-like AlN lattice was then considered [11, 12]. In this configuration, the top AlN layer could serve as an efficient barrier against the diffusion of chemical species towards the silicene surface. The starting configuration corresponds to a compressively strained flat silicene ribbon inserted between two AlN layers, as shown in Fig. 6.2.

During the energy relaxation, the Si–Si bond length increases and reaches its free-standing value of about 2.2 Å. After relaxation, the silicene layer is buckled, with a buckling distance of about 0.21 Å, which is lower than its predicted free-standing buckling distance (about 0.44 Å). In this configuration, the silicene layer weakly interacts with the AlN layers via van der Waals forces, the computation of the partial (Mulliken) atomic charges on the Si, Al, and N atoms indicating no net charge transfer between Si and the Al or N atoms. However, the weak interaction between the Si p_z orbitals and the out-of-plane dipole formed between the Al and N atoms from the bottom and top AlN layers, respectively, is likely responsible for the reduced silicene buckling, as compared to its free-standing configuration. In this van der Waals like AlN/silicene/AlN heterostruture, silicene is predicted to be a gapless semiconductor, as shown in Fig. 6.3, due to the preserved sp^2-sp^3 mixed hybridization of the Si atoms.

The interaction between silicene and layered semiconducting transition metal (di)chalcogenides was reported recently [13–15]. We discuss here the structural and electronic properties of silicene/MoX_2 substrates, with X = S, Se, Tc [14, 15].

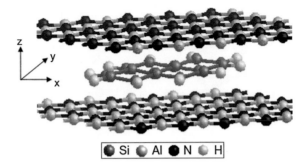

Fig. 6.2 Relaxed atomic configuration of an AlN/silicene/AlN (van der Waals) heterostructure. Reprinted with permission from [11]. Copyright 2010, AIP Publishing LLC

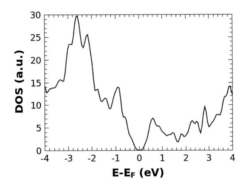

Fig. 6.3 Computed electronic density of states of the AlN/silicene/AlN heterostructure shown in Fig. 6.2. Reprinted with permission from [11]. Copyright 2010, AIP Publishing LLC

The computed in-plane lattice parameters and energy band gap of these substrates are given in Table 6.1.

The initial atomic configuration consists in a supercell with a flat silicene layer on top of "bulk" MoX_2 (which includes 4 atomic layers). The two bottom MoX_2 layers were kept fixed during relaxation and the initial distance between the 2D material and the top MoX_2 layer equals 4.5 Å. The cell parameter was kept to the one of the template, the silicene layer being compressively strained, as compared to

Table 6.1 Computed in-plane lattice parameters and energy band gaps of MoX_2

	In-plane lattice parameter (Å)	Energy band gap (eV)	Silicene buckling distance (Å)	Silicene lattice mismatch (%)
MoS_2	3.16	1.2	1.9	18
$MoSe_2$	3.30	1.1	1.0	14
$MoTe_2$	3.52	1.0	0.7	9

The silicene buckling distance and in-plane lattice mismatch between free-standing silicene and the dichalcogenide substrates are also indicated

the free-standing case. Three different possible arrangements of the silicon atoms with respect to the underlying Mo and X atoms were studied, as shown Fig. 6.4: (a) Si hexagons placed on top of the MoX_2 hexagons—so called AAA stacking like in h-BN, (b) Si hexagons shifted with respect to the MoX_2 hexagons by half a unit cell, and (c) Bernal-like arrangement (ABA stacking), like in graphite.

After energy relaxation, (a) and (c) structures kept their initial configuration, while structure (b) relaxed to configuration (c). The energy difference between the various configurations is typically less than 5 meV/atom, these atomic configurations being thus equally stable (degenerate). This indicates that the interaction between the silicene layer and the dichalcogenide substrate is very weak. On the other hand, the difference in energy between the initial (before relaxation) and the final (after relaxation) structures is more than 0.1 eV/atom. This energy difference is mainly due to the buckling of the silicene layer. In all the three structures, the 2D material indeed buckled after relaxation; the buckling distance is given in Table 6.1 for the silicene/MoX_2 system, together with the in-plane lattice parameter mismatch. From these results, the buckling distance is clearly correlated to the in-plane lattice mismatch. Note that the typical silicene-MoX_2 interlayer distance lies between 3 and 3.5 Å.

The predicted electronic structure of the silicene layer on the MoX_2 template largely depends on the buckling parameter in the 2D material. Highly buckled silicene is predicted to be metallic [14, 15, 23], as illustrated in Fig. 6.5 for the case of silicene on MoS_2.

On the other hand, the low buckled silicene layer on $MoTe_2$ is predicted to preserve its Dirac cones at the K points, as illustrated in Fig. 6.6. Silicene on $MoTe_2$ is thus predicted to be a gapless semiconductor [14], similar to free standing silicene. By increasing the in-plane lattice parameter of the dichalcogenide substrates, it is found out that the buckling distance in the silicene layer can be reduced and the 2D material can eventually preserves its gapless semiconducting behavior, i.e. the partial sp^2 hybridization of the Si atoms is preserved.

The possible growth of silicene on MoS_2 bulk crystals has been recently reported [24]. The STM image of a 0.8 monolayer of Si deposited at 200 °C on a $MoS_2(0002)$ crystal surface is shown in Fig. 6.7a.

(a)   (b)   (c)

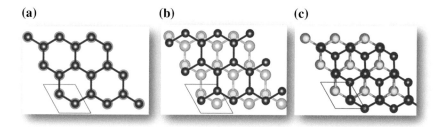

Fig. 6.4 Three different atomic arrangements of the silicene layer on top of MoS_2. *Blue, yellow* and *green* spheres are Si, S and Mo atoms, respectively. **a** AAA stacking. **b** Intermediate position. **c** ABA stacking

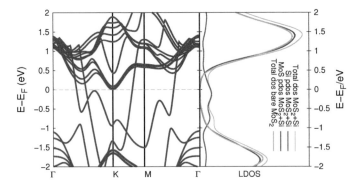

Fig. 6.5 Energy band structure and local density of states (LDOS) of the silicene/MoS$_2$ structure. The LDOS shows that the density of states of the MoS$_2$ substrate still preserve a gap very close to that of the bare MoS$_2$, while all the electronic states close to the Fermi level are due to the contribution of Si atoms, confirming that almost no "interaction" (e.g. hybridization) between Si and Mo/S atomic orbitals is induced

Fig. 6.6 Energy band structure of the silicene/MoTe$_2$ system

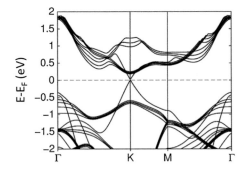

The formation of a uniform 2D Si layer is deduced from this figure, and is corroborated by the well-defined streaks of the reflection high energy electron diffraction pattern of Fig. 6.7b, resembling those of the pristine MoS$_2$ surface and hence indicating the pseudomorphic growth of a Si nanosheet with no apparent reconstruction. Local apertures in the uncompleted Si layer in the STM topography of Fig. 6.7a discriminates the apparently unstructured Si domains from the hexagonal pattern of the underlying MoS$_2$ surface.

This topographic discrepancy suggests a different tip-to-substrate tunneling condition between the two regions, which is related to the characteristic electronic character of the Si ad-layer with respect to the underlying MoS$_2$ surface. An ordered surface pattern can be recognized from the Si covered regions, see Fig. 6.7c, when concomitantly reducing the tip bias and increasing the tunneling current. The magnification of the Si domain reported in the inset of Fig. 6.7c shows well-defined hexagonal rings with a three-fold symmetry, suggesting that Si atoms self-organize

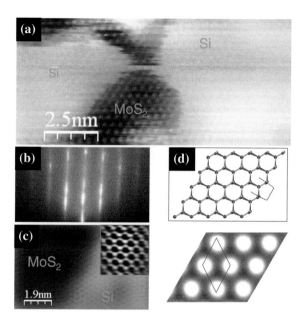

Fig. 6.7 a STM topography of 0.8 Si monolayer grown on a MoS_2(0002) surface at 200 °C (tunneling condition: tip bias 1 V, tunneling current 0.7 nA). Structured and unstructured regions refer to MoS_2 and Si, respectively. **b** Streaky RHEED pattern recorded after Si deposition. **c** STM topography at the boundary between the MoS_2 underlying surface and the Si wetting layer under the tunneling condition: tip bias 0.2 V, tunneling current 2 nA. *Inset* topographic magnification of the Si domain where to extract the structural features of the Si nanosheet lattice. **d** Top view of a minimum energy silicene/MoS_2 configuration showing the coincidence between the two lattices (*upper panel*) and respective STM topography simulation accounting for the alternation of consecutive upper and lower silicene atoms (*lower panel*)

to form a honeycomb lattice under the boundary conditions dictated by the underlying MoS_2 substrate. It should be noticed that the characteristic periodicity of the Si lattice is about 3.2 Å, very close to the MoS_2 lattice constant (3.16 Å). These experimental results confirm the effective role of the MoS_2 surface as a template in determining the atomistic arrangement of the Si nanosheet. This arrangement is consistent with the minimum energy configuration outlined by the DFT simulations discussed above, and corresponds to a highly compressed silicene monolayer supported by a MoS_2 substrate, as illustrated in Fig. 6.7d. According to this picture, the silicene layer naturally puckers as the Si atoms are distributed in two distinct planes whose vertical separation (buckling) is about 2 Å, giving rise to the simulated STM pattern in Fig. 6.7d (bottom image), in full agreement with the experimentally observed STM topography in the inset of Fig. 6.7c. Although the interaction between MoS_2 and the silicene layer is dominated by weak van der Waals interactions, this weak interaction most likely favors the formation of the metastable highly buckled (metallic) silicene phase.

6.3   Silicene/Substrate Interaction: Covalent Bonding

The formation of covalent bonds between silicene and an underlying substrate can result in the partial or complete sp^3 hybridization of the Si atoms, and consequently, in the opening of an energy gap in its electronic structure—like e.g. in silicene functionalized by the adsorption of ad-atoms [25–30]; we discuss here the covalent bonding between silicene and ZnS surfaces [31, 32], as a typical example of a stronger interaction (compared to the weak van der Waals bonding) between silicene and a non-metallic substrate.

ZnS crystallizes in the Wurtzite phase [33, 34] and is a semiconductor, with a direct energy band-gap of about 3.8 eV. Interestingly, its in-plane lattice constant (3.81 Å) is very close to the computed one of free-standing silicene, (about 3.9 Å), ZnS thus appearing as an ideal non-metallic template for the growth of silicene. A (0001) polar ZnS surface is considered here as a possible template for silicene [31]. A slab model with 8 atomic layers (64 atoms) and with a 15 Å vacuum layer was used for the DFT simulations. Displacements of the top and bottom ZnS layers was observed during atomic relaxation, resulting in a surface reconstruction very similar to the one of the non-polar ZnS(1010) surface [35, 36], as discussed in more details in [31, 37]. The reconstructed ZnS(0001) surface is semiconducting, with a computed energy gap of about 2.5 eV, and is predicted to be more stable than the non-reconstructed polar surface for layers up to about 6.6 nm [37]. Note that the polar (non-reconstructed) ZnS surface is metallic, due to the pining of the Fermi level by the anion surface states, like in ZnO [38, 39]. On such a polar surface, silicene is also predicted to be metallic [40].

To study the interaction of silicene with the reconstructed ZnS(0001) surface, a flat silicene layer was placed on top of the surface, followed by atomic relaxation. Different possible arrangements of the Si atoms on the ZnS(0001) surface were considered, as discussed in more details in [31]. The most energetically stable structure is presented in Fig. 6.8, and corresponds to a hexagonal arrangements of the Si atoms placed at intermediate positions between top and hollow sites of the ZnS hexagons. Two Si–S bonds and two Si–Zn bonds are formed, with a charge transfer essentially involving the 3p$_z$ orbitals of the Si atoms and the 4s states of Zn and 3p states of S, the bonded Si atoms thus adopting an sp^3-like character. Four other Si atoms are not bonded to the ZnS surface, two of these atoms lying at about 2.64 Å from the surface (marked "intermediate" on Fig. 6.8) and two other atoms lying at about 3.33 Å from surface (marked "top" on Fig. 6.8). The charge transfer at the silicene/(0001) ZnS interface leads to an excess of negative (Mulliken) charge of about 0.18 e on the top Si atoms, with respect to the intermediate Si atoms, resulting in the formation of a dipole at this interface. The average Si–Si distance (2.30 Å) is very similar to the one of free-standing silicene.

The silicene/(0001)ZnS interface is predicted to be semiconducting, with a computed indirect energy band gap of about 0.7 eV, as shown in Fig. 6.9. The energy gap opening in silicene is due to the charge transfer and partial sp^3 hybridization of the Si atoms bonded to the Zn or S atoms on the surface. The effect

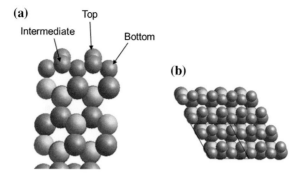

Fig. 6.8 Side view (**a**) and top view (**b**) of the relaxed silicene/(0001) ZnS slab model. *Yellow, gray* and *blue* spheres are S, Zn and Si atoms, respectively

of an out-of-plane electric field on the energy band structure of the system is also illustrated in Fig. 6.9 (dashed lines).

A periodic zigzag potential was applied in the direction perpendicular to silicene/(0001) ZnS interfaces to study the effect of an out-of-plane electric field on the electronic properties of this system [31]. The electric field has a substantial effect on the conduction band near the Γ point, leading to a transition from an indirect (Γ to Y point) to direct (at Γ point) energy band gap in silicene, for an electric field of about 0.5 V/Å, as indicated in Fig. 6.10. The electric-field dependence of the energy band gap of the silicene layer is related to the modulation of the electric dipole at the silicene/ZnS interface [31].

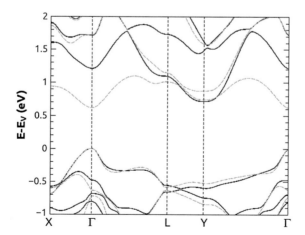

Fig. 6.9 Computed energy band structure of the silicene/(0001) ZnS slab model, without (*solid black lines*) and with (*dashed red lines*) an external electric field of 0.6 V/Å in the direction perpendicular to the interface. The reference (zero) energy level corresponds to the top of the valence band E_V of silicene. Republished with permission of the Royal Society of Chemistry, from [31]; permission conveyed through Copyright Clearance Center, Inc. (Color figure online)

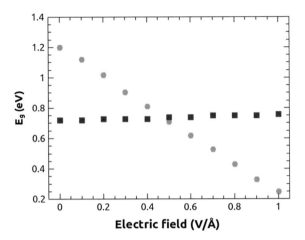

Fig. 6.10 Computed direct (*filled circles*) and indirect (*filled squares*) energy band gaps of the silicene/(0001) ZnS slab model, as a function of the external electric field applied to the system. Republished with permission of the Royal Society of Chemistry, from [31]; permission conveyed through Copyright Clearance Center, Inc.

6.4 Concluding Remarks

The recent progress on the growth and characterization of silicene on different substrates, are very encouraging. However, so far, silicene has been successfully grown on metallic surfaces. Its possible integration in future nanoelectronic devices will require its growth on non-metallic substrates. In this chapter, we have reviewed recent theoretical works on the interaction between silicene and non-metallic surfaces. The weak van der Waals interaction between silicene and e.g. transition metal dichalcogenides can potentially preserves the electronic properties of free-standing silicene, which is predicted to be a gapless semiconductor, like graphene. Very recently, encouraging results have been reported on the possible growth of silicene on MoS_2. However, DFT simulations indicate that silicene on MoS_2 is highly buckled and metallic. Low buckled silicene, with preserved Dirac cones, is predicted to be grown on e.g. $MoTe_2$, a prediction which still needs experimental confirmation. These results could pave the way to so-called silicene-based van der Waals heterostructures for high performances and low-power nanoelectronic applications.

On the other hand, the covalent bonding of silicene on e.g. ZnS surfaces leads to the opening of an energy gap in its electronic structure. Very interestingly, the magnitude and nature (direct or indirect) of this energy gap is predicted to be controlled by an out-of-plane electric field. This theoretical prediction is potentially very interesting for silicene-based logic devices.

Acknowledgments This work has been financially supported by the European Project 2D-NANOLATTICES, within the Future and Emerging Technologies (FET) program of the European Commission, under the FET-grant number 270749, as well as the KU Leuven Research Funds, project GOA/13/011. We are grateful to A. Molle (MDM Laboratory), A. Dimoulas (NCSR Demokritos), G. Pourtois (imec), E. Scalise (Max Planck Institute), B. van den Broek and K. Iordanidou, (KU Leuven) for their valuable contributions to this work and for stimulating discussions.

References

1. P. Vogt, P. De Padova, C. Quaresima, J. Avila, E. Frantzeskakis, M.C. Asensio, A. Resta, B. Ealet, G. Le Lay, Phys. Rev. Lett. **108**, 155501 (2012)
2. B. Feng, Z. Ding, S. Meng, Y. Yao, X. He, P. Cheng, L. Chen, K. Wu, Nano Lett. **12**, 3507 (2012)
3. D. Chiappe, C. Grazianetti, G. Tallarida, M. Fanciulli, A. Molle, Adv. Mat. **24**, 5088 (2012)
4. H. Enriquez, S. Vizzini, A. Kara, B. Lalmi H. Oughaddou. J. Phys. Condens. Matter **24**, 314211 (2012)
5. D. Tsoutsou, E. Xenogiannopoulou, E. Golias, P. Tsipas, A. Dimoulas, Appl. Phys. Lett. **103**, 231604 (2013)
6. P. Moras, T.O. Mentes, P.M. Sheverdyaeva, A. Locatelli, C. Carbone, J. Phys. Condens. Matter **26**, 185001 (2014)
7. A. Fleurence, R. Friedlein, T. Ozaki, H. Kawai, Y. Wang, Y. Takamura, Phys. Rev. Lett. **108**, 245501 (2012)
8. C.C. Lee, A. Fleurence, Y. Yamada-Takamura, T. Ozaki, R. Friedlein, Phys. Rev. B **90**, 075422 (2014)
9. L. Meng, Y. Wang, L. Zhang, S. Du, R. Wu, L. Li, Y. Zhang, G. Li, H. Zhou, W.A. Hofer, M. J. Gao, Nano Lett. **13**, 685 (2013)
10. L. Tao, E. Cinquanta, D. Chiappe, C. Grazianetti, M. Fanciulli, M. Dubey, A. Molle, D. Akinwande, Nature Nanotech. **10**, 227 (2015)
11. M. Houssa, G. Pourtois, V.V. Afanas'ev, A. Stesmans. Appl. Phys. Lett. **97**, 112106 (2010)
12. M. Houssa, G. Pourtois, M.M. Heyns, V.V. Afanas'ev, A. Stesmans. J. Electrochem. Soc. **158**, H107 (2011)
13. Y. Ding, Y. Wang, Appl. Phys. Lett. **103**, 043114 (2013)
14. E. Scalise, M. Houssa, E. Cinquanta, C. Grazianetti, B. van den Broek, G. Pourtois, A. Stesmans, M. Fanciulli. A. Molle, 2D Mater. **1**, 011010 (2014)
15. L.Y. Li, M.W. Zhao, J. Phys. Chem. C **118**, 19129 (2014)
16. J.J. Zhu, U. Schwingenschlögl, ACS Appl. Mat. Interf. **6**, 11675 (2014)
17. L. Linyang, W. Xiaopeng, Z. Xiaoyang, Z. Mingwen, Phys. Lett. A **377**, 2628 (2013)
18. S. Kokott, P. Pflugradt, L. Matthes, F. Bechstedt, J. Phys. Condens. Matter **26**, 185002 (2014)
19. M. Badylevich, S. Shamuilia, V.V. Afanas'ev, A. Stesmans, Y.G. Fedorenko, C. Zhao. J. Appl. Phys. **104**, 093713 (2008)
20. Y.-N. Xu, W.Y. Ching, Phys. Rev. B **48**, 4335 (1993)
21. C.L. Freeman, F. Claeyssens, N.L. Allan, J.H. Harding, Phys. Rev. Lett. **96**, 066102 (2006)
22. P. Tsipas, S. Kassavetis, D. Tsoutsou, E. Xenogiannopoulou, E. Golias, S.A. Giamini, C. Grazianetti, D. Chiappe, A. Molle, M. Fanciulli, A. Dimoulas, Appl. Phys. Lett. **103**, 251605 (2013)
23. S.S. Cahangirov, M. Topsakal, E. Aktürk, H. Sahin, S. Ciraçi. Phys. Rev. Lett **102**, 236804 (2009)
24. D. Chiappe, E. Scalise, E. Cinquanta, C. Grazianetti, B. van den Broek, M. Fanciulli, M. Houssa A. Molle. Adv. Mat. **26**, 2096 (2014)

25. L.C. Lew Yan Voon, E. Sandberg, R.S. Aga, A.A. Farajian. Appl. Phys. Lett. **97**, 163114 (2010)
26. M. Houssa, E. Scalise, K. Sankaran, G. Pourtois, V.V. Afanas'ev, A. Stesmans. Appl. Phys. Lett. **98**, 223107 (2011)
27. R. Quhe, R. Fei, Q. Liu, J. Zheng, H. Li, C. Xu, Z. Ni, Y. Wang, D. Yu, Z. Gao, J. Lu. Sci. Rep. **2**, 853 (2012)
28. Y. Ding, Y. Wang, Appl. Phys. Lett. **100**, 083102 (2012)
29. B. van den Broek, M. Houssa, E. Scalise, G. Pourtois, V.V. Afanas'ev, A. Stesmans. Appl. Surf. Sci. **291**, 104 (2014)
30. T.P. Kaloni, N. Singh, U. Schwingenschlögl, Phys. Rev. B **89**, 035409 (2014)
31. M. Houssa, B. van den Broek, E. Scalise, G. Pourtois, V.V. Afanas'ev, A. Stesmans. Phys. Chem. Chem. Phys. **15**, 3702 (2013)
32. S.S. Li, C.W. Zhang, S.S. Yan, S.J. Hu, W.X. Ji, P.J. Wang, P. Li, J. Phys. Condens. Matter **26**, 395003 (2014)
33. M.J. Weber (ed.) Handbook of Laser Science and Technology (CRC Press, Cleveland, 1986)
34. Y.-N. Xu, W.Y. Ching, Phys. Rev. B **48**, 4335 (1993)
35. J.E. Northrup, J. Neugebauer. Phys. Rev. B **53**, R10477 (1996)
36. A. Filippetti, V. Fiorentini, G. Cappellini, A. Bosin, Phys. Rev. B **59**, 8026 (1999)
37. X. Zhang, H. Zhang, T. He, M. Zhao, J. Appl. Phys. **108**, 064317 (2010)
38. A. Wander, F. Schedin, P. Steadman, A. Norris, R. McGrath, T.S. Turner, G. Thornton, N.M. Harrison, Phys. Rev. Lett. **86**, 3811 (2001)
39. B. Meyer, D. Marx, Phys. Rev. B **67**, 035403 (2003)
40. M. Houssa, B. van den Broek, E. Scalise, G. Pourtois, V.V. Afanas'ev, A. Stesmans. ECS Trans. **53**, 51 (2013)

Part III
Silicene on Substrates

Chapter 7
Silicene on Ag(111): Structure Evolution and Electronic Structure

Noriaki Takagi, Chun Liang Lin and Ryuichi Arafune

Abstract In the last few years, a significant challenge has been to realize the growth of silicene on solid substrates. Silicene can be synthesized by the deposition of Si atoms on Ag(111); this substrate is the most surveyed in the literature both theoretically and experimentally. In this chapter we focus on the geometric and electronic structures of silicene grown on Ag(111), from the sub-monolayer regime to the formation of multilayers, with the focus being on the 4×4 and $4/\sqrt{3} \times 4/\sqrt{3}$ superstructures. Several puzzles that we have solved are discussed, as well as ones that are yet to be solved.

7.1 Introduction

Two-dimensional (2D) materials are of great interest currently because of their exotic properties that could lead to their future applications from electronic and optical devices to green technology devices. Silicene, the 2D honeycomb lattice of Si atoms, has recently emerged as a rising star.

Honeycomb structures have attracted much attention since Ancient times. An Ancient Roman scholar, Marcus Terentius Varro, described the hexagonal form of the bee's honeycomb structure in his book published more than two thousand years ago [1]. He discussed in this book the reason why bees make the hexagonal honeycomb array structure. One explanation was that the hexagonal shape is suitable to accommodate bee's six feet, and the other is that the hexagonal chamber can store honey more efficiently compared to triangular and square ones. This puzzle is

N. Takagi (✉) · C.L. Lin
Department of Advanced Materials Science, Graduate School of Frontier Science,
University of Tokyo, Kashiwa 5-1-5, Chiba 277-8561, Japan
e-mail: n-takagi@k.u-tokyo.ac.jp

R. Arafune
International Center for Materials Nanoarchitectonics (WPI-MANA), National Institute
for Materials Science, 1-1 Namiki, Tsukuba, Ibaraki 304-0044, Japan

© Springer International Publishing Switzerland 2016
M.J.S. Spencer and T. Morishita (eds.), *Silicene*, Springer Series
in Materials Science 235, DOI 10.1007/978-3-319-28344-9_7

known mathematically as classical honeycomb conjecture, that the honeycomb is the best way to partition a surface into regions of equal area with the smallest total perimeter. This conjecture had remained unsolved until a proof was provided by Thomas Hales very recently [1]. In addition to this famous mathematical puzzle, profound physics is hidden in the honeycomb structure [2]. The two-dimensional (2D) honeycomb lattice hosts relativistic particles, i.e., massless Dirac fermions. This hidden property was uncovered in graphene, the one-atom thick honeycomb lattice of carbon atoms, by Andre Geim, Konstantin Novoselov and co-workers in the 2000s [3–6]. They successfully fabricated graphene by peeling off a single layer from graphite using regular adhesive tape (Geim and Novoselov were awarded the Nobel Prize in Physics in 2010 for this groundbreaking discovery). Since then, research exploring the exotic properties of 2D materials has accelerated intensively. Silicene, the 2D honeycomb structure consisting of Si atoms, has emerged along with this research direction.

It was 1994 the first report on the 2D honeycomb lattice of Si atoms was presented. Takeda and Shiraishi (denoted as TS) [7] investigated the thermodynamic stability of freestanding silicene together with freestanding germanene (the 2D honeycomb of Ge atoms) and their electronic structures by density functional theory (DFT) calculations. TS found that silicene and germanene are stable and take buckled honeycomb structures in which two atoms in the unit cell are displaced perpendicular to the basal plane in opposite directions. They also demonstrated that π and π^* bands linearly cross at the Fermi level which is a clear signature of a massless Dirac fermion. Note that TS did not discuss the Dirac fermion in their paper. These results were confirmed later by DFT calculations [8–11]. Since silicene and its bulk counterpart do not exist in nature, in contrast to graphene, it is very challenging to reproduce the theoretical results experimentally. Despite the exciting theoretical predictions, little attention was paid to the DFT results of TS. Recently, silicene has been synthesized on several substrates such as Ag(110) [12–14], Ag(111) [15–40], Au(110) [41], ZrB_2-covered Si(111) [42], Ir(111) [43], and MoS_2 [44]. These works demonstrate that silicene is no longer only a virtual material inside supercomputers. The term, "silicene" was coined by Guzman-Verri and Voon [8]. According to organic chemistry nomenclature, the suffix "ene" is used to describe an organic molecule whose main framework consists of π-double bonds. Thus, strictly speaking, "silicene" should be used to describe a honeycomb lattice made of Si atoms with a π-bonding network. However, it has been widely-used as a term to describe any honeycomb lattice of Si.

In this chapter, the geometric and electronic structures of 2D silicene on Ag(111) are described. As silicene does not take a planar, but a buckled structure (in contrast to graphene), various superstructures appear on Ag(111), depending on the degree of buckling. In the first section, a brief background of silicene is presented. In the second section, the structure evolution of silicene on Ag(111) from the submonolayer regime to multilayers is described. We briefly walk around the phase diagram and then we focus on the atomic structures of 4×4 silicene and $4/\sqrt{3} \times 4/\sqrt{3}$ structure (multilayers) based on the results by Scanning Tunneling Microscopy

(STM), Low Energy Electron Diffraction (LEED) and DFT calculations. Although the structure of multilayer silicene ($4/\sqrt{3} \times 4/\sqrt{3}$ structure) is still under debate, the structure models proposed by several groups (Wu et al., Oshiyama et al. and Hasegawa et al.) are shown and discussed. After the description of the atomic structure, we describe the electronic structures of the 4×4 silicene and $4/\sqrt{3} \times 4/\sqrt{3}$ structures in the third section. Our results acquired by STM (STS under strong magnetic fields and quasi particle interference) and DFT calculations are presented. Finally, we conclude this chapter by discussing the outlook of silicene research in this area.

7.2 Structure Evolution of Silicene on Ag(111)

A large number of papers have been published that report superstructures formed by the adsorption of Si atoms on Ag(111) [15–40]. A simple method is commonly used to synthesize silicene on Ag(111). By heating a piece of Si wafer up to more than 1373 K, Si atoms are deposited onto the clean Ag(111) substrate. A deposition rate of 0.01–0.1 ML/min is usually used, where 1 ML is corresponds to a full coverage of Si atoms on the substrate surface. The precise control of the substrate temperature is key to crystalizing Si atoms to form the honeycomb lattice. The temperature should be kept in the range from 423 to 573 K. Above 600 K, the silicene layer is not formed. Table 7.1 lists the superstructures observed by experimental techniques such as STM and LEED.

Here, these superstructures are described using the conventional nomenclature in surface science where the overlayer structure is described in terms of the underlying substrate unit cell. These structures are sometimes named with reference to the unit cell of the honeycomb lattice. In this case, the 4×4 and $4/\sqrt{3} \times 4/\sqrt{3}$ structures are called 3×3 and $\sqrt{3} \times \sqrt{3}$ structures, respectively. The $\sqrt{3} \times \sqrt{3}R30°$, $\sqrt{7} \times \sqrt{7}$, $2\sqrt{3} \times 2\sqrt{3}R30°$, 3.5×3.5, $\sqrt{13} \times \sqrt{13}R13.9°$, 4×4, $\sqrt{19} \times$

Table 7.1 List of superstructures formed by the deposition of Si atoms on Ag(111)

Superstructure	References
$\sqrt{3} \times \sqrt{3}R30°$	[24]
$\sqrt{7} \times \sqrt{7}$	[19, 21, 22, 39]
$2\sqrt{3} \times 2\sqrt{3}R30°$	[15, 18, 19, 21–23, 28, 30, 33, 38, 39]
3.5×3.5	[24]
$\sqrt{13} \times \sqrt{13}R13.9°$	[16, 18, 22, 24, 26, 33, 34, 37, 39, 40]
4×4[a]	[16–19, 21–24, 26, 27, 29, 31, 33, 37, 39, 40]
$\sqrt{19} \times \sqrt{19}R23.4°$	[24]
$4/\sqrt{3} \times 4/\sqrt{3}$[a]	[19, 20, 24, 25, 27, 32, 33, 35, 36, 39]

[a]The 4×4 and $4/\sqrt{3} \times 4/\sqrt{3}$ structures are sometimes described as 3×3 and $\sqrt{3} \times \sqrt{3}$ structures, respectively, with the reference to the unit cell of honeycomb lattice

$\sqrt{19}R23.4°$ superstructures appear in the monolayer coverage regime, while the $4/\sqrt{3} \times 4/\sqrt{3}$ structure is formed at multilayer coverages. Both the 4×4 and $4/\sqrt{3} \times 4/\sqrt{3}$ structures can be prepared as a single phase, but the other superstructures coexist and it is difficult to distinguish them. This indicates that their thermodynamical stabilities are not greatly different. Although structural models consisting of the honeycomb lattice have been proposed for $\sqrt{7} \times \sqrt{7}$, $2\sqrt{3} \times 2\sqrt{3}R30°$, $\sqrt{13} \times \sqrt{13}R13.9°$, 4×4 and $4/\sqrt{3} \times 4/\sqrt{3}$ structures, the structures are not clear for the $\sqrt{3} \times \sqrt{3}R30°$, 3.5×3.5 and $\sqrt{19} \times \sqrt{19}R23.4°$ overlayers. The emergence of various superstructures originates from the flexibility of the Si–Si bond. The Si–Si bond in silicene has a mixed character of both sp^2 and sp^3 hybridization so that the local configuration has a high degree of freedom in both the bond length and the bond angle as a function of the mixing. Although superstructures are often observed for graphene grown on metal substrates, these structures are not due to the buckling but they are derived from the moiré patterns [45–49].

7.2.1   The 4 × 4 Structure

The 4×4 structure has been surveyed most intensively [16–19, 21–24, 26, 27, 29, 31, 34, 37, 39, 40], partly because it emerges as a single phase with a larger domain size suitable for macroscopic measurements. The STM image of the 4×4 structure shows six protrusions in the unit cell, together with a corner hole, as shown in Fig. 7.1. Since this image looks similar to that of the Si(111) (5×5) structure [50],

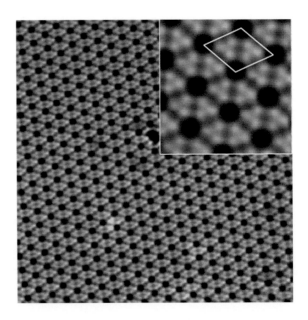

Fig. 7.1 STM topographic image of the 4×4 structure grown on Ag(111) taken at 6 K (sample voltage of $V_S = 0.5$ V, tunneling current of $I_t = 0.3$ nA). The size of the image is 20×20 nm^2. The *inset* shows a high-resolution image. The *white rhombus* is a unit cell of the 4×4 structure (Color figure online)

the STM image alone is insufficient to conclude that the Si atoms construct a honeycomb lattice in the 4 × 4 arrangement.

Two structural models have been proposed using a combination of STM experiments and DFT calculations. One is the model presented independently by Lin et al. [16] and Vogt et al. [17], and the other is the model proposed by Feng et al. [19]. Figure 7.2 shows these models.

The model proposed by Lin et al. [16] and Vogt et al. [17] contains eighteen Si atoms to form the honeycomb network. Six of these Si atoms are shifted upward relatively to the plane composed of the remaining twelve Si atoms. There is a hexagon at the corner of the unit cell. These features nicely explain the STM image consisting of six protrusions together with the darker hole. Meanwhile, Feng et al. [19] presented a different model based on essentially the same STM image that was observed by Lin et al. [16] and Vogt et al. [17]. The model of Feng et al. [19] is composed of hexagons as building blocks, which form a 2D lattice. This model is matched with the one built by removing hexagons at the corner of the unit cell in the model of Lin et al. [16] and Vogt et al. [17]. Thus, there exist Si atoms having dangling bonds and they are terminated with H atoms in the model of Feng et al. [19]. In this regard, the model of Feng et al. [19] can be considered as a sort of 2D Si-hydride. This model also reproduces the STM image according to the DFT calculations of Feng et al. [19]. Since the STM image generally reflects both geometric and electronic structures, it is not straightforward to determine the geometric structure from the image even with the support of DFT calculations. Atomic force microscopy (AFM) has been used to directly visualize the atomic arrangement in the 4 × 4 structure [26, 27]. The AFM image taken in non-contact mode looks very similar to the STM image, and additional information is not provided about the geometric structure. These results indicate that different approaches are required to determine the 4 × 4 structure.

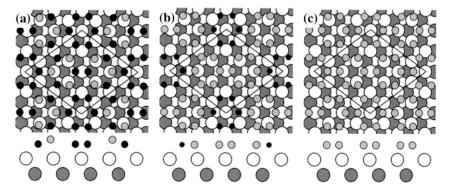

Fig. 7.2 Structural models of the 4 × 4 structure. **a** Model presented by Lin et al. [16] and Vogt et al. [17]. **b** Model presented by Feng et al. [19]. **c** Low-buckled model in which a freestanding silicene is placed on Ag(111) by tuning the lattice constant to match with that of the substrate. Larger and smaller circles are Ag and Si atoms, respectively. The *smaller black circles* in (**b**) represent H atoms which are bonded to the neighboring Si atoms (Color figure online)

Diffraction techniques are the most powerful and reliable for structure determination in a broad range of fields from materials science to biology. In surface
science, LEED has been most successfully used to determine many surface structures from the clean surfaces of single crystals to complicated surfaces accompanied
by long-range reconstructions [51]. The tensor LEED method has also been
established [51–53]. In the structural determination with LEED, the intensities of
the LEED spots are measured as a function of kinetic energy of the incident
electron, which are called the I–V curves. The I–V curves are compared with those
calculated for a model structure by dynamical theory based on multiple scattering of
electrons in condensed matter. To judge how well the calculated I–V curves
reproduce the experimental ones, the reliability factor (R_p) proposed by Pendry is
usually used [52]. Generally, $R_p \leq 0.2$ indicates a good match between them and
that the model structure is convincing [52].

Figure 7.3 shows I–V curves of several LEED spots measured for the 4×4
structure together with the calculated ones for the best-fit structure [31]. Three
structural models were used as initial references. One was a buckled honeycomb
structure similar to the freestanding silicene. Another was the model proposed by
Lin et al. [16] and Vogt et al. [17]. The last one was the model proposed by Feng
et al. [19]. Starting with these models, a structural optimization process was carried

Fig. 7.3 IV curves of several LEED spots measured for the 4×4 structure. The solid lines are measured IV *curves* and the *dashed lines* indicate those calculated for the best-fit structural model ($R_p = 0.17$ with 13 phase shifts). Adapted and reproduced from [31]

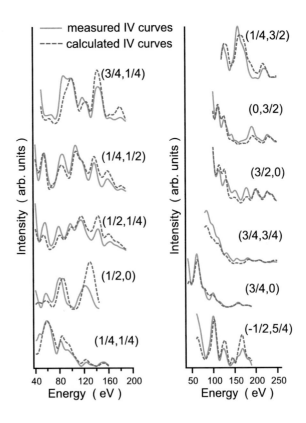

out with ten phase shifts. For the optimized structures, R_p = 0.41, 0.20 and 0.48, respectively. The structural model by Lin et al. [16] and Vogt et al. [17] gave the lowest R_p. When further optimization was carried out with thirteen phase shifts, R_p decreased from 0.20 to 0.17 with Debye temperatures of 250, 180, 200, and 215 K for the silicene layer, and the first, second and bulk layers of Ag, respectively.

Figure 7.4 shows the IV curves calculated for the model of Feng et al. [19]. As this model provided the worst R_p of 0.48, the discrepancy between the measured and calculated IV curves cannot be ignored especially for the (1/4, 1/4), (3/4, 0) and (3/4, 3/4) spots.

Figure 7.5 shows the structure model optimized by the tensor LEED analysis. The positions of the atoms depicted in Fig. 7.5 are listed in Table 7.2 together with those optimized by the DFT calculations [16].

The buckling heights of Si1 and Si2 are 0.77 and 0.74 Å, respectively, and the Si–Si bond lengths range from 2.29 to 2.31 Å. These values are close to the upper limit of the length of Si–Si double bond and are slightly shorter than the Si–Si bond length of 2.35 Å in the diamond crystal structure. The bond lengths of the Si–Si double bond are in the range from 2.14 to 2.29 Å for several molecules [54]. This indicates that the hybridized state of the Si atoms is described as a mixture of sp^2 and sp^3 hybridization. These geometric characteristics demonstrate that the 4 × 4 structure is a buckled silicene.

Fig. 7.4 Comparison of measured IV curves with those calculated for the model of Feng et al. [19] (R_p = 0.48 with 10 phase shifts)

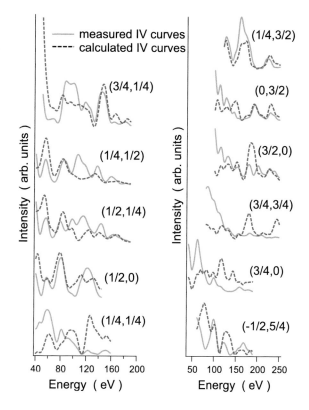

Fig. 7.5 *Top* and *side views* of best-fit structural model for the 4 × 4 structure. The *side view* illustrates the positions of the atoms in a cross section cut along the AA′ line. The *smaller* and *larger circles* represent Si and Ag atoms, respectively

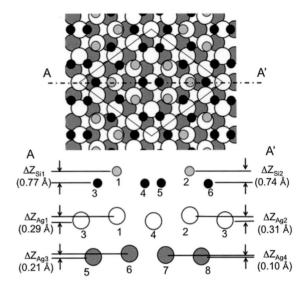

Table 7.2 Positions of the atoms in the 4 × 4 structure specified in Fig. 7.5 as determined by tensor LEED measurements [31]

Atom no.	Lateral displacement (Å)		Height (Å)	
	LEED [31]	DFT [16]	LEED [31]	DFT [16]
Si1			7.67 ± 0.02	7.60
Si2			7.66 ± 0.02	7.59
Si3			6.90 ± 0.03	6.86
Si4			6.91 ± 0.03	6.84
Si5			6.88 ± 0.03	6.83
Si6			6.92 ± 0.02	6.86
Ag1	0.05 ± 0.11	0.04	4.87 ± 0.04	4.89
Ag2	−0.05 ± 0.10	−0.03	4.89 ± 0.04	4.86
Ag3			4.58 ± 0.06	4.59
Ag4	−0.04 ± 0.11	0.01	4.55 ± 0.07	4.55
Ag5	0.03 ± 0.13	0.00	2.30 ± 0.07	2.26
Ag6			2.51 ± 0.10	2.51
Ag7	−0.04 ± 0.14	−0.03	2.43 ± 0.08	2.38
Ag8	0.05 ± 0.14	0.03	2.33 ± 0.06	2.34

The lateral displacements of Ag atoms are the values determined along with the direction of AA′ in Fig. 7.4. The height of each atom is measured from the third Ag layer

Taking a look at the substrate Ag atoms, one can see that the Ag atoms in the first substrate layer (denoted as Ag1 and Ag2 in Fig. 7.5) move upwards, perpendicularly to the surface, by 0.29 and 0.31 Å, respectively. These atoms are located just under the Si atoms that are buckled upwards. The first layer rumples in accordance with the buckling of the silicene layer. The rumpling decays in the second layer, but it still remains. The substrate relaxation plays a crucial role in the structural optimization, and R_p drastically increases to 0.52 without relaxation. The substrate relaxation is also strong evidence of the sizable interfacial coupling of the silicene with the substrate. This helps us figure out the electronic structure of the 4×4 structure discussed below.

It is interesting to compare the structure parameters determined by the tensor LEED with those optimized by the DFT calculations and those determined using reflection high-energy positron diffraction (RHEPD) [29]. Lin et al. [16] and Vogt et al. [17] have optimized the 4×4 silicene by using DFT calculations and determined a buckling of 0.73–0.76 Å [16] and 0.78 Å [17], respectively. The buckling height determined by RHEPD is 0.83 Å. These values are consistent with the buckling heights of 0.74–0.77 Å determined by the tensor LEED. The structural parameters determined independently by different techniques are in good agreement, indicating that the structural model is close to the experimental one.

7.2.2 The $4/\sqrt{3} \times 4/\sqrt{3}$ Structure

The $4/\sqrt{3} \times 4/\sqrt{3}$ structure appears in the high Si coverage regime [19–21, 24, 25, 27, 32, 33, 35, 36, 39]. Figure 7.6a shows an STM image of the $4/\sqrt{3} \times 4/\sqrt{3}$ structure taken at 6 K. The hexagonal array of bright protrusions are well resolved. The distance between the neighboring protrusions is 6.6 Å, which is matched with $4/\sqrt{3}$ times the length of the substrate unit cell. The image shows various triangles separated by straight boundaries. In addition, the $4/\sqrt{3} \times 4/\sqrt{3}$ structure does not wet the Ag surface in contrast to the other monolayer structures such as 4×4 and $\sqrt{13} \times \sqrt{13}R13.9°$. Instead, the $4/\sqrt{3} \times 4/\sqrt{3}$ structure grows perpendicularly to the substrate and the layers consisting of the $4/\sqrt{3} \times 4/\sqrt{3}$ structure pile up into multilayers as shown in Fig. 7.6b [32, 33, 35, 36, 55]. As shown in Fig. 7.6c, the LEED spots are not well resolved and they appear as a part of ring. This suggests that the interaction between the layers is weak and they take different in-plane orientations to each other. Interestingly, for the $4/\sqrt{3} \times 4/\sqrt{3}$ structure a phase transition occurs. Chen et al. [25] reported a phase transition taking place at 40 K based on the STM observation performed in the temperature range from 4 to 77 K. Below 40 K, the triangles separated by the boundaries are observed as shown in Fig. 7.6a, which are converted to the homogeneous hexagonal array of $4/\sqrt{3} \times 4/\sqrt{3}$ periodicity, which appears without the boundaries when the temperature is raised above 40 K.

Several structural models have been proposed for the $4/\sqrt{3} \times 4/\sqrt{3}$ structure. Chen et al. [20, 25] have proposed a single-layer buckled honeycomb structure in which six Si atoms build the unit cell. In contrast, Arafune et al. [24] have

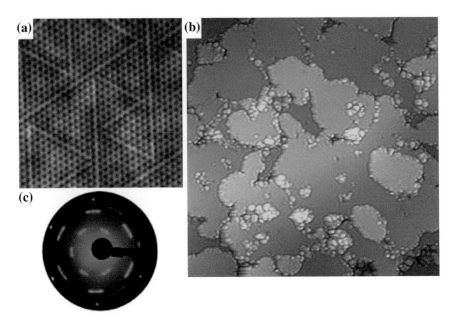

Fig. 7.6 a High-resolution topographic STM image of the $4/\sqrt{3} \times 4/\sqrt{3}$ structure. **b** Topographic STM image of the multilayered $4/\sqrt{3} \times 4/\sqrt{3}$ structure. **c** LEED pattern of the $4/\sqrt{3} \times 4/\sqrt{3}$ structure. The STM images were taken at 6 K with $V_S = -1.0$ V and $I_t = 20$ pA. The image size is **a** 17.0×17.0 nm^2 and **b** 70×70 nm^2. The LEED pattern was measured at room temperature with an electron kinetic energy of 55 eV

suggested a bilayer structure formed by stacking two honeycomb layers from the analysis of the height profile in the STM image. By using DFT calculations, Guo and Oshiyama [56] have reported that the bilayer structure shown in Fig. 7.7a is more favorable than the single-layer silicene proposed by Chen et al. [20, 25]. Pflugradt et al. [57] have found by total energy calculations that the $2\sqrt{3} \times 2\sqrt{3}R30°$ honeycomb bilayer on the $\sqrt{19} \times \sqrt{19}R23.4°$ Ag substrate nicely explains the STM image. Cahangirov et al. [58] have proposed the honeycomb dumbbell silicene in which hexahedrons consisting of 5 Si atoms are arranged to form into a honeycomb lattice. These models basically take on the buckled honeycomb configurations and the modified configurations.

In contrast, an entirely different model has been presented by Shirai et al. [35]. They have proposed an interesting model almost identical to a honeycomb chain trimer (HCT) structure established for the Si(111)$\sqrt{3} \times \sqrt{3}$-Ag surface [59, 60] as shown in Fig. 7.7b. In this model, the substrate Ag atoms segregate in the outermost layer to form a $\sqrt{3} \times \sqrt{3}$ lattice on top of the diamond-like structure of Si. Shirai et al. [35] tested the other models consisting of a honeycomb lattice of Si atoms, but they did not reproduce the I–V curves. The $\sqrt{3} \times \sqrt{3}$ model can also explain the phase transition, because the Si(111)$\sqrt{3} \times \sqrt{3}$-Ag surface shows a phase transition

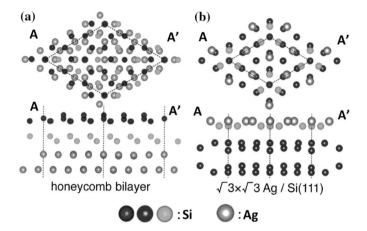

(a) **(b)**

honeycomb bilayer $\sqrt{3}\times\sqrt{3}$ Ag / Si(111)

●●◯ : Si ◯ : Ag

Fig. 7.7 Structural models of the $4/\sqrt{3}\times 4/\sqrt{3}$ structure reproduced from [35, 56]. These models are drawn by VESTA [111]

between a HCT and an inequivalent triangle (IET) structure at 120 K [60]. In addition, a few works have been presented that support this model. Moras et al. [33] have found the growth of a 3D of Si diamond structure when the $4/\sqrt{3}\times 4/\sqrt{3}$ structure is formed. Mannix et al. [36] have pointed out that the STM image of the $4/\sqrt{3}\times 4/\sqrt{3}$ structure is very similar to that of the HCT structure and they also presented a Raman spectrum indicating the growth of a diamond Si crystal. To understand the electronic band structure of the $4/\sqrt{3}\times 4/\sqrt{3}$ structure, it is of great importance to determine the geometric structure. Although the model of Shirai et al. [35] looks better than the other models, more experimental and theoretical analysis is needed.

7.3 Electronic Structure of Silicene on Ag(111)

The most crucial puzzle that needs solving for this system is whether the honeycomb structures of silicene formed on Ag(111) host the Dirac fermions and specifically what evidence provides the most convincing answer? In the research on graphene, we learned that the most convincing evidence is the anomalous quantum Hall effect (QHE) that is essentially different from the QHE observed in a conventional 2D electronic system [6, 61–65]. In contrast to the integer QHE observed in a conventional 2D electron gas, the half-integer QHE is observed for graphene. The difference comes from the fact that an electron in graphene obeys the relativistic Dirac equation while the normal electron obeys the Schrödinger equation.

In general, when an external magnetic field is applied to condensed matter, the orbital motion of the electron is quantized because a parabolic potential well generated by the magnetic field confines the electron. The continuous energy spectrum of a free electron is quantized into discrete Landau levels (LLs). The energy eigenvalues of the LLs for the conventional 2D system are described as:

$$E_n = \hbar\omega\left(n + \frac{1}{2}\right) \tag{7.1}$$

where $\omega = eB/m$ is the cyclotron frequency of an electron with mass m and charge $-e$, $\hbar$ is Plancks constant divided by 2π, B is the external magnetic field perpendicular to the system and n is the quantum number [6]. The energy spectrum is the same as that of a harmonic oscillator and the energy separations between the consecutive LLs are constant with n. On the other hand, the energy spectrum for the Dirac fermion is described as:

$$E_n = \pm v_F\sqrt{2e\hbar Bn} \tag{7.2}$$

where v_F is the Fermi velocity [6]. In contrast to the conventional 2D case, the energy separations between the consecutive LLs are not constant with n because the energy is proportional to $\sqrt{n}$. Thus, the spectral evolution of the LL sequence as a function of magnetic field enables us to distinguish which type of electron exists. Measuring the linear electronic band structure also presents a signature of the Dirac fermion. Angle-resolved photoelectron spectroscopy (ARPES) has been widely used to investigate the electronic band structure. In the ARPES experiments, the kinetic energy of photoelectron is measured as a function of the momentum and then we can determine the band dispersions experimentally. For the graphene, the crossing of the linear bands at the K point in the Brillouin zone near the Fermi level has been observed [66, 67], which gives strong support for the existence of the Dirac fermions. Thus, the observation of the half-integer QHE and/or the unique LL sequence together with the linear band dispersion is desired for the silicene on Ag (111) to confirm the existence of the Dirac fermion.

7.3.1   The 4 × 4 Structure

The electronic band structure of the 4×4 structure on Ag(111) has been investigated by Vogt et al. [17] using ARPES. They observed a linear band near the Fermi level, and concluded that the Dirac fermion survives in the 4×4 structure on Ag (111). As discussed above, measuring the half-integer QHE is the most direct way to prove the existence of the Dirac fermions. It is very difficult to fabricate a Hall bar for the 4×4 structure on Ag(111), however. Lin et al. [69] tried to solve the puzzle by measuring the LL sequence using STS with an external magnetic field. Let us consider the energy spectrum of the LLs for the freestanding silicene. Assuming $B = 7$ T and $v_F = 10^6$ m/s, as expected for freestanding silicene [68], one obtains $E_n = \pm 95\sqrt{n}$ meV. This estimation indicates that the LLs should be clearly observed if the Dirac fermions survive. The domain size is also important for the measurements of the LLs. When the domain size of silicene is small, the electrons are confined by the potential derived from the domain wall to form discrete states. The potential generated by B competes with this confined potential. The magnetic length $\ell = \sqrt{\hbar/eB}$ [6] provides a measure to consider the domain size. For freestanding silicene, $\ell = 14.8$ nm and $\ell = 9.7$ nm at $B = 3$ and 7 T, respectively. Domains that are larger than these values are necessary. In addition, impurities and defects also scatter the electron to hamper the formation of LLs so the measurements should be performed for larger domains free from impurities and defects.

Figure 7.8 shows the spectral evolution of the 4×4 structure as a function of magnetic field [69]. Increasing the magnetic field from $B = 0$ to 7 T, the spectral shape of the 4×4 silicene does not change essentially, and characteristic structures attributed to the LLs are not found. Although each spectrum shows a V-shaped background reminiscent of the Dirac cone, the LLs do not appear. These spectra do not depend on the position where we measured the spectra. Thus, Lin et al. [69] came to the conclusion that the 4×4 structure loses the linear dispersion that is a prerequisite for the Dirac fermion.

The DFT calculations provided a reasonable explanation for the absence of the LLs in the 4×4 structure [69]. Figure 7.9 demonstrates the variation of the band structure for freestanding low-buckled silicene (FLBS), to freestanding 4×4 silicene (F4S) and the 4×4 structure on Ag(111) calculated by Lin et al. [69]. They also considered van der Waals (vdW) interaction between the silicene layer and the Ag(111) substrate, because it has previously been shown to be important for understanding graphene on metal substrates [70]. The calculated results without the vdW interaction were basically the same as those calculated by including the vdW interaction. This indicates that the vdW interaction does not play an important role in determining the electronic structure in the vicinity of the Fermi level for the 4×4 silicene structure on Ag(111).

The Dirac fermion feature appears clearly as a linear dispersion crossing at the Fermi level for the FLBS (Fig. 7.9a). For the F4S, the atomic arrangement, in contrast to the FLBS, opens an energy gap at the Dirac point (Fig. 7.9b). An energy gap opens because the equivalence of the two sublattices inside the honeycomb lattice collapses entirely. One can see that the overall band structure remains similar to that of FLBS except for the part near the Fermi level. After introducing the Ag substrate, the band structure drastically changes, as is seen in Fig. 7.9c. Although the band structure of the 4×4 structure on Ag(111) is very complicated, partly because the bulk Ag bands are folded in the slab model, one can see that neither linear dispersions nor any other features observed in both band structures of the FLBS and the F4S are observed. Definitively, although there are electronic bands that lie around the Fermi level, they are not derived from the Si $3p_z$ orbital but instead the Ag sp states are dominant contributors. On the contrary, the bands derived from the Si $3p_z$ orbital are located below and above 1 eV from the Fermi level. Hence, the electronic states associated with the Si $3p_z$ orbital are strongly

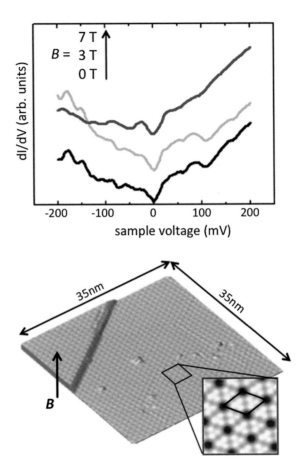

Fig. 7.8 Evolution of STS spectrum with increasing magnetic fields. The magnetic fields generated by a superconducting magnet are applied perpendicularly to the sample surface. The STS spectra were measured at 6 K. The first derivative of I_t was measured as an STS spectrum with lock-in technique. An AC voltage of $V_{rms} = 10$ mV and 312.6 Hz was added to V_S as a modulation voltage in the lock-in measurement. STM image of the 4×4 silicene on Ag(111) used for the measurements ($V_S = 0.70$ V, $I_t = 0.19$ nA and 35×35 nm^2). The *inset* shows a high-resolution STM image of the 4×4 silicene ($V_S = 0.50$ V, $I_t = 0.30$ nA and 3.65×3.65 nm^2). The unit cell is shown by the *rhombus*. Adapted and reproduced from [69]

hybridized with the substrate states. As a result, the wave functions derived from the Si $3p_z$ are not confined inside the silicene layer but instead are delocalized into the substrate. This reasonably explains the absence of the LLs in the STS spectra.

The charge distribution also demonstrates clearly the interfacial coupling between the silicene and Ag substrate. Figure 7.9e shows the differential charge distribution calculated by subtracting the charge distributions of both the isolated Ag substrate and the F4S from that of the 4×4 silicene on Ag(111). The charge redistribution for the 4×4 structure on Ag(111) is observed at the interface of the

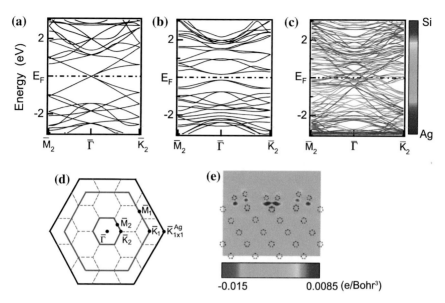

Fig. 7.9 Calculated electronic band structures of (**a**) freestanding low-buckled silicene (FLBS), (**b**) freestanding 4 × 4 silicene (F4S) and (**c**) the 4 × 4 silicene on Ag(111). The *color bar* in (**c**) shows the relative contribution of Si 3p$_z$ orbital for each band. The *red* (*blue*) shows a *higher* (*lower*) contribution. The *blue bands* are basically derived from the substrate Ag. The band structure of the FLBS was calculated for a structure optimized intentionally under the condition that the size of the unit cell is kept the same as that of the 4 × 4 silicene on Ag(111). The total energy, geometric structure and electronic band structure calculated in this manner are not essentially different from those calculated without any constraint. The structure of the F4S is the same as that of the 4 × 4 silicene on Ag(111) except that the substrate is peeled off to pull out the effect of the interface coupling on the band structure. The details about the calculations are described in [69]. **d** Brillouin zones (BZs) corresponding to the unit cells of the 1 × 1 silicene, 4 × 4 silicene on Ag(111) and 1 × 1 Ag(111). The *red* (*blue*) *hexagon* represents the BZ of the 4 × 4 (1 × 1) silicene. The black hexagon indicates the BZ of the 1 × 1 unit cell of Ag(111). The $\overline{K_1}$ point of the blue BZ is matched with the $\overline{\Gamma}$ point of the red BZ in the extended zone scheme, and thus the Dirac cone of FLBS appears at the $\overline{\Gamma}$ point. Note that the 4 × 4 silicene on Ag(111) is also denoted as 3 × 3 silicene referring to the unit cell of honeycomb lattice. **e** Cross section of the differential charge distribution (DCD) for the 4 × 4 silicene. The *black dotted circles* represent Si and Ag atoms. The region where the electron density increases (decreases) is colored *red* (*blue*). Figures (**a–e**) are rendered by VESTA [111]. Figures (**a–e**) are adapted and reproduced from [69]

silicene and Ag substrate. The electron transfer occurs from Ag to Si and the bonding charges are accumulated in between the Si atoms and the Ag atoms underneath (see Fig. 7.9e). The strong coupling at the interface is also supported by the upward displacements of Ag atoms in the first and second layers under the silicene layer, as was discussed in Sect. 7.2.1. These results indicate clearly that the strong coupling accompanied by the charge transfer breaks the symmetry to drastically modulate the band structure of the 4 × 4 silicene on Ag(111). Triggered by the work of Lin et al. [69], a number of works have been carried out to settle the disagreement between the ARPES and the STS results [71–82]. Almost all of them have provided support to the results of Lin et al. [69], and have emphasized that the strong hybridization at the silicene-substrate interface heavily affects the band structure of the 4 × 4 silicene, leading to the absence of the Dirac fermion features. Interestingly, strong coupling is also observed for silicene placed on other metal substrates such as Ir, Cu, Mg, Au, Pt and Al [79]. This indicates that the strong interface coupling is true for various combinations of silicene and metal, and that metal substrates are not adequate as a stage to synthesize silicene hosting Dirac fermions.

We discuss here the linear band reported by Vogt et al. [17] who claimed it as compelling evidence for the Dirac fermion. The origin of the linear band was criticized by several research groups [69, 71, 72, 74, 80, 81], because the linear band is in close resemblance with the dispersion of the Ag bulk sp band and the band width is much larger than that expected in the freestanding silicene. Very recently, Vogt and coworkers [83] have reinvestigated the band structure of the 4 × 4 structure on Ag(111) by using ARPES. They observed two band dispersions. One is the dispersion observed in their previous work, and the other is a new one. They assigned the former and latter dispersions to the sp band of bulk Ag and the 4 × 4-derived band, respectively. The 4 × 4-derived band can be reasonably justified by the DFT calculations of Lin et al. [69]. In Fig. 7.9c, one can see that a band dispersion extends downward from about 0.5 eV below the Fermi level at the $\bar{\Gamma}$ point along both $\bar{\Gamma}\bar{K_2}$ and $\bar{\Gamma}\bar{M_2}$ lines. The main contributor to the band is not the Si $3p_z$ but the Ag sp orbitals. The 4 × 4-derived band observed by Vogt and coworkers [83] is reasonably matched with the band dispersion. The puzzle on the origin of the bands observed by Vogt and coworkers was settled by Mahatha et al. [80] in 2014, who convincingly assigned the bands to be Ag sp bands based on their systematic ARPES measurements and DFT calculations. In addition, Chen and Weinert [81] provided essentially the same explanation on the origin of these bands in the same year. Thus, we can conclude that the electronic structure of the 4 × 4 structure has now been fully rationalized: the 4 × 4 structure does not host the Dirac fermion because the strong interfacial couplings between the silicene layer and the substrate seriously modify the electronic structure. The 4 × 4 structure is a buckled honeycomb lattice without Dirac fermions.

7.3.2 The $4/\sqrt{3} \times 4/\sqrt{3}$ Structure

Chen et al. [20, 25] observed a standing wave pattern in their STM measurement for the $4/\sqrt{3} \times 4/\sqrt{3}$ structure. The standing wave pattern is formed by the interference of incident electron waves with the reflected wave scattered by structural imperfections such as steps and defects. The measurement of the wavelength as a function of voltage applied to the STM junction provides a quasiparticle band dispersion. Chen et al. [20, 25] found a linear band dispersion above the Fermi level by this technique, and claimed that the linear band is a manifestation of the Dirac fermions in the $4/\sqrt{3} \times 4/\sqrt{3}$ structure. Similar measurements were carried out by Arafune et al. [84], who measured the interference patterns in a wider voltage range than that of Chen et al. [20, 25]. Figure 7.10 compares the results of Chen et al. [20, 25] and Arafune et al. [84]. The comparison reasonably shows that the linear dispersion observed by Chen et al. [20, 25] is a part of the parabolic dispersion observed by Arafune et al. [84]. In addition, standing wave patterns are not observed below $V_s = -100$ mV. This implies that the Dirac fermions do not exist. Assuming that the $4/\sqrt{3} \times 4/\sqrt{3}$ structure hosts the Dirac fermions, a pair of linear bands should be observed symmetrically to the Dirac point. Arafune et al. [84] also suggested that the parabolic dispersion comes from the Shockley surface state intrinsic to the clean Ag(111) surface that is modified by the $4/\sqrt{3} \times 4/\sqrt{3}$

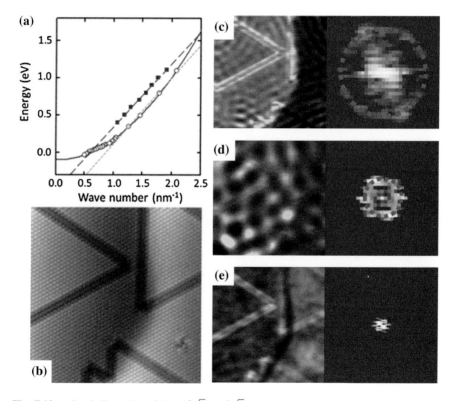

Fig. 7.10 **a** Band dispersion of the $4/\sqrt{3} \times 4/\sqrt{3}$ structure determined by the standing wave patterns. The *filled squares* show the results of Chen et al. [20] and the open circles those of Arafune et al. [84]; **b** STM image of the $4/\sqrt{3} \times 4/\sqrt{3}$ structure used for the measurements ($V_S = 0.5$ V, $I_t = 0.3$ nA and 28.4×28.4 nm^2). The patterns taken at (**c**) $V_S = 500$ mV, (**d**) $V_S = 60$ mV and (**e**) $V_S = -100$ mV as well as the Fourier-transformed images. The measurements were done at 6 K. At $V_S = -100$ mV, clear interference pattern is not observed because the electronic state does not exist. Figures (**a**), (**b**) and (**d**) are adapted and reproduced from [84]

structure. The effective mass determined by the dispersion is $m^*/m = 0.14$, which is smaller than that of the Shockley surface state of Ag(111) ($m^*/m = 0.4$) [85]. The difference remains to be solved.

Shirai et al. [35] have provided an explanation for the inconsistency between the results of Chen et al. [20, 25] and Arafune et al. [84] based on the structural model discussed above.

The electronic band structure of the Si(111)$\sqrt{3} \times \sqrt{3}$-Ag surface shows a parabolic band known as an S_1 surface state. The S_1 state extends from just below the Fermi level to an energy level above the Fermi level [60]. Although this band is essentially parabolic, it is approximated as a linear band away from the Fermi level. Shirai et al. [35] explained that both Chen et al. [20, 25] and Arafune et al. [84] observed the S_1 band with different energy windows. On the other hand, Padova et al. [55] have reported a pair of ∨- and ∧- shaped band dispersions in their ARPES experiments for multilayer silicene. The ∨-shaped band crosses the Fermi level to extend above the Fermi level while the ∧-shaped band is located below the Fermi level and goes down to a deeper energy region. Although the ∨-shaped band can be explained by the S_1 surface band derived from the Si(111)$\sqrt{3} \times \sqrt{3}$-Ag structure, the origins of the ∨- and ∧-shaped bands remain unclear. Comparing the band structure of the Si(111)$\sqrt{3} \times \sqrt{3}$-Ag surface, it is difficult to find the candidates. The electronic band structure of the $4/\sqrt{3} \times 4/\sqrt{3}$ structure remains an open question.

7.4   Summary and Outlook

The current status of our understanding of silicene grown on Ag(111) is summarized here. Various superstructures appear after the deposition of Si atoms on Ag(111); $\sqrt{3} \times \sqrt{3}R30°$, $\sqrt{7} \times \sqrt{7}$, $2\sqrt{3} \times 2\sqrt{3}R30°$, $\sqrt{13} \times \sqrt{13}R13.9°$, 3.5×3.5, 4×4, $\sqrt{19} \times \sqrt{19}R23.4°$ and $4/\sqrt{3} \times 4/\sqrt{3}$. Among these superstructures, structural models consisting of buckled honeycomb lattices are presented for the $\sqrt{7} \times \sqrt{7}$, $2\sqrt{3} \times 2\sqrt{3}R30°$, $\sqrt{13} \times \sqrt{13}R13.9°$ and 4×4 structures. The 4×4 structure has been studied most intensively and the detailed atomic arrangement is determined not only by the combination of STM imaging and DFT calculations but also by diffraction techniques. The unit cell is a buckled honeycomb lattice consisting of eighteen Si atoms, six of which are displaced vertically. The displacement of the Si atoms gives rise to the rumpling of the substrate Ag layers, indicating a strong interface coupling. The coupling severely alters the electronic band structure compared to that of freestanding silicene. The electronic states near the Fermi level have a more dominant Ag character than that from the Si 3p orbitals. As a result, the Landau level sequences expected for the Dirac fermions are not observed. Although the 4×4 structure can be categorized as silicene from a geometric point of view, the structure does not host the Dirac fermion.

The $\sqrt{7} \times \sqrt{7}$, $2\sqrt{3} \times 2\sqrt{3}R30°$ and $\sqrt{13} \times \sqrt{13}R13.9°$ structures are candidates for silicene. The structural models have been proposed using a combination of STM and DFT calculations. A detailed structural analysis, such as using tensor LEED and measuring the electronic structure, has not been carried out for these structures. Since these structures do not appear as a single phase and usually coexist with other structures, it is not easy to investigate their geometric and electronic structure. It has been shown, however, that these structures directly interact with the substrate, similar to the 4×4 structure, suggesting that the strong interfacial couplings can seriously modify the electronic structure compared with that of free-standing silicene.

Several structural models for the $4/\sqrt{3} \times 4/\sqrt{3}$ structure have been presented. They are classified into two types: One is based on the buckled honeycomb structure and the other on the structure consisting of Ag atoms on diamond-like Si layers. There remains, however, a number of intriguing puzzles to be resolved for the electronic structure. For example, the electronic structure is still controversial, i.e., is it linear versus parabolic [84, 86]. The origin of the $\vee$- and $\wedge$- shaped band dispersions is also not clear. In addition to these basic aspects, the $4/\sqrt{3} \times 4/\sqrt{3}$ structure provides pivotal issues regarding the Fermi surface warping [87] and superconductivity [88]. The Femi surface warping indicates that the $4/\sqrt{3} \times 4/\sqrt{3}$ structure is possibly a 2D topological insulator, which is one of the central topics in the material science field, where graphene has been predicted as a 2D topological insulator [89–91]. Freestanding silicene and germanene are the promising candidates and exotic properties have been predicted [68, 92–104]. The $4/\sqrt{3} \times 4/\sqrt{3}$ structure could be a stage to realize these properties. The superconductivity of 'silicene' based materials is also an attractive topic. The STS spectrum shows an energy gap typical for superconductors and the critical temperature estimated from the spectrum is 35–40 K, which is higher than all other single-element superconductors discovered so far. Although the electron-phonon coupling can be a driving force for the pairing, the possibility of chiral superconductivity has been proposed [105]. These results are still under investigation including their ability to be reproduced.

What we have learnt from the 4×4 structure on Ag(111) that metal substrates are not adequate as a stage to realize the exotic silicene because of strong interface couplings. Therefore, we must search for new substrates which satisfy the conditions that (1) the Si atoms grow to wet the substrate two-dimensionally without aggregating three-dimensionally and (2) the couplings of Si with the substrate are enough weak for the honeycomb symmetry to be preserved. Actually, these two conditions look incompatible. Condition (1) requires a strong coupling of Si atoms with the substrates and contradicts Condition (2), and vice versa. Thus, it is not easy to find good substrates experimentally. Thus, theoretical predictions are strongly desired, which could provide a recipe to help realize the Dirac fermion silicene. Several candidates for a substrate have been proposed, such as H-terminated Si (111), boron nitride, graphene, amongst others [71, 106–109]. It has been reported that the intercalation of alkali atoms between the silicene and the Ag(111) substrate reduces the interface coupling, leading to a revival of the Dirac fermions [79]. In these theoretical works, only the stability is discussed based on the total energy calculated at 0 K. Under real growth condition, the effect of finite temperature, growth kinetics, the existence of competing metastable structures including the diamond structure, must be included. More elaborate theoretical considerations are therefore desired, which would help promote further experimental efforts. Although exploring new substrates and growth conditions to synthesize the exotic silicene is a great challenge, epoch-making discoveries are awaiting. Very recently, Tao et al. [110] have opened an avenue to realize the Dirac fermion silicene. They have successfully fabricated a field effect transistor (FET) by peeling off the $2\sqrt{3} \times 2\sqrt{3}R30°$ structure grown on Ag(111) through interface engineering. The Dirac fermion signature is observed in the FET and has similar characteristics to graphene FETs. Although the further identification of geometric and electronic structures of silicene in the FET configuration is necessary, the fabrication of silicene FET is a great leap.

Acknowledgements This work was partially supported by JSPS KAKENHI Grant Number 24241040 and by MEXT KAKENHI Grant Number 25110008. I would like to thank Dr. Noriyuki Tsukahara, Dr. Emi Minamitani, Dr. Tetsuroh Shirasawa, Dr. Yousoo Kim, Prof. Dr. Toshio Takahashi, Prof. Dr. Maki Kawai, Kazuaki Kawahara, Ryo Nagao, Mao Kanno and Naoya Kawakami because this article is based on the collaborative works with them. The DFT calculations were performed by using the computer facilities of the Institute of Solid State Physics (ISSP Super Computer Center, the University of Tokyo) and RIKEN Integrated Cluster of Clusters (RICC) facility.

References

1. T.C. Hales, Discrete Comput. Geom. **25**, 1 (2001)
2. F.D.M. Haldane, Phys. Rev. Lett. **61**, 2015 (1988)
3. K.S. Novoselov, A.K. Geim, S.V. Morozov, D. Jiang, Y. Zhang, S.V. Dubonos, I.V. Grigorieva, A.A. Firsov, Science **306**, 666 (2004)
4. A.K. Geim, K.S. Novoselov, Nat. Mater. **6**, 183 (2007)
5. A.K. Geim, Science **324**, 1530 (2009)
6. A.H. Castro Neto, F. Guinea, N.M.R. Peres, K.S. Novoselov, A.K. Geim, Rev. Mod. Phys. **81**, 109 (2009)
7. K. Takeda, K. Shiraishi, Phys. Rev. B **50**, 14916 (1994)
8. G.G. Guzman-Verri, L.C.L.Y. Voon, Phys. Rev. B **76**, 75131 (2007)
9. S. Cahangirov, M. Topsakal, E. Akturk, H. Sahin, S. Ciraci, Phys. Rev. Lett. **102**, 236804 (2009)
10. M. Houssa, G. Pourtois, V.V. Afanas'ev, A. Stesmans, Appl. Phys. Lett. **97**, 112106 (2010)
11. A. Kara, H. Enriquez, A.P. Seitsonen, C.C.L.Y. Voon, S. Vizzini, B. Aufray, H. Oughaddou, Surf. Sci. Rep. **67**, 1 (2012)
12. C. Léandri, G. Le Lay, B. Aufray, C. Girardeaux, J. Avila, M.E. Davila, M.C. Asensio, C. Ottaviani, A. Cricenti, Surf. Sci. **574**, L9 (2005)
13. G. Le Lay, B. Aufray, C. Leandri, H. Oughaddou, J.-P. Biberian, P. De Padova, M.E. Davila, B. Ealet, A. Kara, Appl. Surf. Sci. **256**, 524 (2010)
14. B. Aufray, A. Kara, S. Vizzini, H. Oughaddou, C. Leandri, B. Ealet, G. Le Lay, Appl. Phys. Lett. **96**, 183102 (2010)

15. B. Lalmi, H. Oughaddou, H. Enriquez, A. Kara, S. Vizzini, B. Ealet, B. Aufray, Appl. Phys. Lett. **97**, 223109 (2010)
16. C.-L. Lin, R. Arafune, K. Kawahara, N. Tsukahara, E. Minamitani, Y. Kim, N. Takagi, M. Kawai, Appl. Phys. Express **5**, 045802 (2012)
17. P. Vogt, P. De Padova, C. Quaresima, J. Avila, E. Frantzeskakis, M.C. Asensio, A. Resta, B. Ealet, G. Le Lay, Phys. Rev. Lett. **108**, 155501 (2012)
18. H. Jamgotchian, Y. Colignon, N. Hamzaoui, B. Ealet, J.Y. Hoarau, B. Aufray, J.P. Biberian, J. Phys. Condens. Matter **24**, 172001 (2012)
19. B. Feng, Z. Ding, S. Meng, Y. Yao, X. He, P. Cheng, L. Chen, K. Wu, Nano Lett. **12**, 3507 (2012)
20. L. Chen, C.-C. Liu, B. Feng, X. He, P. Cheng, Z. Ding, S. Meng, Y. Yao, K. Wu, Phys. Rev. Lett. **109**, 56804 (2012)
21. H. Enriquez, S. Vizzini, A. Kara, B. Lalmi, H. Oughaddou, J. Phys. Condens. Matter **24**, 314211 (2012)
22. D. Chiappe, C. Grazianetti, G. Tallarida, M. Fanciulli, A. Molle, Adv. Mater. **24**, 5088 (2012)
23. G. Le Lay, P. De Padova, A. Resta, T. Bruhn, P. Vogt, J. Phys. D **45**, 392001 (2012)
24. R. Arafune, C.-L. Lin, K. Kawahara, M. Kanno, N. Tsukahara, E. Minamitani, Y. Kim, N. Takagi, M. Kawai, Surf. Sci. **608**, 297 (2013)
25. L. Chen, H. Li, B. Feng, Z. Ding, J. Qiu, P. Cheng, K. Wu, S. Meng, Phys. Rev. Lett. **110**, 85504 (2013)
26. Z. Majzik, M.R. Tchalala, M. Svec, P. Hapala, H. Enriquez, A. Kara, A.J. Mayne, G. Dujardin, P. Jelinek, H. Oughaddou, J. Phys. Condens. Matter **25**, 225301 (2013)
27. A. Resta, T. Leoni, C. Barth, A. Ranguis, C. Becker, T. Bruhn, P. Vogt, G. Le Lay, Sci. Rep. **3**, 2399 (2013)
28. E. Cinquanta, E. Scalise, D. Chiappe, C. Grazianetti, B. van den Broek, M. Houssa, M. Fanciulli, A. Molle, J. Phys. Chem. C **117**, 16719 (2013)
29. Y. Fukaya, I. Mochizuki, W. Maekawa, K. Wada, T. Hyodo, I. Matsuda, A. Kawasuso, Phys. Rev. B **88**, 205413 (2013)
30. A. Acun, B. Poelsema, H. Zandvliet, R. van Gastel, Appl. Phys. Lett. **103**, 263119 (2013)
31. K. Kawahara, T. Shirasawa, R. Arafune, C.-L. Lin, T. Takahashi, M. Kawai, N. Takagi, Surf. Sci. **623**, 25 (2014)
32. E. Salomon, R. El Ajjouri, G. Le Lay, T. Angot, J. Phys. Condens. Matter **26**, 185003 (2014)
33. P. Moras, T.O. Mentes, P.M. Sheverdyaeva, A. Locatelli, C. Carbone, J. Phys. Condens. Matter **26**, 185001 (2014)
34. M.R. Tchalala, H. Enriquez, H. Yildirim, A. Kara, A.J. Mayne, G. Dujardin, M.A. Ali, H. Oughaddou, Appl. Surf. Sci. **303**, 61 (2014)
35. T. Shirai, T. Shirasawa, T. Hirahara, N. Fukui, T. Takahashi, S. Hasegawa, Phys. Rev. B **89**, 241403(R) (2014)
36. A.J. Mannix, B. Kiraly, B.L. Fisher, M.C. Hersam, N.P. Guisinger, ACS Nano **8**, 7538 (2014)
37. Z.-L. Liu, M.-X. Wang, J.-P. Xu, J.-F. Ge, G. Le Lay, P. Vogt, D. Qian, C.-L. Gao, C. Liu, J.-F. Jia, New J. Phys. **16**, 75006 (2014)
38. Z.-L. Liu, M.-X. Wang, C. Liu, J.-F. Jia, P. Vogt, C. Quaresima, C. Ottaviani, B. Olivieri, P. De Padova, G. Le Lay, APL Mater. **2**, 092513 (2014)
39. J. Sone, T. Yamagami, Y. Aoki, K. Nakatsuji, H. Hirayama, New J. Phys. **16**, 95004 (2014)
40. G. Prevot, R. Bernard, H. Gruguel, Y. Borensztein, Appl. Phys. Lett. **105**, 213106 (2014)
41. M.R. Tchalala, H. Enriquez, A.J. Mayne, A. Kara, S. Roth, M.G. Silly, A. Bendounan, F. Sirotti, T. Greber, B. Aufray, G. Dujardin, M.A. Ali, H. Oughaddou, Appl. Phys. Lett. **102**, 083107 (2013)
42. A. Fleurence, R. Friedlein, T. Ozaki, H. Kawai, Y. Wang, Y. Yamada-Takamura, Phys. Rev. Lett. **108**, 245501 (2012)
43. L. Meng, Y. Wang, L. Zhang, S. Du, R. Wu, L. Li, Y. Zhang, G. Li, H. Zhou, W.A. Hofer, H.-J. Gao, Nano Lett. **13**, 685 (2013)

44. D. Chiappe, E. Scalise, E. Cinquanta, C. Grazianetti, B. van den Broek, M. Fanciulli, M. Houssa, A. Molle, Adv. Mater. **26**, 2096 (2014)
45. P. Sutter, J.T. Sadowski, E. Sutter, Phys. Rev. B. **80**, 245411 (2009)
46. M. Gao, Y. Pan, L. Huang, H. Hu, L.Z. Zhang, H.M. Guo, S.X. Du, H.-J. Gao, Appl. Phys. Lett. **98**, 033101 (2011)
47. L. Meng, R. Wu, L. Zhang, L. Li, S. Du, Y. Wang, H.-J. Gao, J. Phys. Condens. Matter **24**, 314214 (2012)
48. S.K. Hämäläinen, M.P. Boneschanscher, P.H. Jacobse, I. Swart, K. Pussi, W. Moritz, J. Lahtinen, P. Liljeroth, J. Sainio. Phys. Rev. B **88**, 201406(R) (2013)
49. W. Moritz, B. Wang, M.-L. Bocquet, T. Brugger, T. Greber, J. Wintterlin, S. Günther, Phys. Rev. Lett. **104**, 136102 (2010)
50. R.M. Feenstra, M.A. Lutz, Phys. Rev. B **42**, 5391 (1990)
51. M.A. Van Hove, Surf. Sci. **603**, 1301 (2009)
52. J.B. Pendry, J. Phys. C **13**, 937 (1980)
53. M.A. Van Hove, W. Moritz, H. Over, P.J. Rous, A. Wander, A. Barbieri, N. Materer, U. Starke, G.A. Somorjai, Surf. Sci. Rep. **19**, 191 (1993)
54. Y. Wang, Y. Xie, P. Wei, R.B. King, H.F. Schaefer III, P.V.R. Schleyer, G.H. Robinson, Science **321**, 1069 (2008)
55. P. De Padova, P. Vogt, A. Resta , J. Avila, I. Razado-Colambo, C. Quaresima, C. Ottaviani, B. Olivieri, T. Bruhn, T. Hirahara, T. Shirai, S. Hasegawa, M.C. Asensio, G. Le Lay, Appl. Phys. Lett. **102**, 163106 (2013)
56. Z.-X. Guo, A. Oshiyama, Phys. Rev. B **89**, 155418 (2014)
57. P. Pflugradt, L. Matthes, F. Bechstedt, Phys. Rev. B **89**, 205428 (2014)
58. S. Cahangirov, V.O. Özcelik, L. Xian, J. Avila, S. Cho, M.C. Asensio, S. Ciraci, A. Rubio, Phys. Rev. B **90**, 035448 (2014)
59. T. Takahashi, S. Nakatani, Surf. Sci. **282**, 17 (1993)
60. H. Aizawa, M. Tsukada, N. Sato, S. Hasegawa, Surf. Sci. **429**, L509 (1999)
61. K.S. Novoselov, A.K. Geim, S.V. Morozov, D. Jiang, M.I. Katsnelson, I.V. Grigorieva, S.V. Dubonos, A.A. Firsov, Nature **438**, 197 (2005)
62. Y.B. Zhang, Y.W. Tan, H.L. Stormer, P. Kim, Nature **438**, 201 (2005)
63. C. Berger, Z. Song, X. Li, X. Wu, N. Brown, C. Naud, D. Mayou, T. Li, J. Hass, A.N. Marchenkov, E.H. Conrad, P.N. First, W.A. de Heer, Science **312**, 1191 (2006)
64. D.L. Miller, K.D. Kubista, G.M. Rutter, M. Ruan, W.A. de Heer, P.N. First, J.A. Stroscio, Science **324**, 924 (2009)
65. G. Li, E.Y. Andrei, Nat. Phys. **3**, 623 (2007)
66. S.Y. Zhou, G.-H. Gweon, J. Graf, A.V. Fedorov, C.D. Spataru, R.D. Diehl, Y. Kopelevich, D.-H. Lee, S.G. Louie, A. Lanzara, Nat. Phys. **2**, 595 (2006)
67. K. Sugawara, T. Sato, S. Souma, T. Takahashi, H. Suematsu, Phys. Rev. B **73**, 045124 (2006)
68. C.-C. Liu, W. Feng, Y. Yao, Phys. Rev. Lett. **107**, 07680 (2011)
69. C.-L. Lin, R. Arafune, K. Kawahara, M. Kanno, N. Tsukahara, E. Minamitani, Y. Kim, M. Kawai, N. Takagi, Phys. Rev. Lett. **110**, 076801 (2013)
70. M. Vanin, J.J. Mortensen, A.K. Kelkkanen, J.M. Garcia-Lastra, K.S. Thygesen, K.W. Jacobsen, Phys. Rev. B **81**, 081408(R) (2010)
71. Z.-X. Guo, S. Furuya, J. Iwata, A. Oshiyama, Phys. Rev. B **82**, 235435 (2013)
72. Y.-P. Wang, H.-P. Cheng, Phys. Rev. B **87**, 245430 (2013)
73. S. Cahangirov, M. Audiffred, P. Tang, A. Iacomino, W. Duan, G. Merino, A. Rubio, Phys. Rev. B **88**, 35432 (2013)
74. P. Gori, O. Pulci, F. Ronci, S. Colonna, F. Bechstedt, J. Appl. Phys. **114**, 113710 (2013)
75. D. Tsoutsou, E. Xenogiannopoulou, E. Golias, P. Tsipas, A. Dimoulas, Appl. Phys. Lett. **103**, 231604 (2013)
76. N. Gao, J.C. Li, Q. Jiang, Chem. Phys. Lett. **592**, 222 (2014)
77. D. Kaltsas, L. Tsetseris, A. Dimoulas, Appl. Surf. Sci. **291**, 93 (2014)

78. M. Houssa, B. van den Broek, E. Scalise, B. Ealet, G. Pourtois, D. Chiappe, E. Cinquanta, C. Grazianetti, M. Fanciulli, A. Molle, V.V. Afanas'ev, A. Stesmans, Appl. Surf. Sci. **291**, 98 (2014)
79. R. Quhe, Y. Yuan, J. Zheng, Y. Wang, Z. Ni, J. Shi, D. Yu, J. Yang, J. Lu, Sci. Rep. **4**, 5476 (2014)
80. S.K. Mahatha, P. Moras, V. Bellini, P.M. Sheverdyaeva, C. Struzzi, L. Petaccia, C. Carbone, Phys. Rev. B **89**, 201416 (2014)
81. M.X. Chen, M. Weinert, Nano Lett. **14**, 5189 (2014)
82. N.W. Johnson, P. Vogt, A. Resta, P. De Padova, I. Perez, D. Muir, E.Z. Kurmaev, G. Le Lay, A. Moewes, Adv. Funct. Mater. **24**, 5253 (2014)
83. J. Avila, P. De Padova, S. Cho, I. Colambo, S. Lorcy, G. Quaresima, P. Vogt, A. Resta, G. Le Lay, M.C. Asensio, J. Phys. Condens. Matter **25**, 262001 (2013)
84. R. Arafune, C.-L. Lin, R. Nagao, M. Kawai, N. Takagi, Phys. Rev. Lett. **110**, 229701 (2013)
85. F. Reinert, G. Nicolay, S. Schmidt, D. Ehm, S. Hüfner, Phys. Rev. B **63**, 115415 (2001)
86. L. Chen, C.-C. Liu, B. Feng, X. He, P. Cheng, Z. Ding, S. Meng, Y. Yao, K. Wu, Phys. Rev. Lett. **110**, 229702 (2013)
87. B. Feng, H. Li, C.-C. Liu, T.-N. Shao, P. Cheng, Y. Yao, S. Meng, L. Chen, K. Wu, ACS Nano **7**, 9049 (2013)
88. L. Chen, B. Feng, K. Wu, Appl. Phys. Lett. **102**, 81602 (2013)
89. C.L. Kane, E.J. Mele, Phys. Rev. Lett. **95**, 146802 (2005)
90. C.L. Kane, E.J. Mele, Phys. Rev. Lett. **95**, 226801 (2005)
91. M.Z. Hassan, C.L. Kane, Rev. Mod. Phys. **82**, 3045 (2010)
92. M. Ezawa, New J. Phys. **14**, 33003 (2012)
93. N.D. Drummond, V. Zólyomi, V.I. Fal'ko, Phys. Rev. B **85**, 075423 (2012)
94. M. Ezawa, Phys. Rev. Lett. **109**, 55502 (2012)
95. M. Ezawa, Phys. Rev. Lett. **110**, 26603 (2013)
96. X.-T. An, Y.-Y. Zhang, J.-J. Liu, S.-S. Li, Appl. Phys. Lett. **102**, 43113 (2013)
97. W.-F. Tsai, C.-Y. Huang, T.-R. Chang, H. Lin, H.-T. Jeng, A. Bansil, Nat. Commun. **4**, 1500 (2013)
98. M. Ezawa, Euro. Phys. J. B **86**, 139 (2013)
99. M. Tahir, A. Manchon, K. Sabeeh, U. Schwingenschloegl, Appl. Phys. Lett. **102**, 162412 (2013)
100. M. Ezawa, Phys. Rev. B **87**, 155415 (2013)
101. H. Pan, Z. Li, C.-C. Liu, G. Zhu, Z. Qiao, Y. Yao, Phys. Rev. Lett. **112**, 106802 (2014)
102. N. Singh, U. Schwingenschloegl, Phys. Stat. Solid-Rapid Research Lett. **8**, 353 (2014)
103. C.J. Tabert, E.J. Nicol, Phys. Rev. Lett. **110**, 197402 (2013)
104. M. Ezawa, Appl. Phys. Lett. **102**, 172103 (2013)
105. F. Liu, C.-C. Liu, K. Wu, F. Yang, Y. Yao, Phys. Rev. Lett. **111**, 066804 (2013)
106. S. Kokott, P. Pflugradt, L. Matthes, F. Bechstedt, J. Phys. Condens. Matter **26**, 185002 (2014)
107. M. Neek-Amal, A. Sadeghi, G.R. Berdiyorov, F.M. Peeters, Appl. Phys. Lett. **103**, 261904 (2013)
108. T.P. Kaloni, M. Tahir, U. Schwingenschloegl, Sci. Rep. **3**, 3192 (2013)
109. M. Kanno, R. Arafune, C.-L. Lin, E. Minamitani, M. Kawai, N. Takagi, New J. Phys. **16**, 105019 (2014)
110. L. Tao, E. Cinquanta, D. Chiappe, C. Grazianetti, M. Fanciulli, M. Dubey, A. Molle, D. Akinwande, Nat. Nanotechnol. **10**, 227 (2015)
111. K. Momma, F. Izumi, J. Appl. Cryst. **41**, 653 (2008)

Chapter 8
Silicene on Ag(111) and Au(110) Surfaces

**Hamid Oughaddou, Hanna Enriquez, Mohammed Rachid Tchalala,
Azzedine Bendounan, Andrew J. Mayne, Fausto Sirroti
and Gérald Dujardin**

Abstract Over the last decade, the existence and stability of silicene has been the subject of numerous studies. Indeed, silicene resembles graphene as it is a two-dimensional material arranged in a honeycomb lattice. Electronically, the main difference between carbon and silicon is the strong preference for $sp3$ over $sp2$ in silicon. It was only in 2010 that researchers presented the first experimental evidence of the formation of silicene on Ag(110) and Ag(111), which has launched a rush for silicene in a similar way as for graphene. This very active field has naturally led to the recent growth of silicene on other substrates such as Ir, ZrB_2 and Au. However, unlike graphene, the existence of silicene as a stand-alone material remains elusive. We present in this chapter the state of the art of silicene growth on Ag(111) and Au(110) substrates.

8.1 Introduction

The discovery of the exotic nature of graphene in 2004 [1–4] has stimulated a growing interest for the silicon and germanium counterparts, namely silicene and germanene. The obvious question was whether silicene and germanene could be

H. Oughaddou (✉) · H. Enriquez · M.R. Tchalala · A.J. Mayne · G. Dujardin
Institut des Sciences Moléculaires d'Orsay, ISMO-CNRS, Bât. 210,
Université Paris-Sud, 91405 Orsay, France
e-mail: Hamid.Oughaddou@u-psud.fr

H. Oughaddou
Département de Physique, Université de Cergy-Pontoise,
95031 Cergy-Pontoise Cedex, France

A. Bendounan · F. Sirroti
TEMPO Beamline, Synchrotron Soleil, L'Orme des Merisiers Saint-Aubin,
B.P. 48, 91192 Gif-sur-Yvette Cedex, France

© Springer International Publishing Switzerland 2016
M.J.S. Spencer and T. Morishita (eds.), *Silicene*, Springer Series
in Materials Science 235, DOI 10.1007/978-3-319-28344-9_8

fabricated and if so would they exhibit similar properties to those of graphene [5–8]. The main difference between carbon and silicon or germanium is the tendency to form sp^2 hybridization for the former and sp^3 for silicon and germanium, as is reflected by the fact that thermodynamically, the most stable structure of bulk Si or Ge is diamond, while for carbon it is a graphitic structure.

Theoretical investigations of free-standing silicene and germanene showed that once these 2D systems are fabricated, they are stable and present properties similar to those of graphene [5–7]. In particular, the free-standing silicene and germanene is expected to present a linear dispersion with a Dirac cone at the Brillouin zone corners and consequently, as in graphene, their charge carriers are expected to behave as massless relativistic particles. Therefore, all expectations held for graphene, such as high-speed electronic nanometric devices based on ballistic transport at room temperature (RT), could be applied to these new materials with the advantage of being directly compatible with existing silicon semiconductor technology. In addition, silicene and germanene have a larger spin-orbit coupling strength than graphene [9], which may lead to a detectable quantum spin Hall effect.

Experimentally, only a few studies have been performed on the Ge/Ag [10–13] and Si/Ag systems [8, 14–17]. Clear evidence for the formation of 2D silicene honeycomb structures for the Si/Ag system has been shown relatively recently [8, 14–16]. Silicene can form either an assembled parallel array of 1D nano-ribbons (NRs) on Ag(110) [14, 17] or a highly ordered sheet on Ag(111) [18–26]. Successful formation of silicene has been also reported on other surfaces including ZrB$_2$ [27] and Ir(111) [28]. These have been discussed at length in a recent review [29], along with the chemical synthesis of silicene and its potential applications in batteries. We present in this chapter the state of the art of silicene growth on Ag (111) and Au(110) substrates.

8.2   Silicene Growth on Ag(111)

Since the discovery of silicene several independent studies of the growth of silicene on the Ag(111) surface have revealed the existence of different ordered phases such as $(2\sqrt{3} \times 2\sqrt{3})R30°$, (4×4), and $(\sqrt{13} \times \sqrt{13})R13.9°$ [18, 20, 21, 26, 30, 31]. These structures are indexed with respect to the Ag(111) unit cell. We have studied in detail these structures and we have shown that these ordered phases correspond to a silicene sheet [25]. The different orientations of the same silicene sheet relative to the Ag(111) surface can be obtained by varying the substrate temperature during the silicon growth [25].

8.2.1 Atomic Structure

8.2.1.1 The $(2\sqrt{3} \times 2\sqrt{3})$R30° Superstructure

A continuous two dimensional (2D) sheet of silicene with a large area of an almost defect free honeycomb structure presenting a $(2\sqrt{3} \times 2\sqrt{3})$R30° superstructure on the Ag(111) surface, was observed for the first time by Scanning Tunneling Microscopy (STM) [16]. Subsequent claims that the STM image of $(2\sqrt{3} \times 2\sqrt{3})$R30° did not correspond to a silicene layer but to a bare Ag(111) surface due to a contrast inversion [32] (a graphically inverted image or to a tip effect) can be ruled out by the STM observations. Figure 8.1(1) and (2) show the STM images obtained, respectively, on the bare Ag(111) substrate, and after the deposition of one silicon monolayer giving rise to a $(2\sqrt{3} \times 2\sqrt{3})$R30° superstructure. The atomic resolved STM images reveal a highly ordered silicon honeycomb lattice structure (Fig. 8.1(2)).

We see clearly that the basal chains of the honeycomb silicene sheet are rotated by an angle of $(30 - 19.1)° = 10.9°$ with respect to the axes of the Ag(111). This matches very well the expected rotation for a $(2\sqrt{3} \times 2\sqrt{3})$R30° superstructure. In the case of an inversion of contrast of the STM image of bare silver, the angle between the observed honeycomb structure, and that of the bare Ag(111) would be 0° and not 10.9°. In Fig. 8.1(3), the line profile along line (A) gives a lateral Si–Si distance of 0.2 nm, and a height difference between neighboring Si atoms (buckling) of 0.02 nm. This can be compared to the corrugation of the bare Ag(111) surface, which is 10 times smaller (0.002 nm).

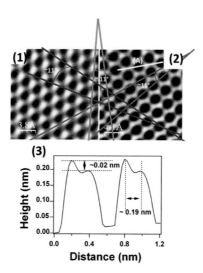

Fig. 8.1 (*1*) Filled-state atomically resolved STM image of the clean Ag(111) surface. (*2*) Filled-state atomically resolved STM image of the same sample (without any rotation) after deposition of one silicon monolayer. (*3*) Line-profile joining neighboring Si atoms along the direction (*A*) indicated in (*2*). Reproduced from [24]

8.2.1.2 The ($\sqrt{13} \times \sqrt{13}$)R13.9° Superstructure

The ($\sqrt{13} \times \sqrt{13}$)R13.9° is obtained after deposition at a substrate temperature around 250 °C [25]. The atomically resolved STM images of the ($\sqrt{13} \times \sqrt{13}$)R13.9° system show two types of structures. The ($\sqrt{13} \times \sqrt{13}$)R13.9° type I and type II superstructures are shown in Fig. 8.2(1) and (2), respectively. We observe that under similar tunneling conditions, the unit cell of the ($\sqrt{13} \times \sqrt{13}$)R13.9° type II is composed of *four* protrusions per unit cell, whereas that of the ($\sqrt{13} \times \sqrt{13}$)R13.9° type I superstructure is composed of only *one* protrusion. This is confirmed by our DFT calculations [24, 26]. Within the unit cell, the relative positions of the Si and Ag atoms are very different between the type I and type II superstructures leading to very different out-of-plane coordinates. A two-dimensional model of the ($\sqrt{13} \times \sqrt{13}$)R13.9° supercell of *the Ag substrate* can coincide with a ($\sqrt{7} \times \sqrt{7}$)R19.1° supercell of *the silicene sheet* for two different relative angles, either 5.2° (i.e. 19.1 − 13.9) or 33° (i.e. 19.1 + 13.9) resulting in the two observed super-structures. Three-dimensional models for the type I and type II superstructures were investigated in our earlier DFT calculations [24, 26]. We find that the type I and type II structures correspond to orientations of 5.2° and 33° relative to the (111) surface, respectively. These results indicate that two different configurations co-exist with different atomic positions. In other words, small changes in the buckling of the silicene modify the observed structure.

8.2.1.3 The (4 × 4) Superstructure

The (4 × 4) superstructure is obtained with a substrate temperature around 200 °C. We have studied this structure for the first time by combining non-contact atomic force microscopy (nc-AFM) and scanning tunneling microscopy [26]. The resulting STM topography and nc-AFM image of the same area are shown on Fig. 8.3(1)

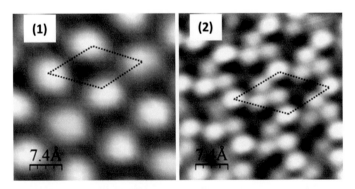

Fig. 8.2 Atomically resolved STM images corresponding to the ($\sqrt{13} \times \sqrt{13}$)R13.9° (V = −1.5 V, I = 0.2 nA) for (*1*) type I, and (*2*) type II. The corresponding unit cells are shown with *dotted lines*. Reproduced from [25]

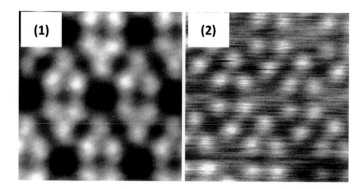

Fig. 8.3 (*1*) Atomically resolved STM image of the (4 × 4) superstructure (4 × 4); (*2*) The same area was mapped in constant height nc-AFM operation. For the latter imaging −17 Hz was used as a feedback set point. Reproduced from [26]

and (2), respectively. The nc-AFM image (Fig. 8.3(1)) was taken with a z-height where the repulsive force over the imaged Si atoms prevails. Therefore, the bright protrusions in the image correspond to sites that had a greater repulsive action on the tip apex.

The STM and nc-AFM images in Fig. 8.3(1) and (2) are remarkably similar, showing protrusions at identical locations in the unit cell. Each triangle represents one half of the elementary surface unit cell. The atomic contrast acquired by nc-AFM and STM consists of 6 characteristic protrusions arranged in 2 triangles within the unit cell. The observation coincides very well with the proposed (4 × 4) model structure [20, 24] as can be seen directly in Fig. 8.3(1) and (2). In particular, the (4 × 4) model structure shows clearly that the Si atoms are relaxed out of the surface plane forming pyramids, whose apices are located at the positions where both the STM and AFM show bright protrusions.

8.2.2 Electronic Structure

Since free-standing silicene is expected to present a Dirac cone [5–8], we can ask the question whether silicene preserves the Dirac cone on a supporting substrate or not. The 4 × 4 silicene structure on Ag(111) has attracted considerable attention in this respect because its atomic structure is well known [20, 30].

Two groups have claimed the existence of Dirac fermions in silicene on the Ag (111) surface [19, 33]. Vogt et al. [19] have studied the electronic structure of the 4 × 4 silicene reconstruction by angular-resolved photoemission spectroscopy (ARPES). They found a linear dispersion near the K point of the Brillion zone of silicene. They attributed the linear dispersion to silicene with the Dirac point 0.3 eV below the Fermi level. In addition, the Fermi velocity was estimated to be 1.3×10^6 ms^{-1}; higher than the value reported for graphene. The second claim of the existence of Dirac fermions in silicene on the Ag(111) surface was reported by Chen et al. [33]. They studied the silicene structure using scanning tunneling spectroscopy (STS). A quasiparticle interference pattern (QPI) at the surface of the silicene was observed, and the Dirac cone was deduced from the linear dispersion curve obtained from the STS curves [33].

However, STM measurements recorded under high magnetic field by Lin et al. [34] pointed to the absence of discrete Landau levels expected in the case of massless carriers. Hence, the authors concluded that the carriers in the 4×4 silicene do not show a Dirac cone.

The interpretation of the single-layer silicene photoemission data [19] has also been questioned on the basis of band structure calculations [35]. Indeed, these theoretical analyses propose that the experimentally reported bands should be attributed to the Ag states. The calculation revealed that the Dirac cone in the 4×4 silicene phase is destroyed as a result of the strong band hybridization between silicene and the Ag surface.

Very recently, Mahatha et al. [36] repeated the same experiment as Vogt et al. [19] in order to confirm, or not confirm, the presence of the Dirac cone. They performed a detailed ARPES study coupled with density functional theory calculations of the 4×4 silicene structure. They found that silicene on Ag(111) does not preserve the Dirac cone at all due to the hybridization of silicene and silver states. Their result contradicts completely the interpretation of the observed linear dispersion by Vogt et al. [19] and Chen et al. [33]. The sp branch of the bare Ag(111) is observed as a linear trend near the K point of 4×4 silicene. The sp states are still observable even after the growth of the 4×4 silicene sheet indicating that the previous observations of the linear dispersion corresponded only to the sp branch of silver. Moreover, a similar *sp*-like linear dispersion of the surface state has been observed in Ag/Pt(111) system [37, 38]. This would suggest that there is no Dirac cone for the 4×4 structure of silicene on Ag(111).

8.2.3 Silicene Multilayers

Overcoming the silicene-substrate interaction is a key challenge at present. One idea is to deposit more silicene to make multilayer silicene. The silicene-substrate interaction should decrease as the thickness of silicene increases. In addition, if we succeed in growing silicene multilayers, this will have a very high impact in the integration of this 2D atomic material into electronic devices.

Very recently, Vogt et al. [39] reported the epitaxial growth of multilayer silicene on Ag(111). Based on STM and low-energy electron diffraction (LEED) measurements, they claimed that they had grown a silicene stack, consisting of a few atomic layers, where the lowest one has the 4×4 structure and all following layers show the $(\sqrt{3} \times \sqrt{3})$ structure [39]. This result was questioned by Acun et al. [40] who used LEEM and micro-LEED experiments. They studied the formation and the stability of silicene layers on a Ag(111) substrate, and they found that silicene layers are intrinsically unstable with respect to the formation of an "sp^3-like" hybridized, bulk-like silicon structure. The claim that the multilayers have diamond-like structure is supported by another LEED study by Shirai et al. [41] on Si multilayers. Their data are consistent with silicon multilayers on Ag(111) that are diamond-like silicon terminated with the Si(111)–$\sqrt{3} \times \sqrt{3}$-Ag surface. They conclude that the $\sqrt{3} \times \sqrt{3}$ structure observed by Vogt et al. [39] is the result of segregation of Ag atoms on top of bulk like silicon surface. The model is similar to the one previously observed in the case of the Ag/Si(111) surface [42, 43]. More experimental and theoretical studies are needed in order to conclude on the evidence of silicene multilayers.

8.3 Silicene Growth on Au(110)

8.3.1 Surface Alloy Formation

Considering that silicene structures form on silver, the growth of silicene was also investigated on gold substrates. Gold presents several common features with silver. First, the unit cell parameters are almost identical (0.2 % difference), which is a critical criterion for the epitaxy of an unstrained and stable silicene structure. Second, the bulk phase diagram of the binary gold-silicon system is a simple eutectic with a large miscibility gap [44], so that one could expect the segregation of silicon at the surface leading to silicene structures.

However, there are also some notable differences with the silicon-silver system, in particular a small solubility (less than 2 at.%) of silicon in solid gold at the eutectic temperature [44]. Metastable bulk Au_xSi_y crystalline phases have been evidenced with different stoichiometry [45–49]. The formation of disordered Au–Si surface alloys was also shown following the room-temperature (RT) deposition of gold on silicon substrates under ultra-high vacuum (UHV) conditions [50–52]. Interestingly, a stable two-dimensional (2D) crystalline Au_4Si_8 gold silicide phase has been obtained through two different methods, namely the surface crystallisation of a eutectic $Au_{82}Si_{18}$ liquid [53] and the annealing of a 3 nm Au layer on a Si(100) substrate above the eutectic temperature in UHV conditions [54].

With this in mind, the deposition of sub-monolayer silicon on an Au(110) substrate was investigated using a combination of STM, LEED and Auger electron spectroscopy under UHV conditions [29, 55]. In contrast to Ag(110), the clean Au (110) substrate undergoes a 2×1 reconstruction, and no ordered structure was observed when the deposition of silicon was performed at room temperature. When the substrate was kept at a temperature above the eutectic temperature (between 360 and 500 °C), new ordered structures were observed as a function of the amount of silicon deposition controlled by Auger electron spectroscopy.

The structural transitions observed as a function of silicon deposition on the substrate at 400 °C are indicated by the evolution of the LEED pattern shown in Fig. 8.4.

The 2 × 1 diffraction spots of the clean Au(110) substrate (Fig. 8.4(1)) start to disappear while a new complex superstructure appears, becoming sharp at a Si coverage of ~ 0.2 monolayer (ML). The LEED pattern of this new superstructure (Fig. 8.4(2)) shows two symmetrical domains with respect to the [001]* and [$\bar{1}$10]* directions of the substrate. The direct-space basis vectors of the two corresponding unit cells are described by the two matrices $\begin{vmatrix} 10 & -1 \\ -2 & 4 \end{vmatrix}$ and $\begin{vmatrix} 10 & 1 \\ 2 & 4 \end{vmatrix}$ [55]. Above this 0.2 ML coverage, the spots of this complex superstructure disappear, and a rectangular superstructure appears becoming sharp at 0.3 ML. The new LEED pattern (Fig. 8.4(3)) presents a very clear 6× periodicity along the [$\bar{1}$10]* direction while the spots along the [001]* direction are not very distinct, showing different spacing, that indicates a variable short-range order along this direction.

The STM topographic measurements were performed at different silicon coverages after the sample had been cooled to RT. The missing-row structure of the STM image of the bare Au(110)–(2 × 1) reconstruction is clearly observed in Fig. 8.5(1). Figure 8.5(2) displays an STM image of the first stage of silicon growth on Au(110) corresponding to a silicon coverage less than 0.1 ML.

On this image, one observes that the clean Au(110)–2 × 1 reconstruction coexists with a superstructure composed of two domains. The oblique unit cells of these two domains are identical to the two unit cells identified in the complex LEED pattern (Fig. 8.4(2)). In addition, the clean Au(110) surface and the new superstructure have the same height in the topography. This suggests the formation of an ordered surface alloy exhibiting chiral domains growing at the expense of the 2 × 1 reconstruction when silicon is deposited at 400 °C. Indeed, at 0.2 ML coverage, the domains of the clean Au(110) surface disappear completely in the STM topography, in agreement with the extinction of the 2× diffraction spots along [001]* in the corresponding LEED pattern (Fig. 8.4(2)). After the deposition of 0.3 ML silicon at 400 °C, the

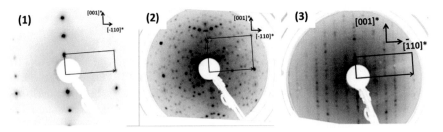

Fig. 8.4 LEED pattern recorded at Ep = 40 eV corresponding to: (*1*) the bare Au(1110)–2 × 1, (2) the $\begin{vmatrix} 10 & -1 \\ -2 & 4 \end{vmatrix}$ reconstruction obtained after deposition of ~0.2 Si ML on Au(110) held at 400 °C, and (*3*) the ×6 superstructure obtained after the deposition of ~0.3 Si ML on Au(110) held at 400 °C. The unit cells are indicated. Reproduced from [55, 59]

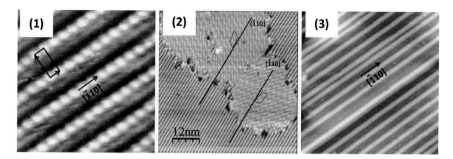

Fig. 8.5 The STM images corresponding to (*1*) the bare Au(110)–2 × 1, (*2*) after deposition of 0.1 ML of silicon, (*3*) after deposition of 0.3 ML of silicon. Reproduced from [55, 59]

STM topography reveals a network of nanoribbons (NRs) not systematically equidistant and all aligned parallel to the [$\bar{1}$10] direction (Fig. 8.5(3)), which corresponds to the rectangular superstructure observed in the LEED (Fig. 8.4(3)), showing no clear periodicity along [001]*.

Figure 8.6 shows an atomically resolved filled-state STM image of one domain of the superstructure at 0.2ML coverage (Fig. 8.6(1)). The proposed atomistic model for this structure in Fig. 8.6(2) produces the simulated STM image in Fig. 8.6(3) which matches very closely the experimental STM image.

The STM topography of the superstructure presents the two-fold symmetry as observed in the LEED pattern. The pattern in the oblique unit cell shown can be described by four similar entities (Fig. 8.6(1)). The STM images display the same contrast at negative and positive sample biases, indicating that we observe the atomic geometry rather than any electronic effect. We therefore propose that each

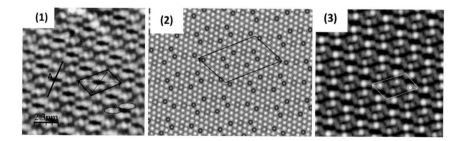

Fig. 8.6 (*1*) Atomically resolved filled-states STM image recorded at $\sim$0.2 Si ML showing the $\begin{vmatrix} 10 & -1 \\ -2 & 4 \end{vmatrix}$ superstructure (V = −1.4 V, I = 2.3 nA). The unit cell and the two-fold symmetry are indicated, (*2*) Proposed model where 8 silicon atoms occupy 4-fold sites and 4 silicon atoms are sitting just below 4 gold atoms. The Au atoms of the 1st and the 2nd monolayers are colored in gold and grey respectively. Si atoms are in *blue*. The Au atoms located on top of Si atoms are colored in *pink*. The unit cell is indicated with *black lines*. (*3*) Simulated STM image corresponding to the model shown in (*2*). The unit cell is indicated with *white lines*. The brightest protrusions are emphasized with circles. Reproduced from [55]

entity is composed of three atoms as highlighted in Fig. 8.6(1), composed of one bright atom and two less bright atoms.

The model for this silicon adsorption superstructure is based on density functional theory (DFT) calculations and STM images simulated using the Tersoff-Hamann method with an s-like tip [56]. Configurations involving 8–12 Si atoms per unit cell were investigated, in agreement both with the Auger electron spectroscopy calibration (~ 0.2 ML), and with the presence of four similar entities in the STM topography. The DFT calculations were performed using a 5-layer thick slab with no reconstruction on the top-layer. A PBE potential was used and the top 3 layers were relaxed [57, 58]. In the different configurations, silicon atoms were adsorbed at different sites (hollow, short-bridge, long bridge and top) into the large unit cell, and ended up on a hollow site after relaxation. However, the calculated STM images for these configurations did not agree with the experimental ones. We have to take into account the possible diffusion of silicon atoms into the subsurface region due to the high temperature of the gold substrate during deposition. Therefore, we considered configurations including Si atoms located below the topmost gold layer. The best agreement was obtained with 12 Si atoms per unit cell, where 8 Si atoms occupy hollow sites and 4 Si atoms sit Immediately below 4 gold atoms, as sketched in the model of Fig. 8.6(3). After relaxation, these 4 gold atoms were located 0.1 nm above the other gold surface atoms, and the corresponding corrugation could explain the bright spots in the entities observed in the experimental STM image (Fig. 8.6(1)). The less bright spots are assigned to the Si atoms in the hollow sites. This is confirmed by the simulated STM image (Fig. 8.6(3)) which reproduces very well the experimental image (Fig. 8.6(1)) for the same bias voltage, based on the atomic configuration shown in Fig. 8.6(2).

Our results indicate that the intermediate superstructure at 0.2 ML is driven by strong interactions between the Si and Au atoms, which is also supported by the thermal stability of this superstructure up to 500 °C. The Si atoms within the unit cell are bound only to Au, and we conclude that the superstructure is made of an ordered surface alloy formed with a stoichiometry of $Au_{38}Si_{12}$.

8.3.2   Silicene Nanoribbons on Au(110)

We now turn to the next structural transition that takes place at the higher coverage of 0.3 ML silicon on Au(110). According to both LEED (Fig. 8.4(3)) and STM topography (Fig. 8.5(3)), the previous surface alloy has completely disappeared and is replaced by a self-assembly of nanoribbons oriented parallel to $[\bar{1}10]$. Only the surface steps limit the length of these nanoribbons. The periodicity along the $[\bar{1}10]$ direction is 1.76 nm (Fig. 8.7(1) and (2)) in agreement with the $6 \times a_{[-110]}$ periodicity observed in the LEED pattern (Fig. 8.4(3)). The transverse line profile (Fig. 8.7(3)) indicates that the ribbons have a width of 1.6 nm, which corresponds to 4 atomic distance along the [001] direction and an atomic height of 0.24 nm.

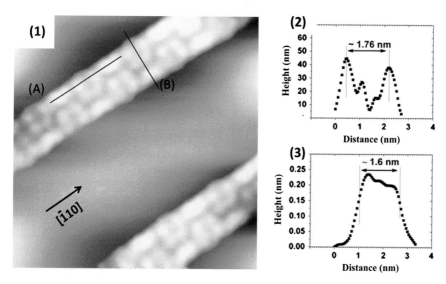

Fig. 8.7 (*1*) High resolution STM image showing the structure of Si NRs on Au(110) (8 × 8 nm², V = −64 mV, I = 3.5 nA). (2) Line profile along A showing the ×6 periodicity. (*3*) Line profile along B showing the asymmetry across the width of the NR. The NR width is 1.6 nm. Reproduced from [59]

Core-level photoemission experiments performed on these silicon NRs [59] showed that the Si 2p spectra measured at normal and at surface sensitive emission angles are almost identical, with the same ratio between the two components (Fig. 8.8(1)). This indicates that all Si atoms are located on the surface and at a similar height, supporting the idea that the NRs observed in STM images are made of silicon. The Si 2p core level spectra are reproduced with two spin-orbit split components with very narrow total widths, indicating a high ordering of the Si atoms within the NRs, with only two Si environments. In addition, the Si 2p core-levels present an asymmetry parameter which indicates the metallic character of the Si NRs. On the other hand, the Au 4f spectra [59] show two components, one corresponding to the bulk environment and the other to the surface. At 0.3 ML silicon coverage, the surface component is shifted to higher binding energies by 1 eV, indicating that the (2 × 1) reconstruction disappears after silicon adsorption, and that only surface Au atoms have a silicon environment.

We proposed a model structure for the NRs composed of a hexagonal silicene lattice with zig-zag edges (Fig. 8.8(2)).

With a lattice unit dimension of 0.346 nm, 5 silicene hexagons correspond to a periodicity of ×6 Au along [110] direction, whereas the ribbon width of 4 silicene rows fits well with 4 Au lattice positions along the [100] direction. The in-plane Si–Si nearest neighbor lateral distance derived from this model is 0.2 nm close to the value found for a silicene sheet on Ag(111). The model presents an asymmetry across the width of the NR in agreement with the STM image. Two silicon

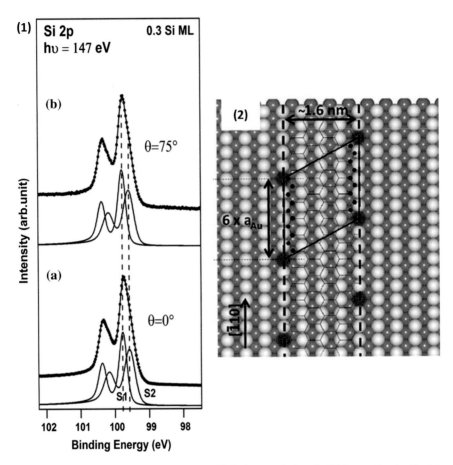

Fig. 8.8 (**1**) Si 2p core levels spectra (*dots*) and their de-convolutions (*solid line* overlapping the data points) with two asymmetric components (*S1*) and (*S2*) recorded after deposition of 0.3 Si ML on Au (110)–(2 × 1) at hv = 147 eV in (*a*) normal emission and (*b*) 75° off-normal. The fitting parameters used are: 150 and 215 meV Gaussian widths for the components (S1) and (S2) respectively, a Lorentzian width of 80 meV, a spin-orbit splitting of 0.605 eV, a branching ratio of 0.52, and an asymmetry parameter of 0.09. (**2**) Proposed model based on the STM image and the LEED pattern. The silicene hexagons are shown in *black*. Reproduced from [59] (Color figure online)

environments can be distinguished, namely the edge atoms of the ribbon and the other atoms in the middle, in agreement with the two components observed in photoemission. However, the internal structure visible in the STM image (Fig. 8.7 (1)) has a larger periodicity by a factor of $\sqrt{3}$. It has been argued that this could be due to the formation of quantum interference patterns generated by electron scattering from the edges of the NRs [60, 61].

The silicon NRs grown on Au(110) present some common features with those obtained on Ag(110), namely a flat structure above the substrate surface, the same width of 1.6 nm, and a metallic character. However, the interaction between silicon and both silver and gold substrates are different, and result in the second case in the existence of an intermediate surface alloy before the transition to a silicon nanoribbon structure. In both cases, a model based on a silicene structure is proposed.

8.4   Conclusion

In this chapter we have presented the recent atomic scale studies of silicene on the Ag(111) and Au(110) surfaces using STM combined with LEED and photoemission measurements along with DFT calculations. Several observations can be made in conclusion: (i) Silicene growth on both surfaces present a variety of surface superstructures which depend on the coverage of the silicon and the temperature of the substrate during growth. (ii) Silicene growth on Au(110) is subtly different from Ag(110); first a surface alloy forms on Au and second, the nanoribbons have a different structure from those on Ag(110). This chapter describes only a small part of the complexity of silicene growth on metals, and yet demonstrates clearly the wealth of physical information that is still to be revealed in further studies.

References

1. K.S. Novoselov, A.K. Geim, S.V. Morozov, D. Jiang, Y. Zhang, S.V. Dubonos, I.V. Grigorieva, A.A. Firsov, Science **306**, 666 (2005)
2. C. Berger, Z. Song, X. Li, X. Wu, N. Brown, C. Naud, D. Mayou, T. Li, J. Hass, A.N. Marchenkov, E.H. Conrad, P.N. First, W.A. de Heer, Science **312**, 1191–1196 (2006)
3. J.C. Meyer, A.K. Geim, M.I. Katsnelson, K.S. Novoselov, T.J. Booth, S. Roht, Nature **446**, 60 (2007)
4. A.H. Castro Neto, F. Guinea, N.M.R. Peres, K.S. Novoselov, A.K. Geim, Rev. Mod. Phys. **81**, 109 (2009)
5. S. Lebègue, O. Eriksson, Phys. Rev. B **79**, 115409 (2009)
6. G.G. Guzman-Verri, L.C.L.Y. Voon, Phys. Rev. B **76**, 075131 (2007)
7. S. Cahangirov, M. Topsakal, E. Akturk, H. Sahin, S. Ciraci, Phys. Rev. Lett. **102**, 236804 (2009)
8. A. Kara, H. Enriquez, A.P. Seitsonen, L.C.L.Y. Voon, S. Vizzini, B. Aufray, H. Oughaddou, Surf. Sci. Rep. **67**, 1–18 (2012)
9. C.-C. Liu, W. Feng, Y. Yao, Phys. Rev. Lett. **107**, 076802 (2011)
10. H. Oughaddou, B. Aufray, J.P. Bibérian, J.Y. Hoarau, Surf. Sci. **429**, 320 (1999)
11. H. Oughaddou, S. Sawaya, J. Goniakowski, B. Aufray, G. Le Lay, J.M. Gay, G. Tréglia, J.P. Bibérian, N. Barrett, C. Guillot, A.J. Mayne, G. Dujardin, Phys. Rev. B **62**, 16653 (2000)
12. H. Oughaddou, C. Léandri, B. Aufray, C. Girardeaux, J. Bernardini, G. Le Lay, J.P. Bibérian, N. Barrett, Appl. Surf. Sci. **1–5**, 9781 (2003)
13. C. Léandri, H. Oughaddou, J.M. Gay, B. Aufray, G. Le Lay, J.P. Bibérian, A. Rangüis, O. Bunk, R.L. Johnson, Surf. Sci. **573**, L369 (2004)
14. B. Aufray, A. Kara, S. Vizzini, H. Oughaddou, C. Léandri, B. Ealet, G. Le Lay, Appl. Phys. Lett. **96**, 183102 (2010)

15. H. Enriquez, S. Vizzini, A. Kara, B. Lalmi, H. Oughaddou, J. Phys.: Condens. Matter **24**, 314211 (2012)
16. B. Lalmi, H. Oughaddou, H. Enriquez, A. Kara, S. Vizzini, B. Ealet, B. Aufray, Appl. Phys. Lett. **97**, 223109 (2010)
17. M.R. Tchalala, H. Enriquez, A.J. Mayne, A. Kara, G. Dujardin, M. Ait Ali, H. Oughaddou, J. Phys. Conf. Ser. **491**, 012002 (2014)
18. C.-L. Lin, R. Arafune, K. Kawahara, N. Tsukahara, E. Minamitani, Y. Kim, N. Takagi, M. Kawai, Appl. Phys. Express **5**, 45802 (2012)
19. P. Vogt, P. De Padova, C. Quaresima, J. Avila, E. Frantzeskakis, M.C. Asensio, A. Resta, B. Ealet, G. Le Lay, Phys. Rev. Lett. **108**, 155501 (2012)
20. H. Jamgotchian, Y. Colignon, N. Hamzaoui, B. Ealet, J.Y. Hoarau, B. Aufray, J.P. Biberian, J. Phys. Condens. Matter **24**, 172001 (2012)
21. B. Feng, Z. Ding, S. Meng, Y. Yao, X. He, P. Cheng, L. Chen, K. Wu, Nano Lett. **12**, 3507–3511 (2012)
22. D. Chiappe, C. Grazianetti, G. Tallarida, M. Fanciulli, A. Molle, Adv. Mater. **24**, 5088–5093 (2012)
23. H. Enriquez, S. Vizzini, A. Kara, B. Lalmi, H. Oughaddou, J. Phys. Condens. Matter **24**, 314211 (2012)
24. H. Enriquez, A. Kara, A.J. Mayne, G. Dujardin, H. Jamgotchian, B. Aufray, H. Oughaddou, J. Phys. Conf. Ser. **491**, 012004 (2014)
25. M.R. Tchalala, H. Enriquez, H. Yildirim, A. Kara, A.J. Mayne, G. Dujardin, M. Ait Ali, H. Oughaddou. Appl. Surf. Sci. **303**, 61–66 (2014)
26. Z. Majzik, M.R. Tchalala, M. Svec, P. Hapala, H. Enriquez, A. Kara, A.J. Mayne, G. Dujardin, P. Jelinek, H. Oughaddou, J. Phys. Condens. Matter **25**, 225301 (2013)
27. A. Fleurence, R. Friedlein, T. Ozaki, H. Kawai, Y. Wang, Y. Yamada-Takamura, Phys. Rev. Lett. **108**, 245501 (2012)
28. L. Meng, Y. Wang, L. Zhang, S. Du, R. Wu, L. Li, Y. Zhang, G. Li, H. Zhou, W.A. Hofer, H.-J. Gao, Nano Lett. **13**, 685–690 (2013)
29. H. Oughaddou, H. Enriquez, M.R. Tchalalaa, H. Yildirim, A.J. Mayne, A. Bendounan, G. Dujardin, M. Ait Ali, A. Kara. Prog. Surf. Sci. **90**, 46–83 (2015)
30. R. Arafune, C.-L. Lin, K. Kawahara, M. Kanno, N. Tsukahara, E. Minamitani, Y. Kim, N. Takagi, M. Kawai, Surf. Sci. **608**, 297–300 (2013)
31. J. Gao, J. Zhao, Sci. Rep. **2**, 861 (2012)
32. G. Le Lay, P. De Padova, A. Resta, T. Bruhn, P. Vogt, J. Phys. D Appl. Phys. **45**, 392001 (2012)
33. L. Chen, C.-C. Liu, B. Feng, X. He, P. Cheng, Z. Ding, S. Meng, Y. Yao, K. Wu, Phys. Rev. Lett. **109**, 056804 (2012)
34. C.-L. Lin, R. Arafune, K. Kawahara, M. Kanno, N. Tsukahara, E. Minamitani, Y. Kim, M. Kawai, N. Takagi, Phys. Rev. Lett. **110**, 076801 (2013)
35. Yun-Peng Wang, Hai-Ping Cheng, Phys. Rev. B. **87**, 245430 (2013)
36. S.K. Mahatha, P. Moras, V. Bellini, P.M. Sheverdyaeva, C. Struzzi, L. Petaccia, C. Carbone, Phys. Rev. B **89**, 201416 (2014)
37. A. Bendounan, K. Aït-Mansour, J. Braun, J. Minár, S. Bornemann, R. Fasel, O. Gröning, F. Sirotti, H. Ebert, Phys. Rev. B **83**, 195427 (2011)
38. A. Bendounan, J. Braun, J. Minár, S. Bornemann, R. Fasel, O. Gröning, Y. Fagot-Revurat, B. Kierren, D. Malterre, F. Sirotti, H. Ebert, Phys. Rev. B **85**, 245403 (2012)
39. P. Vogt et al., Appl. Phys. Lett. **104**, 021602 (2014)
40. A. Acun et al., Appl. Phys. Lett. **103**, 263119 (2013)
41. T. Shirai, T. Shirasawa, T. Hirahara, N. Fukui, T. Takahashi, S. Hasegawa, Phys. Rev. B **89**, 241403 (2014)
42. T. Takahashi, S. Nakatani, Surf. Sci. **282**, 17–32 (1993)
43. H. Aizawa, M. Tsukada, N. Sato, S. Hasegawa, Surf. Sci. **429**, L509–L514 (1999)
44. H. Okamoto, T.B. Massalski, Bulletin of Alloy Phase Diagrams **4**, 190–198 (1983)
45. P. Predecki, B.C. Geissen, N.J. Grant, Trans. Metall. Soc. AIME **233**, 1438 (1965)

46. T.R. Anantharanman, H.L. Luo, W. Klement Jr, Nature **210**, 1040–1041 (1966)
47. A.K. Green, E. Bauer, J. Appl. Phys. **47**, 1284 (1976)
48. A.K. Green, E. Bauer, J. Appl. Phys. **52**, 5098 (1981)
49. R.R. Chromik, L. Zavalij, M.D. Johnson, E.J. Cotts, J. Appl. Phys. **91**, 8992 (2002)
50. S.L. Molodtsov, C. Laubschat, G. Kaindl, A.M. Shikin, V.K. Adamchuk, Phys. Rev. B **44**, 8850 (1991)
51. J.-J. Yeh, J. Hwang, K. Bertness, D.J. Friedman, R. Cao, I. Lindau, Phys. Rev. Lett. **70**, 3768 (1993)
52. Y. Hoshino, Y. Kitsudo, M. Iwami, Y. Kido, Surf. Sci. **602**, 2089–2095 (2008)
53. O.G. Shpyrko, R. Streitel, Venkatachalapathy S.K. Balagurusamy, A.Y. Grigoriev, M. Deutsch, B.M. Ocko, M. Meron, B. Lin, P.S. Pershan, Science **313**, 77 (2006)
54. A.L. Pinardi, S.J. Leake, R. Felici, I.K. Robinson, Phys. Rev. B **79**, 045416 (2009)
55. H. Enriquez, A.J. Mayne, A. Kara, S. Vizzini, S. Roth, B. Lalmi, A.P. Seitsonen, B. Aufray, T. Greber, R. Belkhou, G. Dujardin, H. Oughaddou, Appl. Phys. Lett. **101**, 021605 (2012)
56. J. Tersoff, D.R. Hamann, Phys. Rev. B **31**, 805 (1985)
57. P.E. Blöchl, Phys. Rev. B **50**, 17953–17979 (1994)
58. G. Kresse, D. Joubert, Phys. Rev. B **59**, 1758–1775 (1999)
59. M.R. Tchalala, H. Enriquez, A.J. Mayne, A. Kara, S. Roth, M.G. Silly, A. Bendounan, F. Sirotti, Th Greber, B. Aufray, G. Dujardin, M. Ait, Ali, H. Oughaddou. Appl. Phys. Lett. **102**, 083107 (2013)
60. H. Yang, A.J. Mayne, M. Boucherit, G. Comtet, G. Dujardin, Y. Kuk, Nano Lett. **10**, 943 (2010)
61. C. Park, H. Yang, S. Seo, G. Kim, A.J. Mayne, G. Dujardin, Y. Kuk, J. Ihm, Proc. Natl. Acad. Soc. USA **108**, 18622 (2011)

Chapter 9
Growth of Silicon Nano-ribbons on Ag(110): State of the Art

Bernard Aufray, Bénédicte Ealet, Haik Jamgotchian, Hichem Maradj, Jean-Yves Hoarau and Jean-Paul Biberian

Abstract The adsorption of silicon on Ag(110) forms long nano-ribbons, with very few atomic defects, perfectly aligned along the [$\bar{1}$10] direction of the Ag surface and all with the same width and same height. Despite many experimental and theoretical works, the atomic structure of the nano-ribbons is still debated and up until now, no proposed model agrees with all the experimental characterizations. The same controversy exists for thicker 3D nano-ribbons showing a pyramidal shape: are they stacks of silicene layers or nano-facetting of the substrate? Independently of the atomic structure it is well established that silicon nano-ribbons are hard to oxidize and can be used as templates to grow magnetic or organic nanostructures. In this chapter, we review the growth of silicon nano-ribbons on the Ag(110) surface.

9.1 Introduction

When we started to study the growth of silicon on silver substrates in 2001, in Marseilles, France, obviously our goal was not the synthesis of silicene layers, the new allotropic form of silicon [1–4]. At this time, it was part of the continuation of a

B. Aufray (✉) · B. Ealet · H. Jamgotchian · J.-Y. Hoarau · J.-P. Biberian
Aix-Marseille Université, CNRS, CINaM UMR 7325, 13288 Marseille, France
e-mail: aufray@cinam.univ-mrs.fr

B. Ealet
e-mail: ealet@cinam.univ-mrs.fr

H. Jamgotchian
e-mail: jamgotchian@cinam.univ-mrs.fr

J.-Y. Hoarau
e-mail: hoarau@cinam.univ-mrs.fr

J.-P. Biberian
e-mail: biberian@cinam.univ-mrs.fr

H. Maradj
Laboratoire LSMC, Université d'Oran es-sénia, 31100 Oran, Algeria
e-mail: maradj@cinam.univ-mrs.fr

© Springer International Publishing Switzerland 2016
M.J.S. Spencer and T. Morishita (eds.), *Silicene*, Springer Series
in Materials Science 235, DOI 10.1007/978-3-319-28344-9_9

previous study of germanium on silver, itself being the development of a more general theme on the formation of "surface alloys" [5]. Generally, in the case of metal-on-metal systems, it has been shown that the main dissolution kinetics behavior observed during annealing of a metallic thin film (one or two monolayers thick) deposited on a metallic substrate, can be predicted taking into account using two important parameters: the tendency to stay at the surface of one of the two elements (surface segregation effect) and the tendency for ordering or phase separation between the elements (chemical effect) [6, 7]. This has been studied in details for bi-metallic model systems theoretically [7–9] as well as experimentally [6, 10–13]. In order to expand this approach and to understand the role played by the nature of interatomic binding of the elements in the formation of surface alloys (e.g. covalent/ionic versus metallic bonds), we started to investigate the case of semi-conductors (Si, Ge) on metallic substrates. At this time, surprisingly, very few studies were devoted to the deposition of semiconductors on metallic surfaces and most of them were performed on systems having a tendency towards ordering i.e. forming ordered compounds such as silicide or germanide [14–20]. However, systems with a tendency towards phase separation, which are expected to form a stable semicon-ducting atomic layer on the metal surface, had not been studied at that time.

We started with Ge on a silver substrate (PhD thesis of H. Oughaddou [21]) which is characteristic of phase separation with a large tendency to Ge surface segregation (the surface energy of Ge being lower than that of the silver one). Surprisingly, during the dissolution process of one monolayer of Ge deposited on Ag(100), we observed, instead of a slow dissolution process due to the Ge surface segregation effect (as it was observed for equivalent metal-metal systems [10, 22]), a very rapid Ge dissolution which ceased at an intermediate surface coverage [23, 24]. Using scanning tunneling microscopy (STM), we showed that this phenomenon was due to the formation of stable Ge tetramers self-organized on the surface [24]. In the light of these results, it was then tempting to test silicon on the same substrate (PhD thesis of Léandri [25]). Indeed, the Si-Ag system presents a bulk phase diagram equivalent to the Ge-Ag one but with a stronger tendency to phase separation (i.e. no solubility on both sides of the phases diagram [26]) and a limited surface segregation effect, since silver and silicon have a surface energy very close to each other. We started by studying the growth of Si on Ag(100) at room temperature (RT) using Auger Electron Spectroscopy (AES) and Low Energy Electron Diffraction (LEED) tech-niques [27]. We found a layer-by-layer growth mode but no surface reconstruction in contrast to Ge on the same face. So later, we switched to the Ag(110) face. When we deposited Si on this face at room temperature (RT) it was a great surprise to discover a striking arrangement of massively parallel silicon nano-ribbons [28] (SiNRs) perfectly aligned along the $[\bar{1}10]$ direction of the substrate, as shown in Fig. 9.1. Unambiguously, these first results have been the starting point of a new field of research: the study of the growth of silicon on silver. Indeed, a few years later, by depositing at a higher substrate temperature, silicon nanostructures showing the honeycomb atomic structure, so-called today silicene, were first observed on Ag(100) [29] at 500 K, then on Ag(110) [30] and finally on Ag(111) [31]. Today, it is

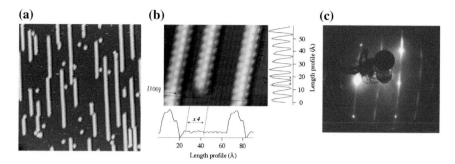

Fig. 9.1 Topographic images and LEED pattern observed after Si deposition on Ag(110) at RT. **a** 42 × 42 nm² filled states STM image showing an overview with SiNRs and nanodots, **b** 10.2 × 10.2 nm² filled-states STM image. The *line* profile indicates the ×2 periodicity of the SiNRs along the [$\bar{1}$10] direction. The *line* profile along the [100] direction shows the SiNRs width and their asymmetric shapes. **c** Typical LEED pattern (Ep = 43 eV) (Adapted from [34])

accepted that the silicon layers form a honeycomb structure (silicene structure) on both Ag(111) and Ag(100), while the atomic structure is not yet elucidated and still under debate on Ag(110).

In the following we recall and discuss the main experimental results and to a less extent, the theoretical approaches devoted to silicon nano-ribbons (SiNRs) grown on the Ag(110) face.

9.2 The Growth Conditions of Silicon Nano-ribbons: Morphologies as a Function of Temperature and Silicon Amount Deposited

9.2.1 Deposition of Silicon

It is important that the deposition of silicon is performed in ultra-high vacuum conditions due to the reactivity of silicon to oxygen. The Ag(110) surface preparation is standard: ion bombardment to clean the surface followed by annealing at 700–750 K for few hours to produce a perfectly clean and flat surface. The silicon deposition can be carried out either by an electron bombardment source or by Joule heating of a small plate of silicon wafer to a temperature between 1500 and 1550 K depending on the deposition rate. It has been shown that the nature of the type of source does not play a role on the superstructures observed [32]. When using a silicon wafer, the evaporator must be placed far from the sample (~ 30 cm) in order to avoid heating the sample by radiation. This can be very important as is has been shown that the substrate temperature is the key parameter for the silicon nanostructure growth [32]. The main discrepancies between different studies are likely due to this radiation effect of the silicon source and more generally to the intrinsic difficulty to determine the exact temperature of the surface. Concerning the amount of silicon deposited, since the atomic structure of the SiNRs is not yet resolved, the knowledge in terms of number of silicon atoms per cm^2 is crucial, and it has been measured only very recently [33]: full coverage of the surface corresponds to about 0.8 monolayer (ML) with respect to the Ag(110) face (1 ML corresponds to the Ag (110) surface atoms density).

Figure 9.1a shows isolated SiNRs obtained after a small amount ($\sim$0.25 ML) of silicon deposited at RT (likely a higher temperature due to radiation heating by the silicon source which was not take into account in this first study) [28]. All SiNRs appear perfectly organized at the atomic scale, and are aligned along the [$\bar{1}$10] direction of the silver substrate. All have the same height ($\sim$0.2 nm), the same width (1.6 nm i.e. $\sim$4 times the distance between two silver dense rows) and different lengths (from 1.5 nm for the shortest to $\sim$30 nm for the longest). It has been assumed that the SiNRs are constructed by self-assembly of individual building blocks [34]. The length of the SiNRs increased after a few minutes of annealing at about 500 K [28]. Figure 9.1b shows a 3D view of SiNRs. The line scan in the [$\bar{1}$10] direction shows a ×2 periodicity corresponding to 2 times the Ag-Ag interatomic distance (0.58 nm). This ×2 periodicity was observed on both sides of the SiNRs but was shifted by one silver interatomic distance giving rise to symmetry breaking relative to a plane perpendicular to the (110) surface through the middle of SiNRs.

The line profile perpendicular to the SiNRs (Fig. 9.1b) also showed a shape asymmetry with a deep depression on the right side looking as if a Ag missing row was produced, which does not exist on the left side. This deep depression is observed on the same side on all SiNRs of large areas. SiNRs with the depression on the left side was also observed on another area of the sample in agreement with the mirror symmetry of the Ag(110) face. Unfortunately, there are no STM images showing the co-existence of both structures. This has been attributed to the very large size of the left- and right-handed domains, which are estimated to be of several microns square [34]. These large domains suggest a surprising and very important cross-talk between isolated SiNRs, due to a strong strain effect in this direction [34].

Fig. 9.2 Filled states STM image (24 nm^2) showing self-assembled SiNRs forming a (5 × 2) superstructure and the corresponding LEED pattern (Ep = 68 eV) (Reproduced from [35])

Figure 9.1c shows the LEED pattern that is characteristic of isolated SiNRs. In addition to the spots of the Ag(110) surface, thin streaks develop through the substrate spots as well as at half-order positions along the perpendicular direction.

When the growth is performed at higher temperature, between 470 and 500 K, the SiNRs self-assemble to form a 1D grating with a periodicity of ~ 2 nm forming a (5×2) superstructure (Fig. 9.2a) [35]. The associated LEED pattern with the spots characteristic of the (5×2) periodicity, is shown in Fig. 9.2b.

Soon after we obtained this first result, the growth of narrower SiNRs (0.8 nm wide) [36–38] locally self-assembled in a (3×2) superstructure and the formation of $(5 \times 4)/(5 \times 2)$ superstructures were reported. In order to clarify these different results Colonna et al. [32] have performed a systematic study of the Si growth on Ag(110) using LEED and STM characterization tools. After an accurate calibration of the sample temperature, it has clearly been shown that temperature and silicon coverage are the key parameters controlling the width and the self-organization of the SiNRs. We discuss in the following their main results that can be considered as references, due to the rigorous methodology used.

The deposition of silicon (~ 0.2 ML) at RT produces mostly isolated SiNRs that are 0.8 nm wide. An increase in the amount of silicon leads to a disordered surface probably linked to low Si atom surface diffusion. At 430 K the silicon deposition (~ 0.6 ML) leads again to SiNRs 0.8 nm wide, but longer and with a tendency to self-assemble to form a (3×2) superstructure. At 490 K the silicon deposition (0.6 ML) leads to SiNRs 1.6 nm wide, that are isolated or locally self-assembled. At 490 K and full coverage (0.8 ML) they form a perfect grating covering the entire surface forming a (5×2) superstructure as shown in Fig. 9.2a. At the same temperature when the amount of silicon exceeds the full coverage (~ 1 ML) the LEED pattern also shows a new superstructure which was indexed as a $c(8 \times 4)$ superstructure. The STM image of this new flat superstructure is shown in Fig. 9.6 of [32].

9.2.2 Discussion on the STM Images Morphologies

Figure 9.3 shows a typical STM image with SiNRs that are 1.6 nm wide and 0.8 nm wide. Both types consist of 4 or 2 lines (0.8 or 1.6 nm wide) of round protrusions in the [$\bar{1}$10] direction of the silver surface, with a distance between the protrusions

Fig. 9.3 Filled-state STM image showing the co-existence of SiNRs 0.8 nm and SiNRs 1.6 nm wide

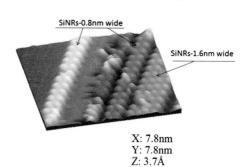

SiNRs-0.8nm wide

SiNRs-1.6nm wide

X: 7.8nm
Y: 7.8nm
Z: 3.7Å

being two times the Ag-Ag interatomic distance (the ×2 periodicity). Independent of the width of the SiNRs, each line is shifted from its neighbor line by one Ag-Ag interatomic distance which induces a symmetry breaking of the SiNRs. Typical STM images do not present the depression on one side, contrary to the ones shown on Fig. 9.1b (SiNRs 1.6 nm wide).

It has been assumed in [32] that the STM images of Fig. 9.1a, b would be due to a resolution loss of the two lines located in the center. This assumption could be correct but it does not explain the depression on one side giving the symmetry break of the profile shown in Fig. 9.1b. Because the SiNRs with a depression on one side were never observed when the SiNRs are self-assembled in (3 × 2) or/and (5 × 2) superstructure, they could be also a signature of an intermediate structure (meta-stable) growing at an intermediate temperature in relation to the formation of the Ag missing row as has been recently shown by Surface X-Ray Diffraction [33].

9.3 The Different Atomic Structures Proposed for the SiNRs

9.3.1 Model 1

The first atomic model of the SiNRs which has been proposed (C. Léandri PhD thesis [25]) was based on the assumption that the perfect alignment of the SiNRs along the [$\bar{1}$10] direction of the Ag substrate is due to the perfect match between four interatomic distances of silver (4 × d_{Ag-Ag} = 1.156 nm) and three unit cells of the (111) surface of silicon (1.152 nm) (Fig. 9.4). Unfortunately this model leads to a ×4 periodicity in the direction of the dense rows of silver in conflict with LEED and STM characterizations. It corresponds to a density of 1.5 ML when forming the (5 × 4) superstructure.

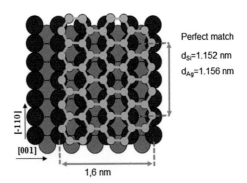

Fig. 9.4 First atomic model proposed of the SiNRs. It is based to the perfect match between four interatomic distances of silver (1.156 nm) and three unit cells of the (111) surface of silicon (1.152 nm). Si atoms are in *grey* and the top most silver atoms are in *dark blue* (Adapted from C. Léandri PhD thesis [25]) (Color figure online)

9.3.2   Model 2

Shortly after our first publication [28], a theoretical model was proposed by He [39] using Density Functional Theory (DFT). Instead of a hexagonal pattern, He shows that the most stable structure under different coverages is a rectangular arrangement of silicon atoms made of two layers: the first one contains four silicon atoms placed in hollow sites and the second one is composed of silicon dimers placed perpendicular to the dense rows of silver. The atomic structure which has been calculated as a (5 × 2) superstructure corresponds to a density 1.2 ML. This model, which takes into account the ×2 periodicity, fails to explain the symmetry breaking observed in all the STM images. It also does not explain the self-limitation of the lateral size which is always a multiple of two times the distance between the dense silver rows (~ 0.8 nm) (Fig. 9.5).

9.3.3   Model 3

Using DFT calculations in a GGA approximation, Kader et al. [30, 40, 41] have proposed an atomic model based on a silicon hexagonal structure (silicene). Starting from an STM image showing the honeycomb structure with four hexagons aligned of $\sim 30°$ with respect to the dense rows of silver (shown in Fig. 9.6a) they have calculated a relaxed atomic structure of the SiNRs (Fig. 9.6b).

After full relaxation of the system, the silicon honeycomb structure is preserved but presents a strong buckling. This buckling gives rise to an asymmetric corrugation in the charge density profile (Fig. 9.6c), in good agreement with STM images (Fig. 9.1b). The side view (Fig. 9.6d) shows that the SiNRs present an arched shape due to silicon atoms that are situated on the edge and are more bound to silver than the silicon atoms located in the middle of the SiNRs. This model, which corresponds to a density of 1.5 ML, seems to be in quite good agreement with the

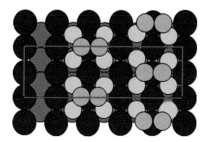

Fig. 9.5 Most stable structure proposed by G. He from a DFT study. Si atoms placed in the hollow sites are in *light blue*, silicon dimers on *top* are in *yellow* and the *top* most silver atoms are in *dark blue* (Adapted from [39]) (Color figure online)

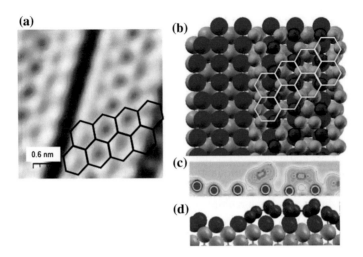

Fig. 9.6 **a** High resolution STM image showing the honeycomb structure. **b, c** charge density in a plane perpendicular to the (110) plane cutting the SiNRs across the width. **d** *Side view* showing the "arch-shaped" of the SiNRs (Si atoms are in *red* and the top-most silver atoms are in *dark blue*) (Reproduced from [30])

different STM images in terms of symmetry and shape. Unfortunately this model also fails in terms of periodicity along the [$\bar{1}$10] of silver which is ×2 and not ×4 as determined by this model. Nevertheless, this has been at the origin of the idea that the growth of silicon on silver substrates could give rise to the possible formation of silicene [30] (See also Chap. 10 for further information on Model 3).

9.3.4 Model 4

Also, based on a high resolution STM image (Fig. 9.7b) showing that the SiNRs 1.6 nm wide are formed by assembling two SiNRs 0.8 nm wide, Masson et al. [42] proposed the model reported in Fig. 9.7c. Compared to He's model (model 3), a row of silicon atoms has been added on the edges which are assumed to be at the origin of the lateral binding between the SiNRs 0.8 nm wide.

This atomic model which takes into account the ×2 periodicity along the [$\bar{1}$10] direction of silver, at the atomic scale does not show the symmetry breaking observed on the STM images. To get around this contradiction it is assumed that the typical protrusions observed (Fig. 9.7a) correspond to a groups of three silicon atoms as shown in Fig. 9.7c. This model is yet to be confirmed by other STM images and theoretical calculations. This model corresponds to a density of ∼ 1.6 ML.

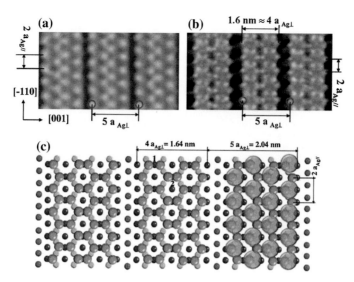

Fig. 9.7 STM image showing two SiNRs grown on Ag(110). Both images correspond to the same SiNRs imaged with different tunneling current and sample bias voltage **a** I_t = 3.2 nA; V_{sample} = 1.5 V and **b** I_t = 3.7 nA; V_{sample} = 86 mV. **c** Proposed atomic arrangement of the two SiNRs shown in Fig. 9.5b (more explanations in [43]). The *circles* on the right-hand of the SiNRs indicate a correspondence between the protrusions visible in (**a**) and groups of the three silicon atoms (Reproduced from [43])

9.3.5 Model 5

Using first principles calculations Lian and Ni [44] studied different geometrical configurations of the SiNRs. They considered SiNRs with zigzag and armchair edges with different widths from 1.1 to 1.9 nm. They found that the most stable structures are always the SiNRs that are 1.6 nm wide. Figure 9.8 shows both structures with zigzag and armchair edges after full structure optimization. For the zigzag structure, they find an atomic structure close to the one of Kara et al. [30] with the same arch shaped profile (side view of Fig. 9.8b). The most stable structure corresponds to the armchair configuration (Fig. 9.8a) where one out of two silver atom is removed on the left edge, producing the chirality of the SiNRs in good agreement with the experiment. Unfortunately, the simulation of the STM images does not show the four characteristic lines or protrusions which must be shifted from each other by one Ag-Ag interatomic distance (see Fig. 4a of [44]). The armchair configuration corresponds to a density of 1.6 ML.

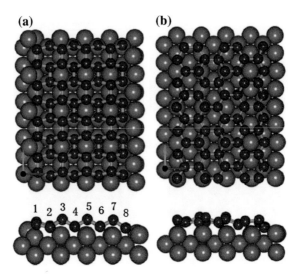

Fig. 9.8 Calculated structures of **a** armchair and **b** zigzag SiNRs on Ag(110) after full optimization. Si atoms are in *blue* (Reproduced from [44]) (Color figure online)

9.3.6 Model 6

Tchalala et al. [45] have proposed an equivalent atomic model of SiNRs with arm-chair edges respecting the ×2 periodicity in the [$\bar{1}$10] direction of silver. Starting from a high resolution STM image but where all silicon atoms are not seen, they propose a honeycomb arrangement of silicon atoms as shown in Fig. 9.9. In comparison with the previous model (Fig. 9.8a), the honeycomb structure seems

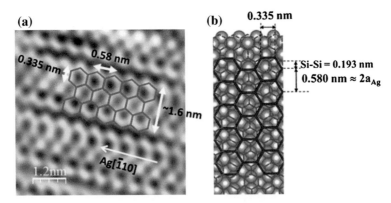

Fig. 9.9 **a** Atomically resolved filled-states STM image of SiNRs grown at 500 K on a Ag(110) surface (6×6 nm^2, V = −0.2 V, I = 1.9 nA). **b** Proposed model where the *grey* and *yellow* atoms correspond to the first and the second silver layers, respectively. The *blue* honeycomb structure corresponds to the internal structure of SiNRs while the *red* one corresponds to the structure observed on the STM image (Reproduced from [45]) (Color figure online)

to be less strained in the [001] direction of silver due to the addition of two extra rows of hexagons. This geometrical atomic model is proposed with the assumption that the Si–Si interatomic distances are preserved by a higher buckling.

This model still needs to be confirmed by other STM images and by theoretical calculations. This model corresponds to a density of ~ 2.2 ML.

9.3.7 Model 7

Finally, Bernard et al. [33] using STM and grazing incidence X-ray diffraction (GISAX) reported an in situ study of the evolution of the silver surface during the growth of the SiNRs. They show that during the growth of the SiNRs there is a release of silver atoms equivalent to one silver atom for two silicon atoms. From GISAX measurements they showed the existence of a missing row reconstruction underneath the SiNRs (one missing row for SiNRs 0.8 nm wide and two missing rows for SiNRs 1.6 nm wide). They do not propose an atomic model but they have accurately measured the silicon density at full coverage i.e. when all the surface is covered by SiNRs self-assembled in the (5 × 2) superstructure, to be 0.8 ML.

9.3.8 Discussion

Amongst all the proposed structures, there is no model which explains all the experimental observations. If the silicon density at full coverage is really 0.8 ML, as determined by Bernard et al. [33], all the atomic structures proposed and evoked above are not valid since they all correspond to a much larger density (1.2 ML for the lowest corresponding to the model 2) and furthermore do not take into account the silver missing rows.

Using the same approach as the one of He [39], we think that a systematic calculation of the structure taking into account the two (one) silver missing rows and the amount of silicon in the (5 × 2) (3 × 2) unit cell should allow one to determine the correct structure. From a general point of view, it is quite surprising that the use of a (111) plane of silicon (more or less buckled) explains all the superstructures observed on Ag(111) [46] and Ag(100) [29] faces, but fails for the Ag(110) face.

There is another possibility that has not been explored so far: the formation of an ordered surface alloy (*silicide*). Indeed, two metastable phases have been reported in rapidly quenched Ag-Si liquid alloys. The first one presents an hcp structure with a = 0.287 nm, c = 0.4528 nm [47]. The second one appears to be quite different from the hcp phase, even though the cooling rates are comparable in both cases [48]. This last intermetallic ordered phase (Lave-type phase) presents an orthorhombic structure with the lattice parameters a = 0.56 nm, b = 0.91 nm and c = 0.85 nm. It is interesting to note that one side of the parallelepiped matches quite well a SiNR 0.8 wide, since the 'a' parameter is close to $2 \times d_{Ag-Ag} = 0.578$ nm (which could explain the ×2 periodicity) and the 'c' parameter is close to the distances between two dense rows (0.81 nm) and could explain the "magic" width of the SiNRs which are always a multiple of two dense rows. Then, the SiNRs could be an ordered surface alloy corresponding to a high temperature metastable phase in 3D but stabilized as a 2D mono-layer at lower temperature by the geometry of the (110) face of silver. This assumption should deserve special attention since there is no strong contradiction with all the experimental result such as the widths observed for the SiNRs, the low silicon density at full coverage, the silver missing rows, the resistance to the oxidation, and the Raman measurements (as will be discussed below).

9.4   Electronic Properties

9.4.1   Experimental Measurement of the Electronic Properties of the Si NRs

The electronic properties of the SiNRs have been studied by high resolution, angle integrated (AI) or angle-resolved (AR) photoelectron spectroscopy (AI-PES; AR-PES) using synchrotron radiation. They have also been studied by scanning tunneling spectroscopy (STS) working at room temperature (RT) and/or low temperature (LT).

After the deposition of silicon (~ 0.4 ML) at RT on the Ag(110) face, angle integrated PES measurements of the valence bands and of the Si 2p core-levels have been performed using synchrotron radiation. A typical Si 2p spectrum is reported in Fig. 9.10a fitted with two peaks using standard procedure. The two components are remarkably narrow which is in line with the perfect atomic order of the SiNRs. The two components indicate the existence of two non-equivalent silicon environments. It has been shown that the shape of the Si 2p core level spectra are very similar for the 0.8 and 1.6 nm wide SiNRs, forming respectively the (3 × 2) and (5 × 2) superstructures [37]. Only small variations in the relative intensities of the different peaks are observed showing that the Si atoms have almost the same atomic environment and share a similar atomic structure independent of the width [37].

The strong asymmetric shape of the Si 2p peaks is a signature of the metallic character of the SiNRs [28, 34]. This metallic character is also revealed by an increase in the density of states at the Fermi level compared to that of the bare silver substrate (Fig. 9.10b). On this curve, one also observes the presence of four new states S4, S3, S2 and S1 at ~ 3.12, ~ 2.37, ~ 1.45 and ~ 0.92 eV from the Fermi level, respectively. These states have been attributed to quantum well states originating from the lateral confinement within the narrow width of the SiNRs [28]. The 1D character of this states has been confirmed by AR-PES showing that all these states disperse along the direction of the SiNRs but not in the orthogonal direction

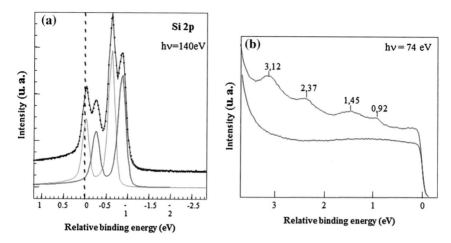

Fig. 9.10 Al-PES measurements of isolated SiNRs **a** Si 2p core level spectrum (*dots*) and its synthesis (*solid line* overlapping the data points) with two asymmetric components (*bottom curves*). **b** Normal incidence valence band spectra limited to the sp region for pristine Ag(110) surface (*black curve*) and after the silicon deposition (*red curve*) (Reproduced from [28]) (Color figure online)

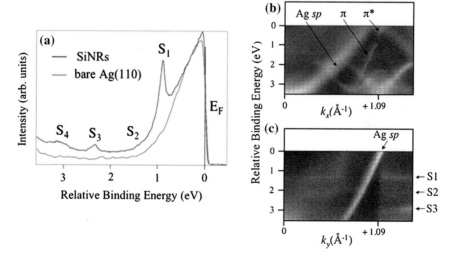

Fig. 9.11 Energy distribution curves for bare Ag(110) and for the array of SiNRs (**a**); band dispersion for the array of SiNRs versus k_x (along the SiNRs) at $k_y = 0.7$ Å^{-1} (**b**) and versus k_y (perpendicular to the SiNRs) at $k_x = 0.35$ Å^{-1} (**c**). The photon energy used was $h_v = 78$ eV (Reproduced from [49])

(Fig. 9.10b, c) [49, 50]. Along the [$\bar{1}$10] direction of silver, we notice a quasi-linear dispersion of the S1 state near the $\bar{X}$ point of the silver surface Brillouin Zone ($k_x = \pm 1.09$ Å^{-1}) (Fig. 9.11b).

On Fig. 3 of [49], by analogy with the behavior of graphene grown on different metallic substrates [51–53] it was tempting to interpret the V-like shape (top) and a Λ-like (bottom) shape, separated by a gap (~ 0.56 eV), as 1D projections of the π^* (upper branch) and π (lower branch) Dirac cones at the K-point of silicene. This interpretation requires that the SiNRs have the honeycomb atomic structure (silicene) oriented with zigzag edges and with 3 hexagons of silicon coinciding with 4 silver interatomic distances (leading to a ×4 periodicity). As mentioned above, this atomic model is probably not correct (mainly due to its wrong periodicity) which means that the linear dispersion observed cannot be the signature of a Dirac cone. Recent theoretical calculations confirmed that the Λ-like shape observed by AR-PES measurements are not due to silicon but to the silver substrate, as an effect of band folding induced by the silicon layer periodicity [54].

The quantum confinement signature (quantum well states) in the SiNRs has been also observed at low temperature by STM/STS in the empty states but not in the filled states [55].

9.4.2 sp^2 Signature of the SiNRs

The nature of the hybridization of the silicon valence band orbitals has been experimentally studied by different groups using different methods. Here too, the conclusions of these different studies are contradictory.

Using reflection electron energy loss spectroscopy, De Padova et al. [56] observed a strong analogy between self-assembled SiNRs grown on Ag(110) and the surface of pyrolytic graphite. The angular dependence of the core electron energy loss at the silicon K-edge (1.840 keV) presents two energy loss structures located at 1.8351 and 1.8398 keV respectively. On a graphite surface, the two equivalent energy loss structures are attributed to the transitions from the carbon 1s to π^* and to σ^* orbitals which are characteristic of the sp^2 hybridization. The two energy loss peaks recorded for the SiNRs which do not exist for Si bulk are then attributed to the transitions from the silicon 1s to π^* and to σ^*, suggesting by analogy with graphite, an sp^2-like hybridization of the Si valence orbitals.

The second study is based on the optical properties of SiNRs experimentally determined by in situ surface differential reflectance spectroscopy [57]. From the strong difference between the reflectance spectrum measured on a self-assembled SiNRs layer and a reflectance spectrum calculated for a silicene layer on silver, the authors concluded that the SiNRs display preferential sp^3 hybridization as in amorphous silicon and not partial sp^2 hybridization as expected for silicene NRs.

An equivalent conclusion has been given by Speiseret et al. [58] who studied by Raman spectroscopy the same (5 × 2) superstructure formed by self-assembled SiNRs. The large difference between the experimental Raman spectra and the theoretical calculations of phonon frequencies of freestanding SiNRs, rule out the assumption that the SiNRs have a honeycomb structure. Furthermore, the Raman signature recorded in this study shows features similar to the ones recorded in the case of small silicon nanoclusters. The authors suggest that the atomic models of SiNRs grown on Ag(110) should take into account this possible formation of silicon nanoclusters with a two Ag missing rows model observed recently.

9.5   3D Nano-ribbons: Stacked Silicene Layers or a Surface Faceting Phenomenon?

During her PhD thesis, C. Léandri [25] observed for temperatures ~ 500 K the growth of extremely long, large and thick SiNRs, perfectly aligned along the [$\bar{1}$10] direction of silver (results mentioned in [28]). Figure 9.12 shows typical 3D nano-ribbons with trapezoidal or triangular cross sections. Their lengths can reach more than several hundred nanometers. Their lateral facets display a stepped structure with a vertical periodicity of 0.3 nm, as revealed by the transverse line profile-1 shown in Fig. 9.12a. On each step we observed a perfect ×2 periodicity in the [$\bar{1}$10] direction of silver (line profile-2 in Fig. 9.12a).

Recently, this kind of 3D nano-ribbon has been interpreted as a multilayer stacked silicene [59] on one hand and in another by a faceting of the (110) surface of the silver substrate [60]. The assumption of a stacking of silicene layers is mainly supported by ARPES measurements taken along the [$\bar{1}$10] direction of silver ($\bar{\Gamma} \rightarrow \bar{X}$ direction) showing linear band dispersions (see Fig. 3a in [59]). The authors use again the analogy with graphene grown on different substrates [51–53] to interpret the V-like shape and the Λ-like shape observed below the Fermi level (see Fig. 3a of [59]). As for SiNRs, this model suggest first that each silicene layer forming the 3D SiNRs is oriented with its hexagon apexes aligned along the silver [$\bar{1}$10] direction, and second, that it has a ×4 periodicity relative to silver along this direction. Because the ×4 periodicity is shown in [59] only in two limited areas on

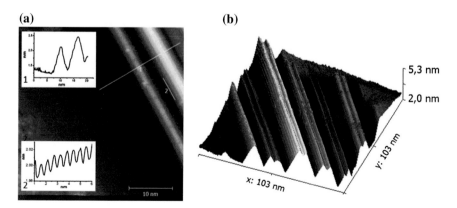

Fig. 9.12 Topographic images of 3D nano-ribbons grown at 500 K as observed by C. Léandri during her PhD thesis [25] (-1.1 V; 2.8 nA). **a** and **b** Correspond to different sample areas

their STM image (Fig. 1b in [59]) and no explanation is given concerning the perfect ×2 periodicity along each step of the 3D SiNRs, we think that the conclusions of this study are highly speculative.

Ronci et al. [60] have performed a more accurate and detailed study at different growth temperatures and with different silicon coverages. They show by STM and LEED that there are two different kinds of 3D nanostructures: nanodikes (NDs) and nanotrenches (NTs) engraved in the Ag(110) substrate. At 490 K these 3D nanostructures coexist with the 2D SiNRs forming the (5 × 2) superstructure, meanwhile at higher temperatures (550 K) only the 3D nanostructures are obtained independently of the silicon deposited. A detailed analysis of the STM images of all NDs and NTs shows that the lateral facets have always the same slope (about 19.5°) with respect to the Ag(110) surface plane and have on each step a ×2 periodicity along the silver [$\bar{1}$10] direction. Furthermore, according to the STM images, on top of NDs and on the base of NTs, bare surfaces of Ag(110) are observed with SiNRs identical to the ones grown on the flat Ag(110) surface (Fig. 4 of [60]). From these observations they propose that the formation of NDs and NTs are a faceting phenomenon i.e. the top and base parts of such nanostructures are Ag(110) planes, while their sides are Ag(221) and Ag(22$\bar{1}$) planes with ×2 periodicity along the [$\bar{1}$10] direction of silver. The atomic model proposed from DFT calculations is in very good agreement with the STM observations (see Fig. 7 in [60]).

9.6 Nano-ribbon Reactivity

9.6.1 Oxidation

The oxidation process of isolated SiNRs has been probed at the atomic scale by STM/STS and PES (using synchrotron radiation) [61]. It has been shown by STM that the oxidation process starts at the extremities of the SiNRs (or by a defect along the NRs) and develops with the oxygen exposure time, along their length like a "wildfire" (called "burning match process" in [61]). This process is well confirmed by the evolution of the Si 2p spectra with increasing oxygen exposure. Five distinct peaks are observed during the oxidation process; two are characteristic of an SiO_2 compound and pure silicon. With increasing exposure to the oxygen, the oxide peak increases while the pure silicon peak decreases. The three others peaks which are characteristic of sub-oxides, remain constant during the oxidation process indicating that they are situated at the interface between the pure silicon and the silicon oxide (SiO_2). Note that the oxidation process of the SiNRs self-assembled in the (5 × 2) grating, starts at higher oxygen exposure (its reactivity is found to be 10^4 less than that of the Si(111) 7 × 7 surface [62]).

9.6.2   Hydrogenation

The reactivity of SiNRs to hydrogen adsorption has been studied using LEED, STM, EELS and PES [37, 63]. It has been shown that contrary to the oxidation process which starts on structural defects and at the extremities of the SiNRs, the attack by hydrogen is not a site selective process. The presence of multihydrides $(Si–H_x)$ phases at the earliest stage of the hydrogen adsorption reveals the formation of covalent bonds between atomic H and the Si atoms. Upon hydrogen adsorption the SiNRs are destroyed into small clusters and lose their metallic character. Eventually, at very high hydrogen exposures, the complete etching of and removal of the SiNRs is observed.

9.6.3   Functionalization (Used as a Template)

The perfect organization of the SiNRs covering uniformly the entire Ag(110) surface and forming the (5×2) superstructure can be used as template. By incorporating a wide variety of elements (organic or inorganic) we can expect the growth of nano-ribbons with new magnetic, electronic, photonic or catalytic properties. In other words the SiNRs forming the (5×2) superstructure can be used as a template to grow new materials with perfect control of the pattern.

Using STM, Salomon et al. [36] have shown that depending on the molecular functional groups, the organic molecules adsorb either randomly on the substrate or preferentially on the SiNRs. In the latter case, chemisorption of suitable organic molecules on the SiNRs leads to a well-defined one-dimensional aggregation and changes the metallic character of the SiNRs to a semi-conducting one.

Concerning the deposition of magnetic elements, at room temperature, the adsorption of Co occurs on top sites of the SiNRs resulting in the formation of nanolines of Co dimers [43, 64]. A nearly layer-by-layer growth mode is observed after increasing the Co coverage resulting in a film composed of 1D Co nanostructures up to 5 ML [65]. Characterization of the magnetic properties of these films has been achieved by X-ray magnetic circular dichroism [66]. The first atomic Co layer (directly on top of Si nanoribbons) presents a weak magnetic response. The magnetization increases for the second Co layer suggesting a ferromagnetic ordering with an in-plane easy axis of magnetization, which is perpendicular to the Co nanolines. For Mn deposition [67], the process seems to be different. The Mn atoms first interact with the side of two adjacent nanowires and grow on the side with increasing coverage. There is no coupling between the Mn atoms as the nanoribbon is metallic. Such results are very promising for using SiNRs as templates.

9.7 Conclusions

In this review, after reminding the reader of the history of the study of the growth of silicon on silver, we have recalled the main experimental results obtained by different groups. Today there are two tendencies of thought in the scientific community: those who believe that the nano-ribbons are pure silicene and those that think that they are not silicene layers. Despite the lack of correct interpretation of the structure of these nanoribbons they show very exiting characteristics: they are highly ordered with large form factors. They have shown quantum electron confinement, a low reactivity to oxidation and are already used as extremely regular templates at the atomic scale. Concerning the atomic structure of the SiNRs, the formation of an ordered 2D silicide, comparable to the metastable 3D silicide, is probably a research direction that needs to be explored.

References

1. G. Guzmán-Verri, L. Lew Yan Voon, Phys. Rev. B **76**, 075131 (2007)
2. S. Lebègue, O. Eriksson, Phys. Rev. B **79**, 115409 (2009)
3. D. Chiappe, E. Scalise, E. Cinquanta, C. Grazianetti, B. van den Broek, M. Fanciulli, M. Houssa, A. Molle, Adv. Mater. **26**, 2096 (2014)
4. A. Kara, H. Enriquez, A.P. Seitsonen, L.C. Lew Yan Voon, S. Vizzini, B. Aufray, H. Oughaddou, Surf. Sci. Rep. **67**, 1 (2012)
5. U. Bardi, Reports Prog. Phys. **57**, 939 (1994)
6. P. Wynblatt, R.C. Ku, Surf. Sci. **65**, 511 (1977)
7. A. Senhaji, G. Tréglia, J. Eugène, A. Khoutami, B. Legrand, Surf. Sci. **288**, 371 (1993)
8. G. Tréglia, B. Legrand, F. Ducastelle, Europhys. Lett. **7**, 575 (1988)
9. J.M. Roussel, G. Tréglia, B. Legrand, Solid State Phenom. **172–174**, 1008 (2011)
10. J. Eugène, B. Aufray, F. Cabané, Surf. Sci. **241**, 1 (1991)
11. J. Eugène, G. Tréglia, B. Legrand, B. Aufray, F. Cabané, Surf. Sci. **251–252**, 664 (1991)
12. M. Lin, X. Chen, X. Li, C. Huang, Y. Li, J. Wang, Appl. Surf. Sci. **297**, 130 (2014)
13. J.Y. Wang, J. Du Plessis, J.J. Terblans, G.N. Van Wyk, Surf. Sci. **423**, 12 (1999)
14. R. Dudde, H. Bernhoff, B. Reihl, Phys. Rev. B **41**, 12029 (1990)
15. J.S. Tsay, a. B. Yang, C.N. Wu, F.S. Shiu, Surf. Sci. **601**, 4265 (2007)
16. C. Polop, C. Rojas, J. Martín-Gago, R. Fasel, J. Hayoz, D. Naumović, P. Aebi, Phys. Rev. B **63**, 115414 (2001)
17. J. Martin-Gago, R. Fasel, J. Hayoz, Phys. Rev. B **55**, 896 (1997)
18. B. Lalmi, C. Girardeaux, A. Portavoce, J. Bernardini, B. Aufray, Defect Diffus. Forum **289–292**, 601 (2009)
19. B. Lalmi, M. Chorro, R. Belkhou, J. Appl. Phys. **114**, 063505 (2013)
20. A. Goldoni, G. Paolucci, C. Rojas, F.J. Palomares, M.F. Lo, **456**, 778 (2000)
21. H. Oughaddou, PhD Thesis, Aix-Marseille Univ. Marseille, Fr. (1999)
22. J. Eugène, B. Aufray, F. Cabané, Surf. Sci. **273**, 372 (1992)
23. H. Oughaddou, B. Aufray, J.P. Biberian, J.Y. Hoarau, Surf. Sci. **429**, 320 (1999)
24. H. Oughaddou, B. Aufray, J.-P. Bibérian, B. Ealet, G. Le Lay, G. Tréglia, A. Kara, T.S. Rahman, Surf. Sci. **602**, 506 (2008)
25. C. Léandri, PhD Thesis, Aix-Marseille Univ. N°22042 (2004)
26. R.W. Oleslnski, A.B. Gokhale, G.J. Abbaschlan, Bull. Alloy Phase Diagrams **10**, 635 (1989)

27. C. Leandri, H. Saifi, O. Guillermet, B. Aufray, Appl. Surf. Sci. **177**, 303 (2001)
28. C. Leandri, G. Le Lay, B. Aufray, C. Girardeaux, J. Avila, M.E. Dávila, M.C. Asensio, C. Ottaviani, A. Cricenti, Surf. Sci. Lett. **574**, L9 (2005)
29. C. Léandri, H. Oughaddou, B. Aufray, J.M. Gay, G. Le Lay, A. Ranguis, Y. Garreau, Surf. Sci. **601**, 262 (2007)
30. B. Aufray, A. Kara, S. Vizzini, H. Oughaddou, C. Léandri, B. Ealet, G. Le Lay, Appl. Phys. Lett. **96**, 183102 (2010)
31. B. Lalmi, H. Oughaddou, H. Enriquez, A. Kara, S. Vizzini, B. Ealet, B. Aufray, Appl. Phys. Lett. **97**, 223109 (2010)
32. S. Colonna, G. Serrano, P. Gori, A. Cricenti, F. Ronci, J. Phys.: Condens. Matter **25**, 315301 (2013)
33. R. Bernard, T. Leoni, A. Wilson, T. Lelaidier, H. Sahaf, E. Moyen, L. Assaud, L. Santinacci, F. Leroy, F. Cheynis, A. Ranguis, H. Jamgotchian, C. Becker, Y. Borensztein, M. Hanbücken, G. Prévot, L. Masson, Phys. Rev. B **88**, 121411 (2013)
34. P. De Padova, C. Quaresima, P. Perfetti, B. Olivieri, C. Leandri, B. Aufray, S. Vizzini, G. Le Lay, Nano Lett. **8**, 271 (2008)
35. H. Sahaf, L. Masson, C. Léandri, B. Aufray, G. Le Lay, F. Ronci, Appl. Phys. Lett. **90**, 263110 (2007)
36. E. Salomon, A. Kahn, Surf. Sci. **602**, L79 (2008)
37. M.E. Dávila, A. Marele, P. De Padova, I. Montero, F. Hennies, A Pietzsch, M.N. Shariati, J.M. Gómez-Rodríguez, G. Le Lay, Nanotechnology **23**, 385703 (2012)
38. P. De Padova, P. Perfetti, B. Olivieri, C. Quaresima, C. Ottaviani, G. Le Lay, J. Phys.: Condens. Matter **24**, 223001 (2012)
39. G. He, Phys. Rev. B **73**, 035311 (2006)
40. A. Kara, S. Vizzini, C. Leandri, B. Ealet, H. Oughaddou, B. Aufray, G. LeLay, J. Phys.: Condens. Matter **22**, 045004 (2010)
41. A. Kara, C. Léandri, M.E. Dávila, P. Padova, B. Ealet, H. Oughaddou, B. Aufray, G. Lay, J. Supercond. Nov. Magn. **22**, 259 (2009)
42. H. Sahaf, F. Dettoni, C. Léandri, E. Moyen, L. Masson, M. Hanbücken, Surf. Interface Anal. **42**, 687 (2010)
43. H. Sahaf, C. Léandri, E. Moyen, M. Macé, L. Masson, M. Hanbücken, EPL **86**, 28006 (2009)
44. C. Lian, J. Ni, Phys. B Condens. Matter **407**, 4695 (2012)
45. M.R. Tchalala, H. Enriquez, A.J. Mayne, A. Kara, G. Dujardin, M.A. Ali, H. Oughaddou, J. Phys: Conf. Ser. **491**, 012002 (2014)
46. H. Jamgotchian, Y. Colignon, N. Hamzaoui, B. Ealet, J.Y. Hoarau, B. Aufray, J.P. Bibérian, J. Phys.: Condens. Matter **24**, 172001 (2012)
47. T.R. Anantharaman., H.L. Luo, W. Klement, Nature **210**, 1040 (1966)
48. C. Suryanarayana, J. Less-Common Met. **35**, 347 (1974)
49. P. De Padova, C. Quaresima, C. Ottaviani, P.M. Sheverdyaeva, P. Moras, C. Carbone, D. Topwal, B. Olivieri, A. Kara, H. Oughaddou, B. Aufray, G. Le Lay, Appl. Phys. Lett. **96**, 261905 (2010)
50. M. A. Valbuena, J. Avila, M.E. Dávila, C. Leandri, B. Aufray, G. Le Lay, M. C. Asensio, Appl. Surf. Sci. **254**, 50 (2007)
51. J. Wintterlin, M.-L. Bocquet, Surf. Sci. **603**, 1841 (2009)
52. A. Grüneis, D.V. Vyalikh, Phys. Rev. B **77**, 193401 (2008)
53. A. Grüneis, K. Kummer, D.V. Vyalikh, New J. Phys. **11**, (2009)
54. P. Gori, O. Pulci, F. Ronci, S. Colonna, F. Bechstedt, J. Appl. Phys. **114**, 113710 (2013)
55. F. Ronci, S. Colonna, A. Cricenti, P. De Padova, C. Ottaviani, C. Quaresima, B. Aufray, G. Le Lay, Phys. Status Solidi **7**, 2716 (2010)
56. P. De Padova, C. Quaresima, B. Olivieri, P. Perfetti, G. Le Lay, Appl. Phys. Lett. **98**, 081909 (2011)
57. Y. Borensztein, G. Prévot, L. Masson, Phys. Rev. B **89**, 245410 (2014)
58. E. Speiser, B. Buick, N. Esser, W. Richter, S. Colonna, A. Cricenti, F. Ronci, Appl. Phys. Lett. **104**, 161612 (2014)

59. P. De Padova, O. Kubo, B. Olivieri, C. Quaresima, T. Nakayama, M. Aono, G. Le Lay, Nano Lett. **12**, 5500 (2012)
60. F. Ronci, G. Serrano, P. Gori, A. Cricenti, S. Colonna, Phys. Rev. B **89**, 115437 (2014)
61. P. De Padova, C. Leandri, S. Vizzini, C. Quaresima, P. Perfetti, B. Olivieri, H. Oughaddou, B. Aufray, G. Le Lay, Nano Lett. **8**, 2299 (2008)
62. P. De Padova, C. Quaresima, B. Olivieri, P. Perfetti, G. Le Lay, J. Phys. D Appl. Phys. **44**, 312001 (2011)
63. E. Salomon, T. Angot, C. Thomas, J.-M. Layet, P. Palmgren, C.I. Nlebedim, M. Göthelid, Surf. Sci. **603**, 3350 (2009)
64. F. Dettoni, H. Sahaf, E. Moyen, L. Masson, M. Hanbücken, EPL **94**, 28007 (2011)
65. L. Masson, H. Sahaf, P. Amsalem, F. Dettoni, E. Moyen, N. Koch, M. Hanbücken, Appl. Surf. Sci. **267**, 192 (2013)
66. L. Michez, K. Chen, F. Cheynis, F. Leroy, A. Ranguis, H. Jamgotchian, M. Hanbücken, L. Masson, Beilstein J. Nanotechnol. **6**, 777 (2015)
67. P. De Padova, C. Ottaviani, F. Ronci, S. Colonna, B. Olivieri, C. Quaresima, A. Cricenti, M.E. Dávila, F. Hennies, A. Pietzsch, N. Shariati, G. Le Lay, J. Phys.: Condens. Matter **25**, 014009 (2013)

Chapter 10
Computational Studies of Silicene on Silver Surfaces

Handan Yildirim and Abdelkader Kara

Abstract Silicon bulk presents a strong preference for sp^3 hybridization and hence will not form 2D materials spontaneously. Standalone silicene does not exist, however, theoretical studies have predicted silicene is stable *once* formed. During the last decade, several experimental groups achieved the synthesis of silicene as sheets and nanoribbons on a number of substrates, with silver being used most. In this chapter we review the computational studies of silicene adsorption on Ag(110) as nanoribbons, and on Ag(111) as nanosheets. On Ag(110), silicene nano-ribbons present a characteristic length of about 16 Å, with an arched shape. On Ag(111), several super-structures exist with different degrees of buckling, resulting in a variety of scanning tunnelling microscopy (STM) images, with remarkable resemblance to those experimentally obtained.

10.1 Introduction

The first theoretical investigation of freestanding silicene was performed using a tight-binding (TB) model where silicon was arranged in a flat honeycomb structure [1]. Following this study, the studies based on density functional theory (DFT) showed that the flat silicon honeycomb structure is meta-stable, presenting imaginary vibrational frequencies [2]. The authors explored a puckered structure where silicon atoms sit on two distinct planes separated by about 0.4 Å. The electronic structure calculations of this puckered structure showed that beside the non-flat geometry, the band structure of silicene presents a linear dispersion with a Dirac cone at the Brillouin zone corners, and consequently, as in graphene, the charge carriers behave as massless relativistic particles [2]. These observations

H. Yildirim
School of Chemical Engineering, Purdue University, Lafayette, IN, USA

A. Kara (✉)
Department of Physics, University of Central Florida, Orlando, FL, USA
e-mail: abdelkader.kara@ucf.edu

© Springer International Publishing Switzerland 2016
M.J.S. Spencer and T. Morishita (eds.), *Silicene*, Springer Series
in Materials Science 235, DOI 10.1007/978-3-319-28344-9_10

sparked a myriad of other investigations of standalone silicene. However, theoretical studies of silicene adsorbed on metal surfaces are scarce [3–12].

The results presented in this chapter are in close connection with the experimental data of the French group of Aufray-Oughaddou [13–27] (see Chaps. 8 and 9). Though historically, silicene nanoribbons (SiNRs) were first studied experimentally, we choose to present the silicene sheets first as these are the most studied ones.

10.2 Silicene Sheets on Ag(111)

Since the first experimental evidence of the synthesis of silicene on Ag(111) [20], several theoretical investigations on different structures of silicene on Ag(111) have been performed [3–12]. All of these studies showed that as silicene is slightly incommensurable with the Ag(111) substrate, several stable configurations co-exist. The consequence of this is a strongly corrugated silicene layer, ~ 1 Å, much larger than the intrinsic buckling of 0.4 Å of the freestanding silicene. In turn, the electronic structure of silicene is strongly altered as compared to that of freestanding silicene. It was found in most cases that the adsorbed silicene is metallic, and the Dirac cone is absent [28–30]. It was shown that the experimentally linear dispersion found experimentally [31, 32] is actually a feature of the Ag(111) electronic structure [28–30].

We have performed detailed studies of the atomic structure, energetics, and electronic structure of Si on Ag (111) and (110) surfaces. We have used a DFT approach by solving the Kohn-Sham equations using the Vienna ab initio simulation package (VASP) [33–35]. Exchange-correlation interactions are included within the generalized gradient approximation (GGA) in the Perdew-Burker-Ernzerhof (PBE) form [36]. The electron-ion interaction is described by the projector augmented wave method in its implementation of Kresse and Joubert [37, 38]. A plane-wave energy cut-off of 250 eV was used for all calculations, and is found to be sufficient for these systems. The bulk lattice constant for Ag is found to be 4.175 using a k-point mesh of $10 \times 10 \times 10$. The slab supercell approach with periodic boundaries is employed to model the surface, and the Brillouin zone sampling is based on the technique devised by Monkhorst and Pack [39]. The slab consists of 5 layers of Ag atoms. The number of Ag atoms per layer and those of silicon atoms depend on the silicene super-structure. The whole system is allowed to relax to the optimum configuration with forces on every atom less than 0.01 eV/Å.

Here, we present some of the structural characteristics of these silicene super-structure configurations on Ag(111); namely (4×4), $(2\sqrt{3} \times 2\sqrt{3})R30°$-I and II, he $(\sqrt{13} \times \sqrt{13})13.9°$ I and II, and $(\sqrt{7} \times \sqrt{7})R9.1°$ calculated using DFT.

We start with the (4×4) structure as shown in Fig. 10.1a (top view), b (side view). The unit cell contains 5 layers of silver each containing 16 silver atoms with 18 silicon atoms adsorbed in a honeycomb structure.

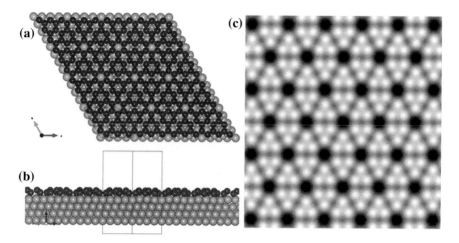

Fig. 10.1 The top (**a**) and side (**b**) views of the (4 × 4) structure, and (**c**) the calculated STM image

We note from Fig. 10.1a that the silicon atoms form a honeycomb structure, and sit at different heights from the surface. The surface atoms of the substrate also experience some changes in their positions, presenting a buckling of the top Ag layer of 0.4 Å. The nearest neighbour Si–Si distance in this structure is 2.35 Å, with a lateral distance of 2.23 Å, which would be measured in an STM image. For this configuration, the average binding energy per silicon atom is 0.46 eV. The Si atoms closest to the substrate bind more strongly than those Si atoms further away from the substrate. The corresponding STM image, shown in Fig. 10.1c, is in very good agreement with that observed experimentally [19].

The DFT optimized (4 × 4) structure is shown in Fig. 10.2 with different colors representing Si atoms with different heights from the surface. Most of the silicon atoms prefer to adsorb as close as possible to the silver substrate resulting in 11 (out of 18) silicon atoms about 2 Å above the substrate, followed by 3 Si atom at 0.55 Å above the first Si atoms, and finally the remaining 4 Si atoms occupy the height at 0.25 Å above the 3 Si atoms. The strong bonding between Si and Ag is also reflected in the corrugation of the top Ag(111) layer amounting to 0.4 Å. This large structural corrugation in the substrate top layer is a result of the relatively strong chemical interaction between silicene and Ag(111). The stacking of silicon atoms into 3 layers induces a corrugation in the surface charge density that leads to bright spots in the calculated STM image (see Fig. 10.1c).

A detailed analysis of the bond-angles can be used to single out the type of hybridization of the bond between inequivalent Si atoms in this system. Note that for the tetrahedral structure found in Si(111) with an sp^3 hybridization would yield a sum of all 3 angles ($\alpha_1 + \alpha_2 + \alpha_3$) of the pyramid equal to 328.4° (see Fig. 10.3); while an sp^2 hybridization like that in graphene would yield a sum of the angles equal to 360°. Our analysis of all bond-angles in the (4 × 4) structure showed a

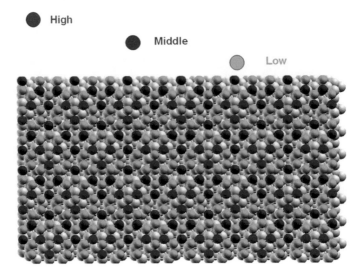

Fig. 10.2 Top-view of the (4 × 4) structure with color-coded atoms at different heights

Fig. 10.3 Top-view of a
pyramidal tetrahedron

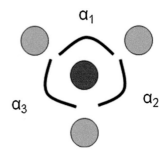

distribution of angles spanning from 320° to 354°. This large range of angles may be the reflection of a mixed sp^2 and sp^3 hybridization.

Let us now move to the second configuration that has been studied, namely the $(2\sqrt{3} \times 2\sqrt{3})R30°$ structure. This structure, unambiguously determined using LEED analysis, presents actually two different signatures in the STM images. We call them $(2\sqrt{3} \times 2\sqrt{3})R30°$- I and II, and they differ by a translation from each other. The first structure is shown in Fig. 10.4a (top), b (side) as top and side views. A structural analysis shows that for this configuration, there is a buckling of the silicene sheet amounting to about 1.5 Å, which is significantly higher than that found for the (4 × 4) structure. Changes in the electronic structure are indeed induced by this large structural corrugation resulting in a change of electronic character of the silicene sheet that is now metallic. Due to the large corrugation in the silicene sheet in this configuration, it was found that the Si–Si nearest neighbour distance is about 2.51 Å, however, due to large buckling, the lateral distance will

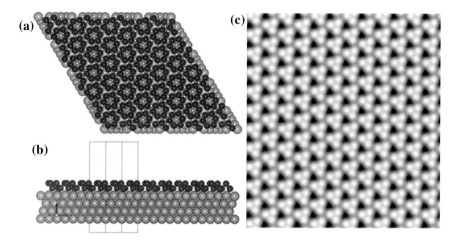

Fig. 10.4 The top (**a**) and side (**b**) views of the $(2\sqrt{3} \times 2\sqrt{3})$R30°-I structure, and (**c**) the corresponding STM image

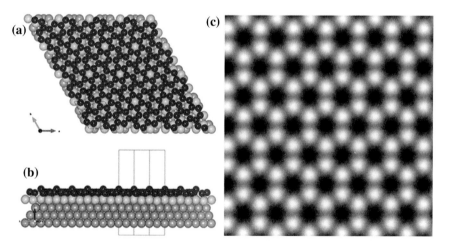

Fig. 10.5 The top (**a**) and side (**b**) views of the $(2\sqrt{3} \times 2\sqrt{3})$R30°-II structure, and (**c**) the corresponding calculated STM image

only be 1.99 Å—the one that is observed by an STM—comparable to that observed by Lalmi et al. [20].

The next superstructure presented here is the $(2\sqrt{3} \times 2\sqrt{3})$R30°-II, a simple translation with respect to the Ag(111) surface of the previous model, shown in Fig. 10.5a (top view), b (side view). Note that in this configuration, the hexagons formed by silicon atoms surround bridge and top sites.

Analysis of the atomic positions of the silicon atoms shows that the buckling in this configuration is about 1.0 Å, which is substantially smaller than that of the

$(2\sqrt{3} \times 2\sqrt{3})$R30°-I structure (about 1.5 Å). This smaller buckling results in a completely different STM image than that of the first structure, which again compares very well with the measured STM image [19]. Due to a smaller corrugation, here the nearest neighbour Si–Si distance varies between 2.28 and 2.37 Å, with a projected lateral distance of 2.0 Å, are again in good agreement with that observed by Lalmi et al. [20]. Finally, one concludes that though these two $(2\sqrt{3} \times 2\sqrt{3})$R30° configurations show the same LEED pattern, they present very different STM images highlighting to the necessity to use multiple tools to come to a conclusion about the atomic structure of silicene adsorbed on Ag(111).

The next super-structure of silicene on Ag(111) is the most observed one after (4×4), namely the $(\sqrt{13} \times \sqrt{13})$R13.9° one. For this structure, as for the previous one, the same LEED pattern is observed for two distinct atomic structures obtained by STM [25]. The first structure is shown in Fig. 10.6a, b for top and side views, respectively. We note from Fig. 10.6b that the silicene sheet is substantially buckled (~ 1.2 Å), with a nearest neighbour Si–Si distance being either 2.35 or 2.43 Å; and the projected lateral distance is 2.34 Å. The corresponding calculated STM image is shown in Fig. 10.6c, and again we notice the striking agreement with the experimental STM images [25].

Next, we present a comparison of this first $\sqrt{13} \times \sqrt{13}$R13.9°-I structure with the similar $\sqrt{13} \times \sqrt{13}$R13.9°-II structure shown in Fig. 10.7. The lowest buckling in the silicene sheet on Ag(111) is found for this structure and it amounts to about 0.7 Å, however the buckling in the substrate surface layer is 0.4 Å, and the nearest neighbour distance here varies between 2.35 and 2.40 Å, with the smallest projected lateral distance being 1.89 Å.

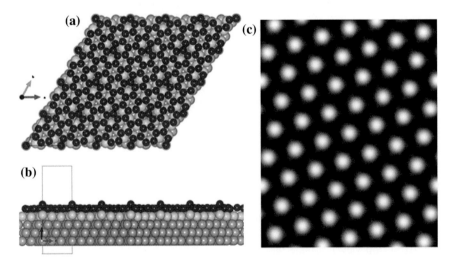

Fig. 10.6 The top (**a**) and side (**b**) views of the $(\sqrt{13} \times \sqrt{13})$13.9°-I structure, and (**c**) the corresponding STM image

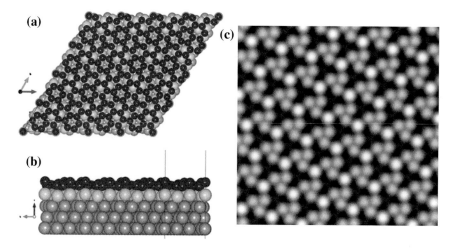

Fig. 10.7 The top (**a**) and side (**b**) views of the ($\sqrt{13} \times \sqrt{13}$)R13.9°-II structure, and (**c**) the corresponding STM image

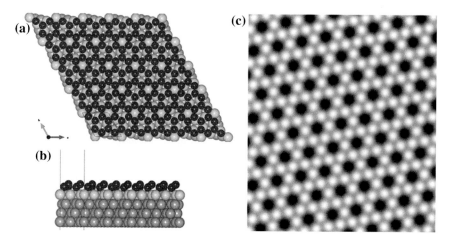

Fig. 10.8 The top (**a**) and side (**b**) views of the ($\sqrt{7} \times \sqrt{7}$)R19.1 structure, and (**c**) the corresponding STM image

Finally, the last silicene superstructure on Ag(111) to be presented here is the ($\sqrt{7} \times \sqrt{7}$)R19.1° one that is presented in Fig. 10.8a (top view), b (side view). This has the smallest unit cell among all those structures presented here. This is the second configuration with a relatively small buckling in the silicene sheet of about 0.7 Å. There are three out of eight silicon atoms that occupy the highest plane, which are the source of the hexagonal protrusions seen in the STM images. For this configuration, the nearest neighbour Si–Si distance ranges between 2.32 and 2.34 Å, while the projected lateral distance (observed by STM) is 2.2 Å.

10.3   Silicene Nanoribbons on Ag(110)

As for graphene, silicene can also grow as nanoribbons, here on Ag(110) with a typical width of 16 Å giving rise to a 2×5 superstructure with respect to the underlying Ag surface and extending tens of nanometers along the channels of the Ag(110) substrate. Using DFT calculations, the first theoretical study of these nanoribbons was done by He [12]. The proposed structure in this study consisted of silicon dimers as building blocks of the SiNRs. This computational study confined the silicon atoms to form a 2×5 rectangular motif, which is not compatible with experimental STM images. As a consequence, we have explored several configurations in order to achieve the best match with the observed STM images [18]. These configurations consisted of silicon nanoribbons containing between 27 and 36 Si atoms per unit cell. The resulting computationally derived STM images were compared to the experimental images. There were several features in the experimental STM images including a square adjacent to a diamond shape protrusion (Fig. 10.9a). We found out that only the Si30 (Fig. 10.9b, c) (a configuration consisting of 4 hexagons and containing 30 Si atoms) provided an excellent resemblance with the experimental image. The protrusions seen in the observed STM images are actually formed by 9 Si atoms out of 30, sitting at a higher position than the other Si atoms. This shows that a corrugated honeycomb structure may present a square-like symmetry.

If the SiNRs maintain their honeycomb structure after full relaxation, their atomic positions experience substantial changes. A side view (Fig. 10.10) shows that SiNRs present a curved shape with the substrate layers experiencing buckling. The substrate surface atoms are more affected on one side than the other giving the SiNRs an asymmetric shape. This asymmetry is also reflected in the electronic structure of the system, as shown in Fig. 10.10, where the charge density, in a plane

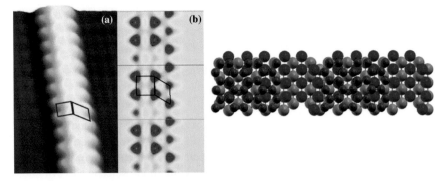

Fig. 10.9 Charge density (*left*) in a plane parallel to (110) containing the topmost Si atoms, shown in the top view (*right*) as *yellow*, with the other Si atoms shown as *red*. The Ag(110) surface layer atoms are shown as *dark blue*

Fig. 10.10 Charge density (*top*, *b*) in a plane perpendicular to the (110) plane cutting the ribbon across the width, and containing some top most Si atoms. A side view (*bottom*) shows the *curved* shape of the nanoribbon (Si atoms are coloured *red*, and Ag(110) surface atoms are shown by *dark blue*)

through the SiNR (perpendicular to the (110) plane) is plotted. This "arch-shaped" configuration explains the "bump" seen in the measured STM images [18].

The electronic structure of these SiNRs adsorbed on Ag(110) present novel features, as a consequence of the substantial changes in the atomic structure of both the SiNRs and the substrate. Due to the strong interaction between the substrate and the SiNRs, the latter becomes metallic, in agreement with a recent computational study [40]. The substrate electronic structure is also altered in such a way that new interface states appear near the Fermi energy, revealing a strong hybridization between the substrate and the SiNRs, in agreement with the experimental observation [17]. These interface states stem from a tandem of confinement-hybridization, with the confinement resulting from the finite lateral size of the NRs.

10.4 Conclusion/Summary

In summary, using DFT calculations we have established that silicene may adsorb on Ag(111) with several superstructures, in a remarkably good agreement with the experimentally observed one. The comprehensive theoretical study shows that the silicene sheet adsorbs strongly on Ag(111), resulting in a substantial change in its electronic structure, i.e. silicene on Ag(111) presents a metallic character.

Silicene NRs have been observed and present mainly a width of 16 Å, nearly four times the lattice constant of silver. The computational studies show that these NRs take an arched-shape due to the very strong chemical bond between the NRs and the substrate. This strong interaction triggers hybridization between the silver substrate and silicene resulting in the appearance of interface states near the Fermi energy.

Acknowledgments This work was partially supported by the U.S. Department of Energy Basic Energy Science under Contract No DE-FG02-11ER16243. This research used resources of the National Energy Research Scientific Computing Center, which is supported by the Office of Science of the U.S. Department of Energy.

References

1. G.G. Guzmán-Verri, L.C.L.Y. Voon, Phys. Rev. B **76**, 075131 (2007)
2. S. Cahangirov, M. Topsakal, E. Akturk, H. Sahin, S. Ciraci, Phys. Rev. Lett. **102**, 236804 (2009)
3. R. Quhe, Y. Yuan, J. Zheng, Y. Wang, Z. Ni, J. Shi, D. Yu, J. Yang, J. Lu, Sci. Rep. **4**, 5476 (2014)
4. Z.-X. Guo, A. Oshiyama, Phys. Rev. B **89**, 155418 (2014)
5. D. Kaltsas, L. Tsetseris, A. Dimoulas, Appl. Surf. Sci. **291**, 93–97 (2014)
6. S. Kokott, P. Pflugradt, L. Matthes, F. Bechstedt, J. Phys. Condens. Matter **26**, 185002 (2014)
7. N. Gao, J.C. Li, Q. Jiang, Chem. Phys. Lett. **592**, 222 (2014)
8. A. Bhattacharya, S. Bhattacharya, G.P. Das, Appl. Phys. Lett. **103**, 123113 (2013)
9. P. Pflugradt, L. Matthes, F. Bechstedt, Phys. Rev. B **26**, 185002 (2014)
10. E. Scalise, E. Cinquanta, M. Houssa, B. van den Broek, D. Chiappe, C. Grazianetti, G. Pourtois, B. Ealet, A. Molle, M. Fanciulli, V.V. Afanas'ev, A. Stesmans, Appl. Surf. Sci. **291**, 113 (2014)
11. H.B. Shu, D. Cao, P. Liang, X. Wang, X. Chen, W. Lu, Phys. Chem. Chem. Phys. **16**, 304 (2014)
12. G.-M. He, Phys. Rev. B **73**, 035311 (2006)
13. A. Kara, H. Enriquez, A.P. Seitsonen, L.C.L.Y. Voon, S. Vizzini, B. Aufray, H. Oughaddou, Surf. Sci. Rep. **67**, 1–18 (2012)
14. H. Oughaddou, B. Aufray, J.P. Bibérian, J.Y. Hoarau, Surf. Sci. **429**, 320 (1999)
15. H. Oughaddou, S. Sawaya, J. Goniakowski, B. Aufray, G. Le Lay, J.M. Gay, G. Tréglia, J. P. Bibérian, N. Barrett, C. Guillot, A.J. Mayne, G. Dujardin, Phys. Rev. B **62**, 16653 (2000)
16. H. Oughaddou, C. Léandri, B. Aufray, C. Girardeaux, J. Bernardini, G. Le Lay, J.P. Bibérian, N. Barrett, Appl. Surf. Sci. **1–5**, 9781 (2003)
17. C. Léandri, H. Oughaddou, J.M. Gay, B. Aufray, G. Le Lay, J.P. Bibérian, A. Rangüis, O. Bunk, R.L. Johnson, Surf. Sci. **573**, L369 (2004)
18. B. Aufray, A. Kara, S. Vizzini, H. Oughaddou, C. Léandri, B. Ealet, G. Le Lay, Appl. Phys. Lett. **96**, 183102 (2010)
19. H. Enriquez, S. Vizzini, A. Kara, B. Lalmi, H. Oughaddou, J. Phys. Condens. Matter **24**, 314211 (2012)
20. B. Lalmi, H. Oughaddou, H. Enriquez, A. Kara, S. Vizzini, B. Ealet, B. Aufray, Appl. Phys. Lett. **97**, 223109 (2010)
21. M.R. Tchalala, H. Enriquez, A.J. Mayne, A. Kara, G. Dujardin, M.A. Ait, H. Oughaddou, J. Phys. Conf. Ser. **491**, 012002 (2014)
22. H. Enriquez, A. Kara, A.J. Mayne, G. Dujardin, H. Jamgotchian, B. Aufray, H. Oughaddou, J. Phys. Conf. Ser. **491**, 012004 (2014)
23. M.R. Tchalala, H. Enriquez, H. Yildirim, A. Kara, A.J. Mayne, G. Dujardin, M.A. Ali, H. Oughaddou, Appl. Surf. Sci. **303**, 61–66 (2014)
24. Z. Majzik, M.R. Tchalala, M. Svec, P. Hapala, H. Enriquez, A. Kara, A.J. Mayne, G. Dujardin, P. Jelinek, H. Oughaddou, J. Phys. Condens. Matter **25**, 225301 (2013)
25. H. Oughaddou, H. Enriquez, M.R. Tchalalaa, H. Yildirim, A.J. Mayne, A. Bendounane, G. Dujardin, M.A. Ali, A. Kara, Prog. Surf. Sci. **90**, 46–83 (2015)
26. H. Enriquez, A.J. Mayne, A. Kara, S. Vizzini, S. Roth, B. Lalmi, A.P. Seitsonen, B. Aufray, Th Greber, R. Belkhou, G. Dujardin, H. Oughaddou, Appl. Phys. Lett. **101**, 021605 (2012)

27. M.R. Tchalala, H. Enriquez, A.J. Mayne, A. Kara, S. Roth, M.G. Silly, A. Bendounan, F. Sirotti, Th Greber, B. Aufray, G. Dujardin, M.A. Ali, H. Oughaddou, Appl. Phys. Lett. **102**, 083107 (2013)
28. R. Arafune, C.-L. Lin, R. Nagao, M. Kawai, N. Takagi, Phys. Rev. Lett. **110**, 229701 (2013)
29. C.-L. Lin, R. Arafune, K. Kawahara, M. Kanno, N. Tsukahara, E. Minamitani, Y. Kim, M. Kawai, N. Takagi, Phys. Rev. Lett. **110**, 076801 (2013)
30. Y.P. Wang, H.P. Cheng, Phys. Rev. B **87**, 245430 (2013)
31. P. Vogt, P. De Padova, C. Quaresima, J. Avila, E. Frantzeskakis, M.C. Asensio, A. Resta, B. Ealet, G. Le Lay, Phys. Rev. Lett. **108**, 155501 (2012)
32. L. Chen, C.-C. Liu, B. Feng, X. He, P. Cheng, Z. Ding, S. Meng, Y. Yao, K. Wu, Phys. Rev. Lett. **109**, 056804 (2012)
33. G. Kresse, J. Furthmuller, Phys. Rev. B **54**, 11169 (1996)
34. G. Kresse, J. Furthmuller, Comp. Mater. Sci. **6**, 15 (1996)
35. G. Kresse, J. Hafner, Phys. Rev. B **47**, 558 (1993)
36. J.P. Perdew, K. Burke, M. Ernzerhof, Phys. Rev. Lett. **77**, 3865 (1996)
37. G. Kresse, D. Joubert, Phys. Rev. B **59**, 1758 (1999)
38. P.E. Blöchl, Phys. Rev. B **50**, 17953 (1994)
39. H.J. Monkhorst, J.D. Pack, Phys. Rev. B **13**, 5188 (1976)
40. C. Lian, J. Ni, Physica B. Cond. Matt. **407**, 4695 (2012)

Chapter 11
Adsorption of Molecules on Silicene

Yi Du and Xun Xu

Abstract In the previous chapters we saw that silicene monolayers can be grown epitaxially on Ag(111) surfaces having various reconstructions and electronic structures. In this chapter, the adsorption of molecules on silicene grown on Ag(111) is reviewed based on theoretical and experimental work. We firstly introduce the thermodynamics and kinetics associated with adsorption on silicene, which is determined by the low-buckled structures. The typical adsorption methods for modifying the physical and chemical properties of silicene, including oxidation and hydrogenation are reviewed in Sect. 11.2. The unique adsorption process corresponding to various types of adatoms, including alkali and metal adatoms is discussed in detail in Sect. 11.3. The effects of adsorption on the electronic structure and chemical properties of silicene are important for the development of its applications, which is discussed in Sect. 11.4. We conclude the chapter with the future research directions.

11.1 Introduction

The last decade has witnessed rapid progress in the application of 2D materials in nano/microelectronic devices such as high-performance field-effect transistors, transparent conductive films and ultrathin solar cells [1–3]. This progress is mainly based on an improved understanding and advanced fabrication of these materials, including graphene, MoS_2 and Bi_2Te_3 [4]. In order to achieve the aim of modulating the electronic properties of 2D materials, the adsorption of molecules on their surface or edge sites has been widely applied. The functionalized 2D materials are, therefore, expected to meet industrial requirements. Since the interaction of

Y. Du (✉) · X. Xu
Institute for Superconducting and Electronic Materials (ISEM), Australian Institute
for Innovative Materials (AIIM), University of Wollongong, Wollongong,
NSW 2522, Australia
e-mail: ydu@uow.edu.au

© Springer International Publishing Switzerland 2016
M.J.S. Spencer and T. Morishita (eds.), *Silicene*, Springer Series
in Materials Science 235, DOI 10.1007/978-3-319-28344-9_11

molecules differs depending on where they adsorb on the material, the molecular arrangements at surface sites may differ from those at edge sites. Different molecules or elements also possess different features after they are attached to 2D materials, and in turn, produce different changes to the chemical and physical properties of these materials.

As a new emerging 2D material, silicene exhibits interesting physical and chemical properties, including a low-buckled structure with Dirac fermion characteristics. Due to large spin-orbital coupling, a band gap intrinsically opens in silicene which makes it more suitable for the development of nanoelectronics [6–10]. In addition, silicene has the advantage of easily being incorporated into silicon-based electronic technology. Because silicene has these remarkable properties that the other 2D materials do not share, functionalization of silicene by adsorption of molecules is expected to be different to the other well-known 2D materials and has been investigated extensively in theoretical and experimental studies.

Surface techniques such as scanning tunneling microscopy (STM) and angle-resolved photoemission spectroscopy (ARPES) enable one to study the behavior of single molecules, adatoms and nanoclusters adsorbed on silicene in terms of their adsorption sites, adsorption dynamics and their influences on the local density of states (LDOS). First principles calculations provide detailed information to compliment the experimental findings and enable one to understand the adsorption mechanism. These techniques have been used in the investigation of the adsorption of molecules on epitaxial silicene layers.

In this chapter, we will focus on studies aimed at obtaining a deeper understanding of how molecular adsorption and interactions occur at the surface of silicene. The effects of adsorption on the electronic structure and chemical properties of silicene are discussed. Related work performed under ultrahigh vacuum conditions will be reviewed in order to demonstrate how different adatoms have different effects on the electronic and chemical properties of silicene. The theoretical simulations on molecular adsorption on silicene are reviewed based on recent breakthrough works. This chapter compliments the studies reviewed in Chap. 5, which focused on theoretical simulations of functionalised silicene.

11.2 Adsorption of Gas Molecules on Silicene

Recent studies in the area of nanoscale physics have aimed at discovering new monolayer materials and revealing what are their properties under different conditions. Among these materials, silicene, was shown to be stable when alternating atoms of hexagons are buckled [1, 2]. More recently, single and multilayer silicene and its derivatives have been grown on the Ag(111) substrate [3–7].

Although both theoretical and experimental studies [8–14] support the stability of silicene, local defects can always exist at finite temperatures. Among these, owing to increased chemical activity, single foreign atoms can adsorb at the defect sites.

Even specific gas molecules can dissociate into constituent atoms. In particular, the dissociation of H is of crucial importance for the hydrogen economy [15–18]. The oxidation of silicene followed by the adsorption of O adatoms is of particular interest because an almost reversible oxidation–deoxidation mechanism leading to a continuous metal–insulator transition is possible [19–22]. The interaction of CO with silicene has also been the subject of interest for fabrication of gas sensors [23]. In this respect, these studies are unique, since the interaction between various molecules and single vacancy sites is examined. Other adsorption properties of atoms on silicene have also been studied theoretically [24–35]. Due to its buckled honeycomb structures, silicene exhibits a much higher chemical reactivity for atom adsorption than graphene with great industrial, environmental and medical applications for new silicene based nano-electronic devices [36]. It is still an open question as to what is the adsorption behavior on silicene of nitrogen-based gas molecules, including NH_3, NO and NO_2, which have been studied via first principles calculations [37].

11.2.1 Fundamental Picture for Silicene Hydrogenation

Hydrogenation was found as an effective chemical method to modify the electronic properties of graphene, where a dramatic band-gap opening was observed upon graphene hydrogenation [38–40]. In contrast to graphene, silicene possesses a hybrid sp^2-sp^3 bonding which should naturally favour a low-barrier hydrogen attach-detach process. Hydrogenation could also be a promising method to modify the properties of silicene. Theoretical calculations have suggested intriguing properties for hydrogenated silicene, for example, a large gap (~3 eV) opening [41], and interesting ferromagnetic [42] and optoelectronic properties [43]. Various hydrogen-adsorption configurations (for example, adsorption on both sides of the silicene or on one-side only on freestanding and substrate-supported silicene) have been studied theoretically.

In particular, Wu's group has published an experimental report on monolayer silicene on an Ag(111) surface by scanning tunneling microscopy (STM) [44]. In contrast to the graphene case, where hydrogen tends to form clusters, hydrogenated silicene exhibits a perfect long-range ordered structure. Combined with the results from first principles calculations, it has been determined that there are seven hydrogen atoms in each (3 × 3) unit cell and that the buckling configuration of the Si atoms in silicene spontaneously rearranges upon hydrogenation. Moreover, by annealing the sample to a moderate temperature, about 450 K, dehydrogenation occurs and a clean silicene surface is recovered. This work provides a clear and fundamental picture for silicene hydrogenation. Such a uniformly ordered, reversible hydrogenation can be useful to control the electronic properties of silicene for potential applications.

Amongst these phases of silicene that can be formed on the Ag(111) surface, the (4 × 4) is the simplest and most well understood, in which a (3 × 3)-Si supercell is placed commensurate with a (4 × 4)-Ag supercell, and is chosen as the model system for hydrogen adsorption. For convenience, it will be referred to as (3 × 3) with respect to the pristine silicene-(1 × 1) in the following.

Figure 11.1a shows a typical high-resolution STM image of our silicene-(3 × 3) film, with a characteristic hexagonal arrangement of triangular structures around dark centers. Each (3 × 3) unit cell (UC) is composed of two triangular half unit cells (HUC). Figure 11.1b is the structural model, with the red balls indicating upper-buckled Si atoms, which are roughly on top of the Ag atoms. The yellow balls represent lower-buckled Si atoms. Thus, in the total 18 Si atoms in each (3 × 3) UC, only six are upper buckled. These six Si atoms correspond to the six protrusions observed in the STM images.

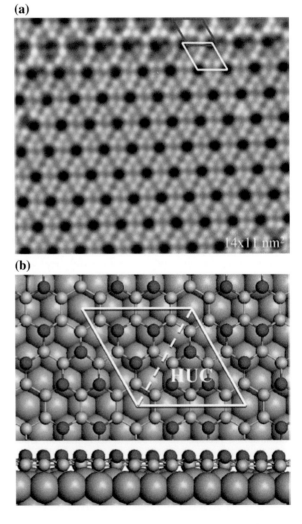

Fig. 11.1 STM image and structural model of clean silicene-(3 × 3) on Ag(111) [44]. **a** A typical STM image (14 × 11 nm) of clean silicene-(3 × 3). In the *upper-right* part of the image, there is a small area consisting of a metastable β-(3 × 3) phase. The *white rhombus* marks a (3 × 3) unit cell and the *red rhombus* is the metastable β-(3 × 3) unit cell. **b** Structural model of silicene-(3 × 3). Each unit cell consists of six upper-buckled Si atoms and the two HUCs are mirror symmetric (Color figure online)

Typical changes induced by the hydrogenation of silicene-(3 × 3) are illustrated in Fig. 11.2. Upon exposure of 900 L hydrogen at room temperature, a perfectly ordered structure with the same (3 × 3) periodicity can be observed, as shown in Fig. 11.2a. Further increases to the hydrogen dosage do not induce further changes, indicating that the hydrogen adsorption is saturated. A high-resolution image of the hydrogenated structure manifests two inequivalent HUCs, one with six bright spots while the other has only one bright spot in the middle, as shown in Fig. 11.2b. The distance between the nearest bright spots is about 3.8 Å, corresponding to the lattice constant of silicene-(1 × 1).

If there are regions that are not fully hydrogenated, it was possible to find clean silicene-(3 × 3) (symmetric HUCs) coexisting with the hydrogenated, symmetry-broken area. An example is shown in Fig. 11.2c, where clean silicene-(3 × 3) is found in the left part of the image whereas the right part is hydrogenated.

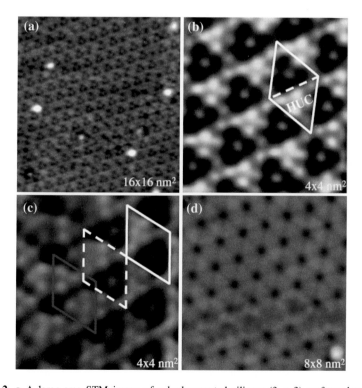

Fig. 11.2 **a** A large-area STM image of a hydrogenated silicene-(3 × 3) surface showing an ordered (3 × 3) structure [44]. **b** Enlargement of a STM image of the hydrogenated (3 × 3) phase. The *white rhombus* marks an apparent unit cell of the structure. There are six bright protrusions in one HUC and one protrusion in the other HUC. **c** STM image showing the comparison between the position of apparent UCs of clean and hydrogenated silicene-(3 × 3). The *red* and *white rhombuses* correspond to clean (3 × 3) UC and the hydrogenated (3 × 3) UC, respectively. A translation of the white UC (*dotted line*) does not match the *red* one. **d** The clean silicene-(3 × 3) surface is fully recovered after annealing the surface at 450 K

The apparent unit cells of clean and hydrogenated silicene-(3 × 3) are indicated by the red and white solid rhombuses, respectively, in Fig. 11.2c. Surprisingly, the two sets of (3 × 3) UCs do not overlap after translation. Instead, they are shifted along the Si–Si bond direction by a distance of one Si–Si bond length. There can be only two possibilities. First, the whole silicene film may be laterally shifted. This is, however, unlikely since the monolayer silicene film fully covers the Ag(111) substrate and there is no room for a lateral shift. The second possibility is that the buckling configurations of Si atoms may be changed after hydrogenation, resulting in a shift of the apparent UCs. It is remarkable that the fully hydrogenated silicene sheet can be completely restored to its original state by annealing the sample to a moderate temperature of about 450 K. As is shown in Fig. 11.2d, a regular monolayer silicene-(3 × 3) structure and domain boundary are completely restored after hydrogen desorption. The adsorption-desorption cycle can be repeated many times without degradation of the silicene film if the UHV system is clean enough.

The relatively lower desorption temperature is consistent with the lower binding energy of H on silicene (~ 2.67 eV per H atom) as compared with that on graphene (6.56 eV per H atom) [10]. In the case of graphene, the desorption temperature is as high as 1100 K [13]. The easily reversible hydrogenation of monolayer silicene suggests that silicene may be useful for controllable hydrogen storage. The work provides a comprehensive picture for silicene hydrogenation, and paves the way for further investigation of the electronic properties of hydrogenated silicene, such as band-gap controlling and magnetism that have been predicted by theoretical calculations [41, 42].

The dissociative adsorption of a H_2 molecule on silicene with different tensile strains was investigated by DFT calculations [45]. From Fig. 11.3, it was found that

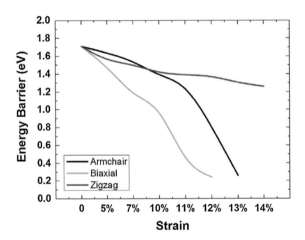

Fig. 11.3 Energy barrier of the dissociative adsorption of a H_2 molecule on silicene under increasing tensile strain [45]. The *black, red* and *blue lines* stand for the results under biaxial tensile strain, uniaxial tensile strain along the armchair direction and uniaxial tensile strain along the zigzag direction, respectively (Color figure online)

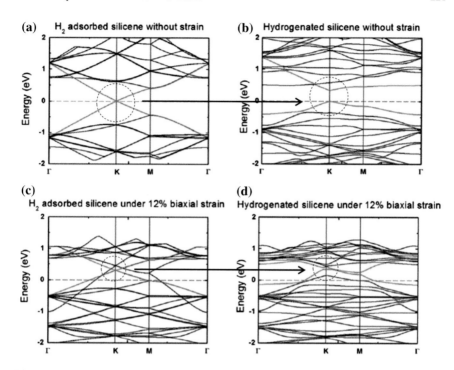

Fig. 11.4   Band structures of pure and hydrogenated silicene without strain (**a**) and (**b**) and under 12 %
biaxial strain (**c**) and (**d**) [45]. The *red lines* are the bands near the Dirac point (Color figure online)

the energy barrier of the dissociative adsorption of a H_2 molecule on silicene could
be reduced significantly by applying biaxial tensile strain or uniaxial tensile strain
along the armchair direction, while the biaxial strain had an improved effect at
reducing the energy barrier. The energy barrier also decreased under uniaxial strain
along the zigzag direction, but the effect was not so obvious.

From Fig. 11.4, under 12 % biaxial strain, the energy barrier drops from 1.71 eV
to about 0.24 eV, which can greatly reduce the reaction time from 8.06×10^{16} to
1.68×10^8 s. Thus, the hydrogenation of silicene can be facilitated efficiently under
the strain, which is essential to tune its electronic properties for application in
electronic devices.

11.2.2   Effects of Oxygen Adsorption on Silicene

Epitaxial silicene, which is composed of a single layer of silicon atoms packed in a
honeycomb structure, demonstrates a strong interaction with the substrate that
dramatically affects its electronic structure. The reconstructions and hybridized
electronic structures of epitaxial (4 × 4) silicene on Ag(111), were revealed by

scanning tunneling microscopy and angle-resolved photoemission spectroscopy. X-ray photoemission spectroscopy confirmed the decoupling of Si–Ag bonds after oxygen treatment as well as the relative oxygen resistance of the Ag(111) surface, in contrast to (4 × 4) silicene [with respect to Ag(111)]. First-principles calculations have confirmed the evolution of the electronic structure of silicene during oxidation [46]. Silicene monolayers grown on the Ag(111) surface have a band gap that is tunable by oxygen adatoms, from semimetallic to semiconducting. With the use of low-temperature scanning tunneling microscopy, the work demonstrates the feasibility of tuning the band gap of silicene with oxygen adatoms, which, in turn, expands the properties of two-dimensional electronic materials for devices that is hardly achieved with graphene oxide [47].

Figure 11.5 shows 15 × 15 nm^2 STM topographical images of the Ag(111) surface and a 4 × 4 silicene monolayer epitaxially grown on the substrate. Quantum-interference patterns are clearly visible on the Ag(111) surface, as shown in Fig. 11.5a. Electrons in the two dimensional surface states can be scattered by surface point defects, leading to periodic spatial oscillations of the electronic local density of states (LDOS) [48]. The quantum-interference pattern (QPI) with a period of several tens of angstroms reflects the nature of the 2D electron wave in the Ag(111) surface-state band. In Fig. 11.5b, 4 × 4 silicene on a 1 × 1 Ag(111) surface exhibits a lattice constant of 1.06 nm. The low-buckled configuration can be verified by the different heights of the Si atoms at the edges. The height of buckling is 0.86 Å in 4 × 4 silicene, which is different from the calculated value for free-standing 1 × 1 silicene [49, 50].

No quasiparticle interference pattern (QPI) as observed in the as-grown silicene layer, which suggests the absence of Dirac fermion characteristics in epitaxial 4 × 4 silicene. The strong coupling, accompanied by the charge transfer, leads to a modulation of the electronic structure of silicene on Ag(111). In order to reveal the nature of the hybridization state, oxygen molecules were introduced onto the 4 × 4 silicene surface by a leak valve in precise doses at 77 K. Figure 11.5c, d shows typical STM images of silicene layers exposed to 10 Langmuir (L) and 600 L O$_2$, respectively. The topmost Si atoms in the buckled silicene are defined as "top-layer" (TL), and the other atomic layers are defined as "bottom-layer" (BL). At the low oxygen dose, the oxygen adatoms prefer to reside on a bridging site that forms a Si(TL)–O–Si(BL) configuration. The Si–O bonds significantly modulate the surface metallic band in silicene on Ag(111). As shown in Fig. 11.5c inset, a gapped electronic state was identified in scanning tunneling spectroscopy (STS) measurements carried out at oxygen adatom sites on the silicene layer. The surface metallic band is therefore tuned to a semiconducting-like characteristic. When the oxygen dose is increased to 600 L, the silicene layer is oxidized and forms a disordered structure, as shown in Fig. 11.5d. Some areas of the bare Ag (111) substrate were exposed. Interestingly, the QPI pattern again appears on the Ag(111) surface with the same oscillating periothe Ag(111) substrate acts as an inert material compared to 4 × 4 silicene in the process of oxidation.

Figure 11.6 shows the ARPES results of occupied states along the Γ-M$_{Ag}$ and Γ-K$_{Ag}$ directions of 4 × 4 silicene/Ag(111) before and after oxidation. Figure 11.6d

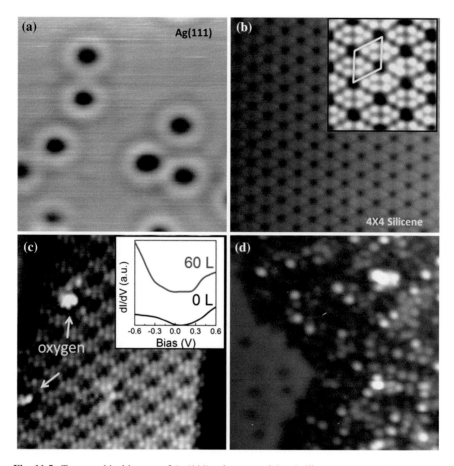

Fig. 11.5 Topographical images of Ag(111) substrate and 4 × 4 silicene grown on Ag(111) [46]. **a** STM topographical image of clean Ag(111) substrate, which shows a clear quantum-interference pattern due to point defects (scanning area 15 nm × 15 nm, V_{bias} = −0.2 V, I = 4 nA). **b** STM topographical image of 4 × 4 silicene on Ag(111) (scanning area 15 nm × 315 nm, V_{bias} = 20.8 V, I = 2 nA). *Inset* is an enlarged view of an area 4 nm × 34 nm in size. **c** STM image of silicene layer oxidized by an oxygen dose of 10 L. O adatoms prefer to reside at bridge sites. The *inset* contains STS spectra of silicene and silicene oxide samples, indicating that there is gap opening due to oxidation. **d** STM image of the 4 × 4 silicene oxidized under 600 L O_2. The bare Ag(111) surface can be seen at the *bottom left* of (**d**)

shows the reciprocal space Brillouin zones (BZ) of 1 × 1 Ag(111) (blue hexagon), free-standing silicene (dashed red hexagon), and 4 × 4 epitaxial silicene with respect to Ag(111) (or 3 × 3 silicene with respect to 1 × 1 silicene) (orange hexagons). Note that the Γ_{Ag} and K_{Ag} points of Ag(111) coincide with the Γ_{Si} and K_{Si} points of 4 × 4 silicene in the BZ. Figure 11.6a displays the Shockley surface state (SSS) of Ag(111) substrate at the BZ center Γ point ($\kappa = 0$ Å^{-1}).

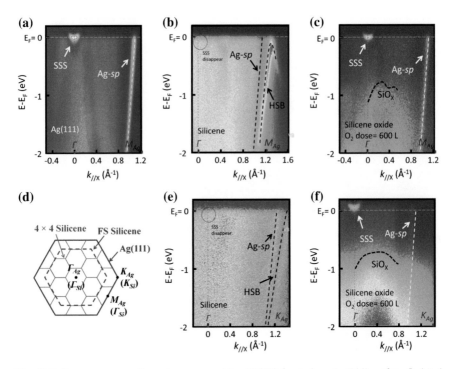

Fig. 11.6 Energy versus κ dispersion measured by ARPES for **a** clean Ag(111) surface, **b** 4 × 4 silicene grown on Ag(111) along the Γ-M$_{Ag}$ direction, and **c** oxidized silicene on Ag(111) along the Γ-M$_{Ag}$ direction, respectively [46]. SSS in (**a**) and (**b**) denotes the Shockley surface state. HSB in (**b**) denotes the hybrid surface band. **d** Schematic diagram of the BZ for 4 × 4 silicene grown on Ag(111): *red*, *blue* and *orange* honeycomb structures correspond to free-standing (FS) silicene, Ag (111) and 4 × 4 silicene with respect to Ag(111) (or 3 × 3 silicene with respect to 1 × 1 silicene), respectively. **e** 4 × 4 silicene grown on Ag(111) along the Γ-K$_{Ag}$ direction, and **f** oxidized silicene on Ag(111) along the Γ-K$_{Ag}$ direction, respectively

The SSS arises primarily from surface states of nearly free electrons and is associated with the special boundary conditions introduced by the metal/vacuum interface. The typical bulk *sp*-band of Ag lies across the Fermi level at κ = 1.15 Å^{-1}. As the coverage of silicene increases, the SSS and *sp*-band of Ag become faint, and the SSS eventually disappears when the Ag(111) surface is fully covered by the silicene layer, as shown in Fig. 11.6b. The weak Ag *sp*-band is still observable, which indicates that this band remains stable upon Si deposition. There is a clear new "∩"-shaped state with a top point at the M$_{Ag}$ point. As shown in Fig. 11.6e, the "∩"-shaped state along the Γ-K$_{Ag}$ direction exhibits a variation, where the band traverses the Fermi surface at the K$_{Ag}$ (K$_{Si}$) point. The results of band structures along both the Γ-M$_{Ag}$ and the Γ-K$_{Ag}$ directions are consistent with previous reported works [51], indicating that the "∩"-shaped state is attributable to a hybridization of Si and Ag orbitals that resembles the *p*-band dispersion in graphene. The apex of the state at κ = 1.28 Å^{-1} in Fig. 11.6b is about 0.15 eV below the Fermi level, which is the saddle point of the surface state and at the middle point between two adjacent K_{Ag} (K_{Si}) points. It should be noted that this feature is absent in the clean Ag(111) spectra and has been associated with a Ag(111)-related surface band that appears only after Si deposition.

Figure 11.6c, f shows ARPES results of an oxidized 4 × 4 silicene/Ag(111) sample under an oxygen dose of 600 L along the Γ-M_{Ag} and Γ-K_{Ag} directions, respectively. The signature metallic HSB has disappeared. The two ARPES results show similar features to each other. In despite of the disappearance of the metallic HSB, the well-defined SSS bands are revived in the oxidized silicene/Ag(111) sample. The intensities of the Ag(111) states are weak because the Ag(111) surface is still partially covered by silicene oxide. Moreover, an asymmetric band with the highest energy level at about 20.6 eV can be observed in Fig. 11.6c. It is worth noting that SSS in metal is extremely surface sensitive, so that it can reflect modifications to the surface atomic and electronic properties [52]. In ARPES results, the revived SSS in the oxidised sample demonstrates that oxygen would preferentially react with Si rather than Ag(111) when a low oxygen dose is present. As a result, the surface states of Ag(111) can be chemically protected against oxygen molecules by 4 × 4 silicene. The disappearance of the HSB in Fig. 11.6c, f implies that the hybridization between Si and Ag is broken due to oxidation.

A detailed X-ray photoelectron spectroscopy (XPS) characterisation of the chemical bonds in the samples was carried out. Figure 11.7a, b shows Ag 3d core level XPS spectra for 4 × 4 silicene deposited on an Ag(111) sample before and after oxygen treatment, respectively. The experimental data points are displayed with black dots, while the fitted curves are red lines. For the bare Ag(111) substrate, the Ag $3d_{3/2}$ and $3d_{5/2}$ peaks at 371.5 and 365.5 eV have originated from A–Si bonds. A downward energy shift ($\sim$0.7 eV) for the Ag 3d orbital is observed after the deposition of silicene, where the Ag–Si chemical bond forms, indicating that the chemical activity of silicene is higher than for the pure Ag–Ag bond arising from Ag^0. Peak splitting of the Ag 3d line was observed after exposure to 600 L oxygen in Fig. 11.7b. The peaks could be decomposed into two contributions, coming from the Ag–Ag bond component and the Ag–Si bond component, respectively. The dramatic fall in intensity of the Ag–Si state and the recovery of Ag–Ag bonds indicates that Ag–Si bonds are broken after the oxygen treatment. Moreover, no Ag–O chemical structure is present in the XPS spectrum, which implies that oxygen adatoms most likely only form bonds with silicon atoms, which supports the resurgence of QPI patterns on Ag(111) after oxidation, as shown in Fig. 11.5d.

Figure 11.7c, d shows Si 2p core level spectra for the sample before and after oxygen treatment, respectively. The fitting results for the Si 2p line, as shown in Fig. 11.3c, make it clear that there are two groups of bonding components, labelled as Si1 and Si2, respectively. The energy gap of the two peaks in each group is a constant value, indicating that the two fitting peaks in one group are related to two Si $2p_{3/2}$ and $2p_{1/2}$ peaks, respectively. The Si2 peaks at a binding energy around 98.8 eV are related to the Si–Si bonds in silicene, consistent with the report [53].

The Si1 group is attributed to Si–Ag bonding, since there are no other elements induced in the process of deposition, combined with the fitting results on Ag–Si

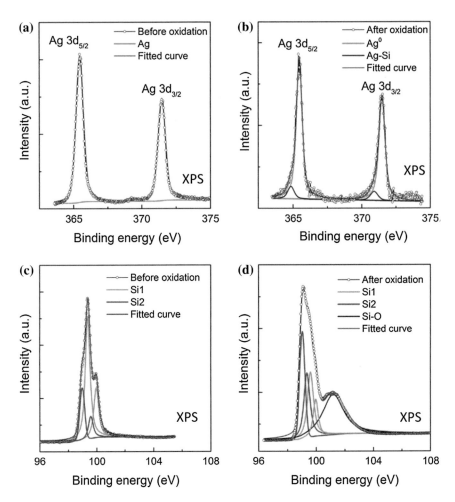

Fig. 11.7 Representative Ag 3*d* core level XPS spectra of 4 × 4 silicene on Ag(111) **a** before and **b** after oxidation [46]. Si 2*p* core level XPS spectra of 4 × 4 silicene on Ag(111) **c** before and **d** after oxidation. Si1 and Si2 represent Si–Ag and Si–Si, respectively. The spectra indicate that the 4 × 4 silicene layer is oxidized and decoupled from Ag(111) when the oxygen dose is as high as 600 L

bonding in Fig. 11.7a, b. The Si–O peaks are clearly present after oxygen treatment. The binding energy value (101.6 eV) is lower than the peak position for the SiO_2-like binding energy (102.3 eV) [54], indicating that the valence states of Si–O bonds are lower than Si^{4+}. Therefore, silicene was not fully oxidized to SiO_2, which is consistent with the report [47] in which the oxygen adatoms are the most energetically favored on the surface of silicene. The intensity of the peaks related to the Si–Ag bonds is significantly reduced with the emergence of the Si–O peak. The variation of the peak intensity demonstrates that oxygen adatoms prefer to decouple the Si–Ag bonds rather than the Si–Si bonds. The XPS results are in a good agreement with the STM and ARPES results, and confirm the decoupling of Si–Ag bonds after oxygen treatment, as well as the relatively high oxygen resistance of the Ag(111) surface, in contrast to 4 × 4 silicene [55–57].

The density functional theory (DFT) calculations were carried out to investigate the revived SSS on Ag(111) and to confirm the origins of the VB in silicene oxide, as shown in Fig. 11.8. The first step in the calculations was to determine the superstructure of silicene grown on Ag(111). One layer of silicene was put on top of 5 layers of 4 × 4 Ag(111). The simulated structure shows the same reconstruction as in the above STM results, and as displayed in Fig. 11.8a. The Ag d-state and the Si p-state make the heaviest contributions to the density of states (DOS) at the Fermi level (E_F), which indicates that the metallic HSB is indeed contributed by the p_z

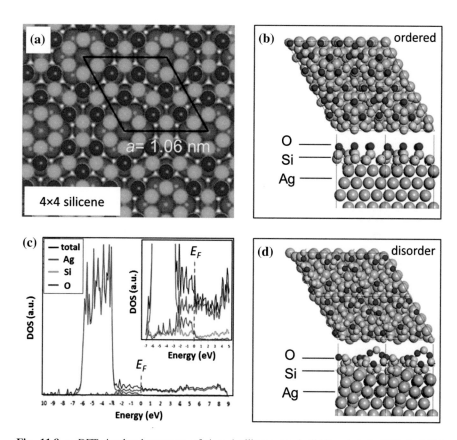

Fig. 11.8 a DFT-simulated structure of 4 × 4 silicene on Ag(111) substrate [46]. **b** Initial adsorption sites of oxygen adatoms on silicene. **c** Calculated DOS of oxidized silicene with oxygen coverage of 0.5 ML. The *inset* is an enlarged view of the DOS near the Fermi level. **d** The energy favoured stable adsorption site after running molecular dynamics for 7 ps. It indicates that oxygen adatoms prefer to form Si–O–Si bonds at bridge sites in the 4 × 4 silicene surface. *Red, yellow,* and *blue balls* in (**b**) and (**d**) represent oxygen, silicon, and silver atoms, respectively

electrons of the Si atoms and the $4d$ electrons of the Ag(111) substrate after putting 0.5 monolayer (ML) oxygen on the stabilized silicene surface. After running molecular dynamics for 7 ps, all the singly coordinated Si atoms have moved to bridge sites, indicating that there is a low energy barrier that needs to be overcome for the other O atoms to move to more highly coordinated sites on silicene. Meanwhile, the Si atomic layer becomes disordered, demonstrating that the silicene oxide layer has started to decouple from the underlying Ag(111) substrate.

The disordering of the Si overlayer induced by the oxygen adatoms is in excellent agreement with the STM observations on the disordered nature of silicene oxide. Figure 11.8c shows the calculated DOS on 4 × 4 silicene covered by 0.5 ML oxygen. The deep level (~ 22 eV) is mainly contributed by Ag d-states. The DOS near E_F, however, consist of Ag, Si, and O orbitals, as shown in the inset of Fig. 11.8c. It should be pointed out that the Si $3p$ states and O $2p$ states form a new band below E_F, although Si and O also contribute some partial DOS at E_F. The top of this band is at 20.4 eV, which matches well with the asymmetric band (20.6 eV) shown in the ARPES results (see Fig. 11.6c, f). Thus, it was confirmed that this shallow band is the VB of partially oxidized silicene. It should be noted that 0.5 ML oxygen is not enough to oxidize the whole silicene layer, so that the hybridisation between Si and Ag still exists in some regions. For areas of silicene oxide, the Shockley surface state would be revived due to decoupling between the silicene overlayer and the Ag(111). The Ag state at the Fermi level could be composed of both the metallic HSB and the revived Shockley surface state.

11.2.3 The Adsorption of NH₃, NO, and NO₂ on Silicene

It is still an open question as to what is the adsorption behavior of nitrogen-based gas molecules, including NH_3, NO and NO_2, which are all of great practical interest for industrial, environmental and medical applications. In the recent report, the adsorption behaviors and electronic properties of NH_3, NO and NO_2 on silicene were studied via first principles calculations, and indicated that NH_3, NO and NO_2 prefer to chemically adsorb on silicene with high adsorption energies. Charge is transferred from silicene to the molecules, resulting in a p-type doping of silicene with tunable band gap opening at silicene's Dirac point. The calculated adsorption energies indicated that the silicene can be used as a reusable molecule sensor for NH_3 and NO, and a disposable molecule sensor for NO_2 [37].

All the calculations performed were based on density functional theory (DFT) as implemented in the VASP package [58]. The generalized gradient approximation of Perdew, Burke, and Ernzerhof (GGA-PBE) [59] with van der Waals (vdW) correction proposed by Grimme (DFT-D2)86 was chosen due to its good description of long-range vdW interactions. As a benchmark, DFT-D2 calculations give a good bilayer distance of 3.25 Å and binding energy of −25 meV per carbon atom for bilayer graphene, in full agreement with experimental [60, 61] and theoretical [62, 63] studies. The energy cutoff was set to be 500 eV and the surface Brillouin zone was sampled with a 3 × 3 regular mesh and 240 k points for calculating the small band gaps at the Dirac points of silicene. All the geometry structures were fully relaxed until the energy and forces were converged to 10^{-5} eV and 0.01 eV $\mathring{A}^{-1}$, respectively. Dipole corrections were employed to cancel the errors of electrostatic potential, atomic forces and total energy caused by periodic boundary condition [64]. The electronic band structures of NH_3, NO and NO_2 adsorbed on silicene are shown in Fig. 11.9 [37]. NH_3 adsorption on silicene in supercells of different sizes was considered. When the supercell size was 3N × 3N unit cells (i.e. 3 × 3 and 6 × 6), negligible energy band gaps were opened at two different kinds of silicene's Dirac points, as shown in Fig. 11.9c, f. Recent theoretical simulations have also reported similar results for atom-doped graphene and silicene [65–67], in which two kinds Dirac points for grapheme were produced by band folding in the 3N × 3N or $\sqrt{3}N \times \sqrt{3}N$ supercells. In particular, these Dirac points were easily distorted with the introduction of dopants or defects due to the break of bond symmetry in graphene and silicene [65–68]. When NH_3, NO or NO_2 molecules were chemisorbed on

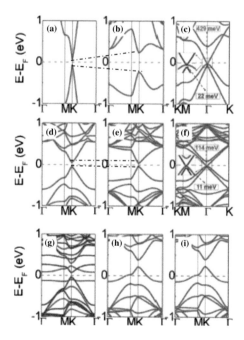

Fig. 11.9 Electronic band structures of NH_3, NO, and NO_2 adsorption on silicene [37]. NH_3 adsorption on silicene in the **a** 1 × 1, **b** 2 × 2, **c** 3 × 3, **d** 4 × 4, **e** 5 × 5 and **f** 6 × 6 supercells, **g** NO, **h** and **i** NO_2 (N–Si and O–Si) adsorption on silicene in the 4 × 4 supercell. The band gaps at two different Dirac points of silicene adsorbed NH_3 in the 3 × 3 and 6 × 6 supercells are shown in the *inset*. The red and blue color lines represent spin-up and spin-down states, respectively. The Fermi level is set to zero and marked by *green dotted lines* (Color figure online)

silicene with the supercell size of 4×4, sizable band gaps at the level of hundreds of meV were opened at the Dirac point of silicene. These values are significantly larger (by tens of meV) than those for common gas molecules physically adsorbed on graphene with similar adsorbate concentrations and thermal fluctuations (25 meV) at room temperature. Hence, the energy band gap of silicene depends sensitively on the molecules and their adsorption concentrations.

The band gap values increased with the increasing molecule concentrations. Moreover, the band gaps values could be further tuned via other external methods, such as electric fields [69] and atom doping [70], implying potential applications for silicene-based FETs. When NO_2, NH_3 and NO molecules adsorb on the silicene, electrons are transferred from silicene to the molecules, resulting in a p-type doping of silicene itself. For the adsorption of these molecules on silicene with a 4×4 unit cell-sized supercell, the calculated charge transfer ranges from 0.14–0.82 e for silicene, distinctly larger than on graphene (−0.027, −0.018 and 0.099 e, respectively for NH_3, NO and NO_2) [71]. For NH_3- and NO chemisorbed on silicene, the Fermi energy levels remain in their induced gaps at the Dirac points of silicene, showing semiconducting behavior with small amounts of electrons transferring (as represented by ρ) from silicene to the gas molecules ($\rho(NH_3) = 0.14$ e and $\rho(NO) = 0.46$ e). For NO_2- chemisorbed silicene, however, a large transfer of electrons occurs from silicene to NO_2 in two configurations ($\rho(NO_2, N–Si) = 0.69$ e and $\rho(NO_2, O–Si) = 0.82$ e), which move the Dirac points of silicene to 0.31 and 0.30 eV above the Fermi levels. The dominant charge transfer confirms the strong adsorption of NO_2 molecules on silicene, resulting in heavier p-type doping of silicene than that on graphene with NO_2 physisorption [71]. It can, therefore, be calculated that NO_2-chemisorption on silicene with large adsorption energies and charge transfer can effectively enhance the hole conductivity in silicene, opening up new applications for nano-electronic devices.

Figure 11.10 shows the total and partial density of states (DOS) of NH_3, NO, NO_2 adsorbed on silicene and graphene. From Fig. 11.10a, the frontier orbitals (highest occupied molecular orbital (HOMO) and lowest unoccupied molecular orbital (LUMO) of NH_3, NO and NO_2 are very close to the Dirac point of silicene compared to other common gas molecules (an example of N_2 is given in Fig. 11.10a). Thus, these nitrogen-based gas molecules have a high reactivity to silicene. The differences of chemical reactivity between silicene and graphene can also be found from Fig. 11.10b, c. Graphene is a carbon sp^2 hybridized monolayer with a stable planar honeycomb structure.

However, silicene is a silicon hybrid sp^2 and sp^3 hybridized monolayer with stable buckled honeycomb structures, which possesses chemically rich active electronic states at the top of the silicon atoms. Moreover, the corresponding electronic state peaks are closer to silicene's Fermi level compared to that of graphene. Therefore, silicene exhibits a higher reactivity than graphene for adsorption of atoms and molecules, leading to potential applications of silicene in gas sensors and nano-electronic devices.

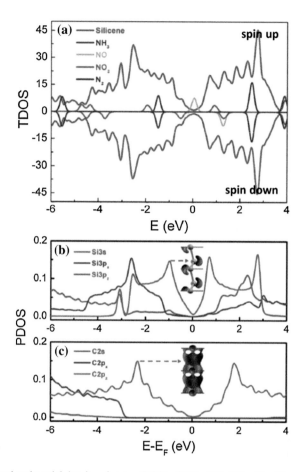

Fig. 11.10 Total and partial density of states (DOS) of NH_3, NO, NO_2, N_2, silicene and graphene [37]. **a** TDOS of NH_3, NO, NO_2, N_2 and silicene in the 4 × 4 supercell. The positive and negative values represent spin-up and spin-down states, respectively. The Fermi level of silicene is set to zero and all the energy levels are referenced to the vacuum level of silicene. **b** and **c** PDOS of silicene (Si3s, Si3px = Si3py and Si3pz) and graphene (C2s, C2px = C2py and C2pz) in the unit cell. Partial charge densities with an isosurface value of 0.1 e Å^{-3} of silicene and graphene close to their Fermi levels are shown in the *insets*. The *yellow* and *gray balls* denote silicon and carbon atoms, respectively. The Fermi levels of silicene and graphene are set to zero (Color figure online)

11.2.4 Environmental Gas Adsorption on Silicene Nanoribbons

Because sensors are typically exposed to environmental gases such as N_2, O_2, CO_2, and H_2O, the effect of these gases on the conductance of silicene nanoribbons was studied by Osborn and Farajian [72]. Their calculations show that N_2 interacts with silicene via physisorption with an energy of 0.42 eV/N_2 and is most energetically stable on the edge of the nanoribbon. This N_2 adsorption does not lead to significant deformation of the nanoribbon. O_2, on the other hand, interacts strongly with the pristine silicene nanoribbon with an adsorption energy of 2.96 eV/O_2. Upon relaxation, the O_2 molecule dissociates in favor of individual Si–O bonds. This large adsorption energy seems to indicate that pristine silicene nanoribbons would easily oxidize under ambient conditions.

To explore the effects of environmental gases on the conductance of the nanoribbon, the conductance was calculated before and after the adsorption of O_2 and N_2 molecules. The conduction curves depicted in the top panel of Fig. 11.11 [72] confirm the inert behavior of N_2 (blue curve), showing conductance nearly identical to that of the pristine nanoribbon. For oxygen, we see that conductance is significantly reduced (red curve) while the 0.09 eV band gap is preserved. These results indicate that although the CO sensing capability of silicene nanoribbons may diminish in an oxygen containing atmosphere, the capability is preserved in a nitrogen-containing atmosphere. The effects of CO_2 and H_2O adsorption were also investigated. The conductance results are shown in the middle panel of Fig. 11.5 and the structures are presented in the bottom panel.

For H_2O adsorption, the minimum energy configuration results from water splitting [73] and subsequent attachment of H and OH at the edge (with a binding energy of 1.62 eV) while CO_2 adsorbs via physisorption (with a binding energy of 0.46 eV). Similar to the case of oxygen adsorption, owing to the destructive effect on the nanoribbon structure, water molecules should also be removed from the

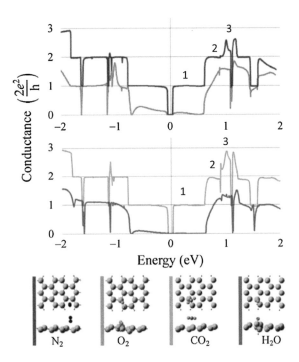

Fig. 11.11 Quantum conductance modulation resulting from environmental gas molecules adsorption on nanoribbon [72]: N_2 (*blue*), O_2 (*red*), CO_2 (*green*) and H_2O (*purple*). The middle of the gap is set at zero (Color figure online)

environment for proper CO sensor functionality. Overall, these results indicate that long silicene nanoribbons could provide a unique nano-sensor capable of single molecule resolution.

11.3 Adsorption of Alkali and Transition Metal Adatoms on Silicene

The adsorption of metal atoms on both monolayer and multi-layer silicene have been studied extensively over the past few years [24, 29, 32, 70, 74]. The electronic structures and physical and chemical properties associated with 2D materials, specifically the Dirac fermions that are now near-synonymous with graphene, have motivated studies to gain a fundamental understanding of and hence technological applications of silicene-based materials. Adsorption of metal clusters on 2D materials is expected to be one of the most promising ways to modify the electronic structure and in turn, to tune such properties in 2D materials like silicene. The various experimental and theoretical reports of silicene adsorbed on, or interacting with, metallic substrates also indicates that the interaction between metal clusters and silicene layers determines the structure and properties of the silicene. In this section, the progress of investigations examining adsorbed metal adatoms and clusters on silicene layers is discussed. In particular, the results are discussed in context of how their electronic structures and properties are affected. While the calculations are performed free-standing silicene, and not on silicene supported on a substrate, the findings yield important information that is likely to be similar for supported silicene.

The adsorption characteristics of metal adatoms on silicene have been simulated by first principles calculations [24, 29, 32]. The adsorption behavior is determined by the chemical activity of the adatoms, for example, alkali metals possess chemical activity that increases from Li to K. The characteristic bonding geometry of alkali atoms is depicted in Fig. 11.12. After full geometry optimization, all alkali atoms Li, Na, and K, favor bonding on the hollow site of the silicene layer. The adsorption of alkali atoms does not yield any significant distortion or stress on the silicene lattice. The valley site on the low-lying silicon atoms is the next favorable site. However, bridge site adsorption of alkali atoms is not possible on a silicene lattice.

Therefore, adsorption on a bridge site is a type of transition state between the top and valley sites. The band structure of silicene after adsorption of alkali atoms demonstrates typical metallic features. In contrast, alkaline-earth adatoms, such as Be and Mg, prefer to be adsorbed at valley sites, resulting in a band gap opening. Alkaline-earth metals have two valence electrons in their outermost orbital and compared to alkali metals have a smaller atomic size, higher melting point, higher ionization energy and larger effective charge. The strong interaction between alkaline-earth metals and silicon atoms favors silicon etching and surface engineering techniques. Therefore, one can expect strong bonding of alkaline-earth

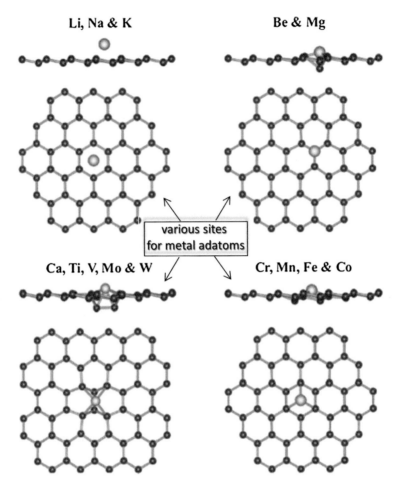

Fig. 11.12 Side and top view for characteristic adsorption geometries for alkali, alkaline-earth, and transition metal atoms [47]

atoms to a silicene surface. This hypothesis was also theoretically tested by Sahin and Peeters [29] by using the first principles calculation. Adsorption of transition metal adatoms that have $3d$, $4d$ and $5d$ electrons was systematically investigated in terms of their adsorption sites on silicene and their effects on the electronic structure of silicene. They included the elements Ti, V, Cr, Mn, Fe, Co, Mo, and W. Since the outermost s orbitals of these transition metals are fully filled, distinct adsorption behavior for different transition metal adatoms means that transition metal adatoms can easily bind to the silicene lattice. Nevertheless, it appears that the adsorption of alkali atoms does not induce any significant change in [29], only Cr, Mn, Fe, and Co, similar to alkalis, adsorb on the hollow site. However, in contrast to the alkali elements, the transition metals studied had quite strong binding (3.20, 3.48, 4.79, and 5.61 eV for Cr, Mn, Fe, and Co, respectively) to three uppermost Si atoms. Therefore, instead of being adsorbed on top of a hollow site like alkali metals, transition metals are almost confined within the silicene plane. Notably, bonding of Ti to silicene, with a binding energy of 4.89 eV, results in a significant lattice distortion whereas the adsorption of Ti occurs on the hollow site of graphene without disturbing the planar lattice structure. For group VI elements, such as Mo and W, adsorption on the bridge site becomes preferable due to their larger atomic radii. It can be concluded that the effect of the atomic radii can also be seen even for elements in the same row: adatoms Ti and V, which have a covalent radius larger than that of Cr, are relaxed to the bridge site. Thus, this indicates that only transition metal adatoms having a covalent atomic radii larger than ~1.50 Å favor bridge site adsorption [24, 29].

Several dimer adsorption configurations were also studied for transition metal elements. Seven possible configurations were proposed:

(1)   hollow site-hollow site [h-h],
(2)   atom site-hollow site [a-h],
(3)   atom site-hollow site [b-h],
(4)   vertically stacked at the hollow site [h-h*],
(5)   vertically stacked at the a-atom site [a-a*],
(6)   vertically stacked at the b-atom site [b-b*], and
(7)   adjacent and directly opposite atop bridge site with a hexagon [m-m].

It was found from the DFT calculations that only five configurations were stable, being the h-h, b-h, h-h*, b-b*, and m-m dimers. As shown in Fig. 11.13, the Ni h-h* configuration transforms to a b-h configuration after relaxation. Similarly, the Fe b-b* configuration relaxes to the b-h configuration. All a-h and a-a* dimers, just like the corresponding adatom adsorption, became b-h and b-b* dimers, respectively [75]. Co and Ni dimers adsorbed most strongly in the h-h configuration;

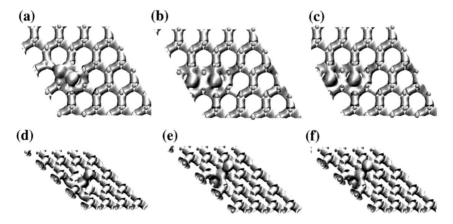

Fig. 11.13 Electron density isosurfaces of the most stable (**a–c**) and least stable (**d–f**) Fe, Co and Ni dimer adsorption configurations, respectively [75]. **a** b-h Fe. **b** h-h Co. **c** h-h Ni. **d** h-h* Fe. **e** b-b* Co. **f** b-b* Ni

while the Fe dimer preferred the b-h configuration. In the h-h adsorption configuration, each Fe atom has an electron configuration $3d^{7.1}4s^{0.4}$, however, it prefers $3d^{6.9}4s^{0.6}$ when the Fe atom adsorb in a b-h configuration. The lower average inter-configurational s-d transfer of 0.15 charge between the Fe atoms when adsorbed in the b-h configuration was sufficient to offset the energy released from being adsorbed in a higher coordination site on the underlying Si lattice than when adsorbed in the h-h configuration. It was found that the 3d-4s inter-configurational energy for Fe was greater than Co and Ni. Consequently, there was substantial electron density between the Fe atoms when adsorbed as a dimer in the b-h configuration while there was little electron density between the Co atoms when adsorbed as dimers in the h-h configuration. The Co and Ni dimer adsorption is, therefore, more akin to separate adatom adsorption. In contrast, the m-m dimer configuration is weakest in all cases. The reason for this is the strong metal–metal interaction breaks down the metal–Si coordination.

11.4 Electronic Structure Tuning of Silicene by Adsorption

Theoretical and experimental studies of silicene showed that the presence of various adatoms and dimers within the silicene structure can tune its electronic structure. For example the band gap in epitaxial silicene has been tuned by oxygen adatoms from a zero-gap-type to semiconductor-type material [47]. With a tunable band gap, specific electronic components could be made-to-order for applications that require specific band gaps. The band gap could be brought down to 0.1 eV, which is considerably smaller than the band gap (0.4 eV) found in traditional field effect transistors (FETs). This prospective aim makes modulation of the electronic structure in silicene extremely important.

The electronic structures of silicene layers with metal adatoms, in terms of spin-polarized electronic band dispersion and DOS, have been investigated by the first principles calculation. Because the alkaline-earth and transition metal adsorption significantly disturb the hexagonal lattice symmetry, electronic band dispersion along the high symmetry points of perfect silicene may not represent the real electronic properties of the whole structure. In order to overcome this drawback, a density of states plot covering a large number of high symmetry points in the Brillouin zone (BZ) is more convenient for a reliable description of the electronic structure. As shown in Fig. 11.14 [29], the band structures of alkali metals adsorbed on silicene show little difference to each other. In contrast to pure silicene, a gap of 6–31 meV opens up in all cases for alkaline elements adsorbed on silicene. As the chemical activity of the element increases, the doping level of the silicene increases from 0.2 to 0.25 and 0.27 eV for Li, Na and K adatoms, respectively. The Dirac cone is shifted from the K point to the Γ point in the BZ due to band folding of the silicene superlattice. As a result of adsorption of an alkali atom, semimetallic

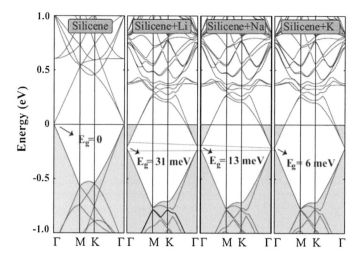

Fig. 11.14 Band structures for perfect, Li, Na, and K adsorbed silicene [29]. Fermi level is set to zero

silicene becomes metallic due to the electron doping (~ 0.8e charge) from the metal adatom to the silicene conduction band. The adatoms which are stabilized at hollow sites show a significant charge transfer that resembles ionic bonding. In addition, the band dispersion of pure silicene becomes distorted because nature. This is verified by increased the doping level. It is seen that all the alkali metal atom adsorption bands formed by the hybridization of adatom s states with the silicene p states appear in the vicinity of 0.4 eV [24, 29, 32].

Transition metal adatoms show different effects on the electronic structure of silicene, in comparison to alkaline adatoms. Since their d states have comparable energy values to that of their s states, the d shell of these atoms is partially filled. Although the d-shell electrons are located close to the nucleus, like the characteristic core electrons, they can spread out much further like valence electrons. Therefore, transition metal adatoms will relax to different sites on silicene and result in various electronic structures. The spin-polarized DOS of different transition metals adsorbed on silicene are shown in Fig. 11.15 [29].

Ti adsorption on silicene shows quite different characteristics compared to the other transition metals. The strong bonding to silicene results in a significant structural distortion. The Ti-d states are completely hybridized with Si atoms. The metallic bands of the structure originate only from Ti d orbitals. Meanwhile, silicene becomes a half-metal when Ti adatoms are adsorbed. V adsorption on silicene significantly contributes to the valence band maximum due to the V-d states. The Si-p states mainly contribute to the conduction band. Cr adsorption also turns the semimetallic silicene into a half-metallic material with similar characteristics to the V adatoms. However, different to Cr, the adsorption of Mo and W that occurs on bridge sites results in a nonmagnetic ground state. Mo-adsorbed silicene becomes metallic while W-adsorbed silicene is a nonmagnetic semiconductor with

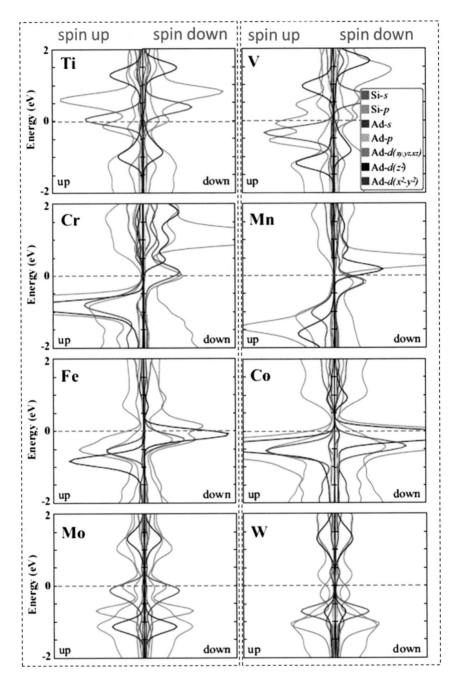

Fig. 11.15 Partial DOS for Ti, V, Cr, and Mn adsorbates and nearest silicon atoms. The Fermi level is set to zero [24]

a band gap of 0.02 eV. The adsorbate-induced magnetic moments in the silicene layer lead to an exchange-splitting especially in d eigenstates. The calculated value of the exchange splitting is 0.4, 2.0, 2.1, 1.9, 0.5 and 0.4 eV for Ti, V, Cr, Mn, Fe and Co, respectively.

It has been predicted that the band structure of silicene can be tailored to show different characteristics from a semimetal, to a semiconductor or insulator, by chemical functionalization and specifically by oxidation [55, 56]. STM and STS results on partially oxidized silicene layers on Ag(111) are shown in Fig. 11.16. This displays a series of spectra taken along lines cut across the oxygen adatoms on the three silicene superstructures. The magnitude of the gap shows significant variation corresponding to the different superstructures. Another common characteristic is that the gap is larger at oxygen adatom sites and becomes smaller in the locations away from the absorption sites. Despite that, the gap still exists at a lateral distance of 3 nm around the oxygen adatoms, which indicates that the oxygen adatoms could affect the electronic structure of silicene over a large area. Since the average distances between neighboring oxygen adatoms on silicene in each structure are less than 3 nm, this suggests that the gap is opened over the whole silicene surface due to adsorption of oxygen adatoms even with a low oxygen dose of 20 L. It should be noted that the oxygen adatoms do not show an ordered structure, which might lead to variations in the size of the gap value at different sites on the oxidized silicene surface.

In 4×4 silicene, the gap varies from 0.18 to 0.30 eV after an oxygen dose of 20 L. The most typical gap is about 0.18 eV, as shown in Fig. 11.16b, while the $\sqrt{13} \times \sqrt{13}$ and $2\sqrt{3} \times 2\sqrt{3}$ structures show a band gap of 0.11–0.14 and 0.15–0.18 eV, respectively, as shown in Fig. 11.16a, c. These values of the band gaps are in qualitative agreement with DFT calculations. Because pure silicene in each structure exhibits a characteristic semimetal zero gap, this clearly demonstrates that there is a band gap opening associated with the number of oxygen adatoms.

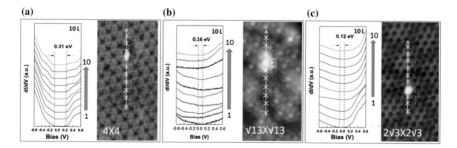

Fig. 11.16 Spatial evolution of the electronic states measured on **a** $2\sqrt{3} \times 2\sqrt{3}$, **b** 4×4, and **c** $\sqrt{13} \times \sqrt{13}$ silicene exposed under oxygen dose of 20 L [47]. Tunneling spectra (dI/dV curves) were obtained along a line denoted by the *arrows* in the corresponding STM topographic images on the *right*. The *dashed lines* in each STS result illustrate the value of band gap. STM images were obtained at $V_{bias} = -0.8$ V, I = 0.6 nA. The oxygen adatoms appear as bright protrusions on the silicene layers

The band gap is increased with increasing oxygen dose. The band gap of oxidized silicene is homogeneous when the oxygen dose is greater than 30 L. The band gaps are 0.18, 0.9, and 0.22 eV for the $\sqrt{13} \times \sqrt{13}$, 4×4, and $2\sqrt{3} \times 2\sqrt{3}$ structures for oxidation with an oxygen dose of 60 L, respectively. These values are well below the width of the semiconducting band gap in bulk silicon. While the gap opens homogeneously for oxidized silicene (oxygen dose >30 L), small differences in the occupied and unoccupied states can be observed, which are most likely due to the inhomogeneous local density of states induced by disordered oxygen adatoms. According to DFT simulations [56], the conduction band of partially oxidized silicene is mainly composed of the Si p-orbital and O p-orbital, and the valence band originates from the O p-orbital. The width of the band gap is predominantly influenced by the adsorption sites of the oxygen adatoms. Since the valence band of silicene oxide mainly originates from the p-orbital of O, the dangling bonds of TL Si in oxidized 4×4 silicene are fully saturated by oxygen adatoms, which results in the largest gap in the oxidized silicene among the three structures. Unpaired electrons in oxidized $\sqrt{13} \times \sqrt{13}$ and $2\sqrt{3} \times 2\sqrt{3}$ silicene layers, however, contribute to the narrow gap under low oxygen doses. So, by varying the oxygen dose, it is found that the band gaps are indeed tunable and dominated by the coverage of oxygen adatoms. In contrast to graphene [57], oxygen adatoms prefer to be accommodated at the surface of silicene rather than the edge, which is most likely because the dangling bonds on the Si edge atoms are passivated by the Ag(111) surface. The band gap can be modulated from semimetallic to semiconducting type, which can very well overcome the zero-gap disadvantage of silicene. In fully oxidized silicene, the buckled silicene structure vanishes, with subsequent crumpling of the sample and exposure of bare Ag(111) surface areas.

11.5 Summary and Future Prospects

Modification of the electronic structure of silicene though the adsorption of small molecules or adatoms has significant effects on the potential applications of silicene as nanoelectronics. The electronic structure of functionalized silicene is highly dependent on the type of adsorbed molecule or adatom. The adsorbate-silicene interactions can occur in a number of ways, either by modifying the local silicene buckling structures or through charge transfer to the silicene. The strong hybridization between adatoms and silicene can tune the material from a metal to a semimetal and further to a semiconductor. The multitude of studies of molecular adsorption on silicene are very useful for understanding the influence of the environment on silicene and pave the way for a better design of silicene-based devices.

Acknowledgement The authors would like to acknowledge the Australian Research Council (ARC) for the financial supports through Discovery Project (DP 140102581), LIEF grants (LE100100081 and LE110100099).

References

1. S. Cahangirov, M. Topsakal, E. Akturk, H. Sahin, S. Ciraci, Phys. Rev. Lett. **102**, 236804-4 (2009)
2. E. Durgun, S. Tongay, S. Ciraci, Phys. Rev. B **72**, 075420-10 (2005)
3. H. Nakano, T. Mitsuoka, M. Harada, K. Horibuchi, H. Ozaki, N. Takahashi, T. Nonaka, Y. Seno, H. Nakamura, Angew. Chem. Int. Ed. **45**, 6303–6306 (2006)
4. B. Aufray, A. Kara, S. Vizzini, H. Oughaddou, C. Leandri, B. Ealet, G. Le Lay, Appl. Phys. Lett. **96**, 183102 (2010)
5. P. Vogt, P. De Padova, C. Quaresima, J. Avila, E. Frantzeskakis, M.C. Asensio, A. Resta, B. Ealet, G. Le Lay, Phys. Rev. Lett. **108**, 155501–155505 (2012)
6. S. Cahangirov, V.O. Özçelik, L. Xian, J. Avila, S. Cho, M.C. Asensio, S. Ciraci, A. Rubio, Phys. Rev. B **90**, 035448-4 (2014)
7. S. Cahangirov, V.O. Özçelik, A. Rubio, S. Ciraci, Phys. Rev. B **90**, 085426 (2014)
8. O.V. Yazyev, S.G. Louie, Phys. Rev. B **81**, 195420–195427 (2010)
9. A. Hashimoto, K. Suenaga, A. Gloter, K. Urita, S. Iijima, Nature **430**, 870–873 (2004)
10. A.V. Krasheninnikov, F. Banhart, Nat. Mater. **6**, 723 (2007)
11. V.H. Crespi, L.X. Benedict, M.L. Cohen, S.G. Louie, Phys. Rev. B **53**, R1330 (1996)
12. Y. Kim, J. Ihm, E. Yoon, G.D. Lee, Phys. Rev. B **84**, 075445-5 (2011)
13. R. Singh, P. Kroll, J. Phys.: Cond. Matter **21**, 196002–196007 (2009)
14. R. Faccio, L.F. Werner, H. Pardo, C. Goyenola, O.N. Ventura, A.W. Mombru, J. Phys. Chem. C **114**, 18961–18971 (2010)
15. S. Dag, Y. Ozturk, S. Ciraci, T. Yildirim, Phys. Rev. B **72**, 155404–155408 (2005)
16. T. Yildirim, S. Ciraci, Phys. Rev. Lett. **94**, 175501–175504 (2005)
17. E. Durgun, S. Ciraci, W. Zhou, T. Yildirim, Phys. Rev. Lett. **97**, 226102–226104 (2006)
18. C. Ataca, E. Akturk, S. Ciraci, H. Ustunel, Appl. Phys. Lett. **93**, 043123 (2008)
19. C.C. Liu, W. Feng, Y. Yao, Phys. Rev. Lett. **107**, 076802 (2011)
20. D.A. Dikin, S. Stankovich, E.J. Zimney, R.D. Piner, G.H. Dommett, G. Evmenenko, S.T. Nguyen, R.S. Ruoff, Nature **448**, 457–460 (2007)
21. Z. Wei et al., Science **328**, 1373–1376 (2010)
22. M. Topsakal, H.H. Gurel, S. Ciraci, J. Phys. Chem. C **117**, 5943 (2013)
23. H. Hakan Gurel, V. Ongun Ozçelik, S. Ciraci, J. Phys. Chem. C **118**, 27574 (2014)
24. X. Lin, J. Ni, Phys. Rev. B **86**, 075440 (2012)
25. R. Quhe, R. Fei, Q. Liu, J. Zheng, H. Li, C. Xu, Z. Ni, Y. Wang, D. Yu, Z. Gao, J. Lu, Sci. Rep. **2**, 85 (2012)
26. T.H. Osborn, A.A. Farajian, J. Phys. Chem. C **116**, 22916 (2012)
27. N. Gao, W.T. Zheng, Q. Jiang, Phys. Chem. Chem. Phys. **14**, 257 (2012)
28. J. Sivek, H. Sahin, B. Partoens, F.M. Peeters, Phys. Rev. B **87**, 085444 (2013)
29. H. Sahin, F.M. Peeters, Phys. Rev. B **87**, 085423 (2013)
30. R. Friedlein, A. Fleurence, J.T. Sadowski, Y. Yamada-Takamura, Appl. Phys. Lett. **102**, 221603 (2013)
31. F. Zheng, C. Zhang, S. Yanb, F. Li, J. Mater. Chem. C **1**, 2735 (2013)
32. G.A. Tritsaris, E. Kaxiras, S. Meng, E. Wang, Nano Lett. **13**, 2258 (2013)
33. J. Wang, J. Li, S.-S. Li, Y. Liu, J. Appl. Phys. **114**, 124309 (2013)
34. C. Li, S. Yang, S.-S. Li, J.-B. Xia, J. Li, J. Phys. Chem. C **117**, 483 (2013)
35. B. Huang, H.J. Xiang, S.-H. Wei, Phys. Rev. Lett. **111**, 145502 (2013)
36. A. Kara, H. Enriquez, A.P. Seitsonen, L.C.L.Y. Voon, S. Vizzini, B. Aufray, H. Oughaddou, Sci. Rep **67**, 1 (2012)
37. W. Hu, N. Xia, X. Wu, Z. Li, J. Yang. Phys. Chem. Chem. Phys. **16**, 6957–6962 (2014)
38. J.O. Sofo, A.S. Chaudhari, G.D. Barber, Phys. Rev. B **75**, 153401 (2007)
39. D.C. Elias, R.R. Nair, T.M.G. Mohiuddin, S.V. Morozov, P. Blake, M.P. Halsall, A.C. Ferrari, D.W. Boukhvalov, M.I. Katsnelson, A.K. Geim, K.S. Novoselov, Science **323**, 610 (2009)

40. R. Balog, B. Jørgensen, L. Nilsson, M. Andersen, E. Rienks, M. Bianchi, M. Fanctti, E. Lægsgaard, A. Baraldi, S. Lizzit, Z. Sljivancanin, F. Besenbacher, B. Hammer, T.G. Pedersen, P. Hofmann, L. Hornekær, Nat. Mater. **9**, 315 (2010)
41. L.C. Lew Yan Voon, E. Sandberg, R.S. Aga, A.A. Farajian, Appl. Phys. Lett. **97**, 163114 (2010)
42. F.B. Zheng, C.W. Zhang, Nanoscale Res. Lett. **7**, 422 (2012)
43. B. Huang, H.X. Deng, H. Lee, M. Yoon, B.G. Sumpter, F. Liu, S.C. Smith, S.H. Wei, Phys. Rev. X **4**, 021029 (2014)
44. J. Qiu, H. Fu, Y. Xu, A.I. Oreshkin, T. Shao, H. Li, S. Meng, L. Chen, K. Wu, Phys. Rev. Lett. **114**, 126101 (2015)
45. W.C. Wu, Z.M. Ao, C.H. Yang, S. Li, G.X. Wang, C.M. Lie, S. Lia, J. Mater. Chem. C **3**, 2593–2602 (2015)
46. X. Xu, J. Zhuang, Y. Du, H. Feng, N. Zhang, C. Liu, T. Lei, J. Wang, M. Spencer, T. Morishit, X. Wang, S.X. Dou, Sci. Rep. **4**, 7543 (2014)
47. Y. Du, J. Zhuang, H. Liu, X. Xu, S. Eilers, K. Wu, P. Cheng, J. Zhao, X. Pi, K.W. See, G. Peleckis, X. Wang, S.X. Dou, ACS Nano **8**, 10019–10025 (2014)
48. M.F. Crommie, C.P. Lutz, D.M. Eigler, Science **262**, 218–220 (1993)
49. J.A. Yan et al., Phys. Rev. B **88**, 121403 (2013)
50. E. Scalise et al., Nano Res. **6**, 19–28 (2013)
51. D. Tsoutsou et al., Appl. Phys. Lett. **103**, 231604 (2013)
52. T. Andreev, I. Barke, H. Hvel, Phys. Rev. B **70**, 205426 (2004)
53. A. Molle et al., Adv. Funct. Mater. **23**, 4340–4344 (2013)
54. F.J. Himpsel et al., Phys. Rev. B **38**, 6084 (1988)
55. D. Chiappe, C. Grazianetti, G. Tallarida, M. Fanciulli, A. Molle, Adv. Mater. **24**, 5088–5093 (2012)
56. R. Wang, X. Pi, Z. Ni, Y. Liu, S. Lin, M. Xu, D. Yang, Sci. Rep. **3**, 3507 (2013)
57. Z. Liu, K. Suenaga, P.J.F. Harris, S. Iijima, Phys. Rev. Lett. **102**, 015501 (2009)
58. G. Kresse, J. Hafner, Phys. Rev. B **47**, 558 (1993)
59. J.P. Perdew, K. Burke, M. Ernzerhof, Phys. Rev. Lett. **77**, 3865 (1996)
60. Y. Baskin, L. Mayer, Phys. Rev. **100**, 544 (1955)
61. R. Zacharia, H. Ulbricht, T. Hertel, Phys. Rev. B **69**, 155406 (2004)
62. R.E. Mapasha, A.M. Ukpong, N. Chetty, Phys. Rev. B **85**, 205402 (2012)
63. W. Hu, Z. Li, J. Yang, J. Chem. Phys. **138**, 054701 (2013)
64. G. Makov, M.C. Payne, Phys. Rev. B **51**, 4014 (1995)
65. P. Lambin, H. Amara, F. Ducastelle, L. Henrard, Phys. Rev. B **86**, 045448 (2012)
66. Y.-C. Zhou, H.-L. Zhang, W.-Q. Deng, Nanotechnology **24**, 225705 (2013)
67. M. Farjam, H. Rafii-Tabar, Phys. Rev. B **79**, 045417 (2009)
68. R. Quhe, R. Fei, Q. Liu, J. Zheng, H. Li, C. Xu, Z. Ni, Y. Wang, D. Yu, Z. Gao, J. Lu, Sci. Rep. **2**, 85 (2012)
69. N.D. Drummond, V. Zolyomi, V.I. Falko, Phys. Rev. B **85**, 075423 (2012)
70. S. Li, C. Zhang, W. Ji, F. Li, P. Wang, S. Hu, S. Yan, Y. Liu, Phys. Chem. Chem. Phys. **16**, 15968 (2014)
71. O. Leenaerts, B. Partoens, F.M. Peeters, Phys. Rev. B **77**, 125416 (2008)
72. T.H. Osborn, A.A. Farajian, Nano Res. **7**, 945–952 (2014)
73. R. Konečný, D.J. Doren, J. Chem. Phys. **106**, 2426–2435 (1997)
74. V. Bui, T. Pham, H.S. Nguyen, H.M. Le, J. Phys. Chem. C **117**, 23364 (2013)
75. H. Johll, M.D.K. Lee, S.P.N. Ng, H.C. Kang, E.S. Tok, Sci. Rep. **4**, 7594 (2014)

Chapter 12
Epitaxial Silicene: Beyond Silicene on Silver Substrates

Antoine Fleurence

Abstract The growth and physical properties of epitaxial forms of silicene formed on several conductive substrates such as $ZrB_2(0001)$, $ZrC(111)$, $Ir(111)$ and $Au(110)$ are described in detail in this chapter. The exploratory investigation of the structure and electronic properties of two-dimensional forms of silicon on conductive substrates by ab initio calculations is also addressed.

12.1 Introduction

Whereas the pioneering observation of silicene on Ag(110) and Ag(111) surfaces demonstrates that a graphitic form of silicon can exist, the experimental observation of silicene on other substrates provides evidence that the stability of silicene is not specific to silver surfaces. However, the number of substrates on which silicene was reported to form remains limited, and all of them are $_2$made of conductive materials. The first section of this chapter is dedicated to the particular form of silicene that spontaneously crystallizes on the surface of ZrB(0001) thin films grown on Si(111). In the second section of this chapter, the growth, by silicon deposition, of two-dimensional silicon layers on Ir(111), ZrC(111) and Au(110) will be presented. Section 12.4 will give an overview of the ab initio calculation works aimed at finding other templates capable of stabilizing silicene.

A. Fleurence (✉)
Japan Advanced Institute of Science and Technology, 1-1 Asahi-dai, Nomi, Ishikawa 923-1292, Japan
e-mail: antoine@jaist.ac.jp

© Springer International Publishing Switzerland 2016
M.J.S. Spencer and T. Morishita (eds.), *Silicene*, Springer Series in Materials Science 235, DOI 10.1007/978-3-319-28344-9_12

12.2 Epitaxial Silicene on ZrB$_2$(0001)

In this first section, the state-of-the-art of the research on silicene on ZrB$_2$(0001) will be presented in detail. The structural and electronic properties of the single form of silicene that does not need to be grown, were investigated thoroughly by a wide range of experimental techniques and with the support of theoretical calculations. Owing to the ease of preparation and high reproducibility, silicene spontaneously formed on ZrB$_2$(0001) thin film on Si(111) is a perfect benchmark to study the chemical properties of silicene or its functionalization by adsorption of foreign species.

12.2.1 ZrB$_2$: Bulk and Surface Properties

Transition-metal diborides are a conductive material with high stiffness and high melting temperature used for instance as a hard coating layers, thin film resistors or diffusion barriers in microelectronic devices. These materials generally crystallize in the AlB$_2$-type simple hexagonal structure (space group P6/mmm) made of stacking along the c-axis of close-packed metal layers (for instance Mg, Nb, Hf, Ta, Ti, Zr, etc.) and B honeycomb layers (Fig. 12.1a). The bonds between boron atoms have a strong covalent character and those in the metal layers are metallic. The bonding between the metal and boron layer is of both ionic and covalent nature. For

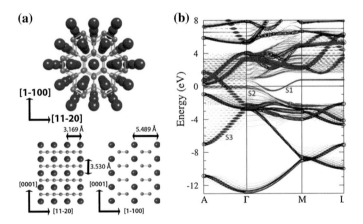

Fig. 12.1 Zirconium diboride (ZrB$_2$). **a** Crystal and **b** computed band structure. Zr and B atoms are *red-* and *blue-colored*, respectively. The contribution of the outermost Zr atoms is *red-colored* and the total contribution of B and Zr atoms is *black-colored*. Circle radius is proportional to the spectral weight. S1–S3 denotes the surface states of the bulk ZrB$_2$(0001) surface. **a** Reproduced from [21] with permission from John Wiley and sons. **b** © IOP Publishing. Reproduced with permission from [3]. All rights reserved

zirconium diboride (ZrB_2), the a-axis lattice parameter is 3.169 Å and that of the c-axis is 3.530 Å. The calculation of the band structure of ZrB_2 [1–3] (Fig. 12.1b) shows that the Fermi level lies in a depression in the density of states (DOS) conferring a semi-metal character to ZrB_2 [1, 4]. At the Fermi level, the density of states is dominated by the contribution of Zr $4d$ orbitals [4, 5].

The growth of ZrB_2 single-crystals by an RF-floating zone method is possible [6, 7] and although limited by the high melting temperature (3220 °C) of ZrB_2 [8] a mirror-polished oxide-free $ZrB_2(0001)$ substrate can be produced [9, 10]. As for other group 5-transition metal diborides, the (0001) surface is metal-terminated and has the same periodicity as that of the bulk [1, 11]. The band structure of (0001)-oriented slabs features surface bands [1–3] originating predominantly from the $4d$, $5s$ and $5p$ orbitals of the Zr topmost atoms [1].

12.2.2 ZrB₂ Thin Film Growth on Si(111)

As their respective a-axis lattice parameters and thermal expansion coefficients [12] are very close to each other, ZrB_2 is a perfect template for the growth of group III-nitrides [13], which are wide band gap semiconductors used for instance in ultraviolet light emitting diodes (LEDs). The growth of ZrB_2 buffer layers on silicon substrates opened the way to the integration of group III-nitride with silicon-based technologies. A ZrB_2 buffer layer is a perfect growth template for a number of reasons; it is an inter-diffusion barrier; it is an optical reflector preventing the adsorption of the ultraviolet light; it is a back contact allowing for the high efficiency vertical LED design [14–17].

ZrB_2 thin films can be grown on Si(111) (but also on other substrates such as $Al_2O_3(0001)$ [18]) by ultra high vacuum—chemical vapour epitaxy (UHC-CVE) using the thermal decomposition of zirconium borohydride ($Zr[BH_4]_4$) following the reaction: $Zr(BH_4)_4$ $_{(g)}$ → B_2H_6 $_{(g)}$ + $5H_2$ $_{(g)}$ + ZrB_2 $_{(s)}$ [19]. The exposure to Zr $(BH_4)_4$ of clean and oxide-free Si(111) held at temperatures in the 900–1000 °C range induces the nucleation of crystalline epitaxial ZrB_2 islands which then progressively coalesce until formation of a continuous thin film [20]. The epitaxial relationship of the thin film with the substrate is predominantly $ZrB_2(0001)//Si(111)$ and $ZrB_2[11\bar{2}0]//Si[1-10]$ owing to the 5:6 magic mismatch between the $ZrB_2(0001)$ and Si(111) lattices [20]. Cross-sectional transmission electron microscopy (XTEM) shows an abrupt interface between the thin film and the substrate [20] (Fig. 12.2a).

Misoriented ZrB_2 crystallites are locally observed [21], but their parasitic growth can be kinetically hindered [22]. The $ZrB_2(0001)$ thin film is strained as the a-axis lattice parameter is slightly larger (3.186 Å) than that of the bulk [23, 24].

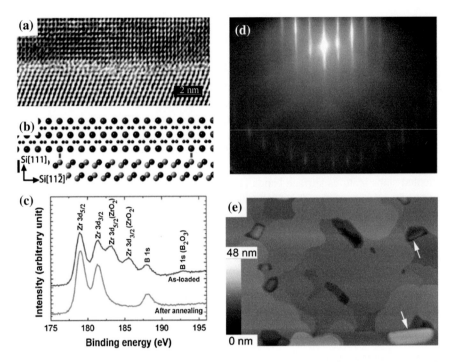

Fig. 12.2 ZrB$_2$ thin film on Si(111). **a** XTEM image of the interface. **b** Sketch of the 5:6 magic mismatch at the interface between the ZrB$_2$(0001) and Si(111) lattices. **c** XPS spectra in B *1s* and Zr *3d* regions recorded with Al Kα before and after annealing in UHV of a ZrB$_2$//Si(111) sample exposed to ambient air. The different valences states of oxidized Zr and B are indicated. **d** RHEED pattern recorded with the electron beam parallel to ZrB$_2$⟨11$\bar{2}$0⟩ of the oxide-free ZrB$_2$ surface. **e** STM image (1000 nm × 700 nm, $V = 1$ V, $I = 48$ pA). **a** Reprinted from [20] with permission from Elsevier, **b** and **e** Reproduced from [22] with permission from Elsevier, **c** and **d** Adapted with permission from [17]. Copyrighted by the American Physical Society (Color figure online)

12.2.3 Spontaneous Segregation of Silicon on Oxide-Free ZrB$_2$(0001) Thin Films Grown on Si(111)

X-ray photoemission spectroscopy (XPS) (Fig. 12.2c) shows that exposing the ZrB$_2$(0001) surface to ambient air leads to the formation of a mixed B and Zr oxide layer which can be removed by annealing at 800 °C [17]. The resulting oxide-free surface is (2 × 2)-reconstructed as shown on the reflecting high energy electron diffraction (RHEED) pattern of Fig. 12.2d. This (2 × 2) reconstruction turns reversibly into (1 × 1) when the temperature is raised above 660 °C [17]. On large scale scanning tunneling microscopy (STM) images, the surface appears atomically flat on few hundred nanometer-wide terraces (Fig. 12.2e) [17]. The ZrB$_2$(0001)- (2 × 2) reconstruction, that is not observed in the case of the bulk material, is associated with the presence of silicon atoms segregating on the surface, as

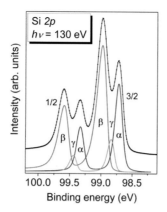

Fig. 12.3 Surface-sensitive core-level Si *2p* PES spectrum recorded in normal emission. Its decomposition into three components labeled α, β and γ is shown. Reprinted with permission from [26]. Copyright 2014, AIP Publishing LLC (Color figure online)

evidenced by the core-level Si *2p* photoemission spectroscopy (PES) spectrum recorded with a surface-sensitive photon energy hv = 130 eV (Fig. 12.3) [25, 26].

The Si *2p* PES spectrum can be decomposed into three well-defined spin–orbit split doublets [25, 26] giving evidence for the ordered character of the segregated silicon layer. They are labeled α, β and γ and their respective relative weights are 32, 55, and 13 %. The binding energies of the α and γ components are respectively shifted by 260 and 140 meV with respect to the prominent β component. The Si *2p* spectrum of the segregating Si layer is shifted by about 700 meV towards a lower binding energy with respect to that of bulk silicon [26]. The work function of the Si-rich $ZrB_2(0001)$ surface, determined to be 4.61 eV [24] is very close to that of bulk $ZrB_2(0001)$ (4.58 eV) [27].

12.2.4 Silicon Deposition on Bulk $ZrB_2(0001)$

The $ZrB_2(0001)$-(2×2) reconstruction can also be realized by depositing silicon on $ZrB_2(0001)$ bulk susbtrates [28]. The (2×2) structures formed by deposition of silicon and resulting from the spontaneous segregation of Si atoms is suggested by the reversible (2×2) to (1×1) phase transition occurring at approximately the same temperature [17, 28].

For growth temperatures high enough, the amount of silicon on the surface saturates once the coverage corresponding to the completion of the (2×2) reconstruction is reached. It suggests a steep decrease in the sticking probability at the saturation coverage. Any excess Si deposited at room temperature beyond the saturation coverage is easily desorbed by subsequent heating.

12.2.5 Structure of Silicene on ZrB₂(0001)

12.2.5.1 Large-Scale Domain Structure

STM images show that the silicon layer segregating on $ZrB_2(0001)$ thin films grown on Si(111) is textured with one-dimensional arrays of stripe-shaped (2 × 2)-reconstructed domains [25]. The domains are aligned along one of the three equivalent $\langle 11\bar{2}0 \rangle$ directions of $ZrB_2(0001)$ and the periodicity along the perpendicular $\langle 1\bar{1}00 \rangle$ direction is 2.7 nm (Fig. 12.4a). The (2 × 2) reconstruction appears as an array of protrusions with a single protrusion per unit cell. The domain boundaries result from shifts by about one unit cell of $ZrB_2(0001)$ along the two $\langle 11\bar{2}0 \rangle$ directions rotated by 120° and 240° with respect to the axis of the domains. The direction of the shift between domains alternates, further pointing out the large-scale ordering of the spontaneously segregating silicon layer. In contrast to epitaxial silicene on Ag(111) where different phases with different orientations and epitaxial relationships may coexist [29–31], the (2 × 2)-reconstructed structure solely covers the entire $ZrB_2(0001)$ surface. The silicon layer grown on bulk $ZrB_2(0001)$ also features this large-scale ordering as evidenced by the splitting of the (2 × 2) streaks in the RHEED pattern [28].

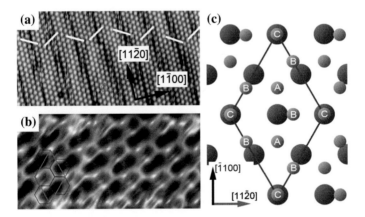

Fig. 12.4 Silicon honeycomb lattice on $ZrB_2(0001)$. **a** STM image (20 nm × 5 nm, $I = 55$ pA, $V = 700$ mV) of the silicene domain structure. The *arrows* indicate the crystallographic direction of the $ZrB_2(0001)$ surface. *White lines* show the direction of the shift between domains at the boundaries. **b** STM image (4.2 nm × 2 nm, $I = 600$ pA, $V = 100$ mV) of the silicon honeycomb structure. The ($\sqrt{3} \times \sqrt{3}$) unit cell and the honeycomb structure are emphasized. **c** Model of the interface between silicene (*light gray*) atoms and the uppermost Zr atoms (*dark gray/red* online). *A*, *B* and *C* denote the atoms sitting on hollow, bridge and on-top sites of the Zr layer, respectively. **a**, **b** and **c** Modified from [25] (Color figure online)

12.2.5.2 A Si Honeycomb Lattice on a Zr-Terminated ZrB₂(0001) Surface

In the higher resolution STM image (Fig. 12.4b), a finer structure is visible between the protrusions. It can be identified as a honeycomb structure whose lattice parameter (3.68 Å) is in such a way that the ($\sqrt{3} \times \sqrt{3}$) unit cell of this honeycomb lattice is commensurate with the (2 × 2) unit cell of $ZrB_2(0001)$ [25].

The observation by angle-resolved ultraviolet photoemission spectroscopy (ARPES) (see Sect. 12.2.6.1) of intact $ZrB_2(0001)$ surface bands, gives evidence for the crystallization of the silicon honeycomb structure on a bulk-like $ZrB_2(0001)$ surface with a relatively weak interaction between the silicon layer and the substrate. The sole possible positioning of the silicon atoms capable of reproducing the ratios 2:3:1 of the α, β and γ components of the Si $2p$ spectrum is depicted in Fig. 12.4c. Among the six Si atoms of the ($\sqrt{3} \times \sqrt{3}$) unit cell of the Si honeycomb lattice, two atoms are sitting on hollow sites of the Zr topmost layer, three are sitting near the bridge sites and one single atom is sitting on top of a Zr atom. The α, β and γ components of the Si $2p$ spectrum are thus assigned to the atoms in the three distinct environments, which are labeled A-, B- and C-site atoms. In this structure the lattice parameter of the silicon honeycomb structure is smaller than that expected for free-standing silicene (in the 3.8–3.9 Å range) [32–34].

12.2.5.3 Insights into the Structure of Epitaxial Silicene on ZrB₂(0001) by PES

Due to diffraction effect of the photoelectron, the ratio between the intensities of the different components in the PES spectra recorded with a given photon energy may not reflect the exact quantities of the atoms they originate from. To determine more accurately the real weights of the different components identified in the Si $2p$ spectrum of Fig. 12.3, PES spectra were recorded at several photon energies ranging from 120 to 240 eV [26]. The relative normalized weights of the α, β and γ components (Fig. 12.5a) are oscillating around average ratios (2.3:3.4:0.3) deviating from those expected for the model of the Fig. 12.4c (2:3:1). This deviation can be attributed to a small misfit between the Si honeycomb lattice and the $ZrB_2(0001)$ surface in the direction perpendicular to the ribbon-shaped domains. As a consequence of this, the slight displacement of the C-site atoms away from the on-top position, results in the under-representation of the γ component corresponding to those atoms.

The buckling of the silicon structure was investigated by photoelectron diffraction (PhD) using a surface-sensitive photon energy of 130 eV [25]. As shown in Fig. 12.5b, a decay of the α component with respect to the β component occurs when the electron analyzer is tilted away from the direction of normal emission towards the direction of the Si–Si bonds. In contrast, no change occurs when the perpendicular direction is probed. This anisotropic behavior is attributed to the buckling of the silicene structure. The comparison of the α/β and γ/β ratios suggests

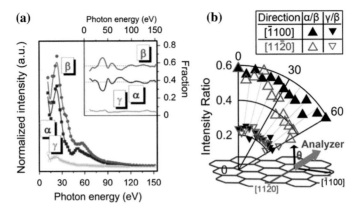

Fig. 12.5 Insights on the structure of silicene on $ZrB_2(0001)$ from PES. **a** Intensities of the α, β and γ components of the Si $2p$ spectrum as a function of the photon energy. The *inset* shows the variation of the relative weights of the three components. **b** Intensity ratios α/β and γ/β as a function of the polar photoelectron emission angle. The photon energy was set to 130 eV. **a** Reprinted with permission from [26]. Copyright (2014), AIP Publishing LLC. **b** Modified from [25] (Color figure online)

a structural difference between the corresponding A and C atoms. Although those atoms belong to the same sub-lattice, they may be sitting at different heights with respect to the topmost Zr atoms.

12.2.5.4 Structure of Epitaxial Silicene on $ZrB_2(0001)$ from DFT Calculations

Density functional theory (DFT) calculations were performed within the generalized gradient approximation (GGA) with the OpenMX code [35] based on norm-conserving pseudopotentials generated with multi-reference energies and pseudoatomic basis functions.

The binding energies of a single Si atom adsorbed in on-top, hollow and bridge sites are 5.05, 6.08 and 5.79 eV, respectively and the distances of the Si atom to the topmost Zr surface are respectively 2.57, 1.97 and 2.07 Å [36]. Hollow sites are thus the most favorable adsorption sites whereas the on-top sites are the least favorable. In light of the site-dependence of the binding energies for a single atom, the positions of the Si atoms within the $(\sqrt{3} \times \sqrt{3})$ unit cell appears as a result of the tradeoff between the energy cost represented by the atoms sitting on the on-top sites (C-site atoms) and the gain of energy represented by the atoms sitting on the hollow sites (A-site atoms) and in close proximity to the bridge sites (B-site atoms).

Two different structures (Fig. 12.6) were found to be stable in the calculation [25, 36]. They differ from each other by their buckling and by the length of the Si–Si bonds. The most stable structure (Fig. 12.6a) resembles that found for the $(\sqrt{3} \times \sqrt{3})$-reconstructed silicene phase on Ag(111) [37] and for other epitaxial forms of silicene (described in Sect. 12.3). The binding energy per Si atom, 1.14 eV, is

Fig. 12.6 Computed structures of epitaxial silicene on $ZrB_2(0001)$. **a** Planar-like and **b** regularly-buckled-like structures. Adapted with permission from [38]. Copyrighted by the American Physical Society

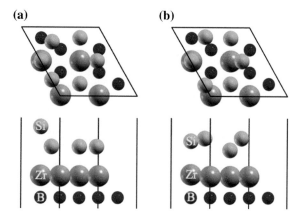

much smaller than that of a single Si atom due to the strong binding between Si atoms. This "planar-like" structure has all but one Si atoms sitting at almost the same height (2.3 Å) above the topmost Zr atoms. The Si atom sitting above a Zr atom, is protruding at a height of 3.9 Å in agreement with the unfavorable character of this position.

A metastable structure (Fig. 12.6b), with a binding energy per Si atom of 0.5 eV higher, was also found. This so-called regularly-buckled-like structure deviates from that of free-standing silicene by a larger height difference between the two sub-lattices (0.9 Å instead of 0.44 Å) and by the z-position of the atom sitting on top of a Zr atom (C-site), which is at a higher position than the A-site atoms belonging to the same sub-lattice (2.68 Å instead of 1.96 Å). Of note, the two structures also differ by their bond lengths, which are 2.31–2.36 Å in the planar-like structure and approximately 2.25 Å in the regularly-buckled-like structure [36].

12.2.5.5 Phonon Dispersion of Epitaxial Silicene on Zrb₂(0001)

The phonon dispersion of epitaxial silicene formed by deposition of silicon on $ZrB_2(0001)$ was measured using high resolution electron loss spectroscopy (HREELS) along both the $\Gamma-K$ and $\Gamma-M$ directions (Fig. 12.7a). Modes in the 90–100 meV range and in the 60–70 meV range were assignable to the boron modes of the substrate from comparison with the phonon dispersion of bulk $ZrB_2(0001)$ [9]. In contrast, the modes between 43 and 60 meV, can be attributed to silicene phonon modes.

The phonon dispersion of the planar-like structure was calculated using two different methods [28, 39]. The resulting dispersions (Fig. 12.7b) do not feature any imaginary frequencies, pointing out the overall stability of this epitaxial form of silicene. Except at energies around 40 meV, the phonon modes of silicene are decoupled from those of the $ZrB_2(0001)$ slab. The phonon partial density of states (PDOS) in the silicene layer shows that the vertical modes have frequencies between 15 and 30 meV, and the longitudinal modes lie in the 45–60 meV range [28]. At lower energies, the vibration mode have both characters.

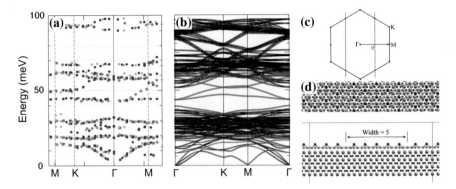

Fig. 12.7 Phonon dispersion in epitaxial silicene on ZrB$_2$(0001). **a** Phonon dispersion measured by HREELS represented in the Brillouin zone of the silicene-($\sqrt{3} \times \sqrt{3}$) unit cell. The different colors correspond to different primary electron energies ranging from 9 to 27 eV. **b** Computed phonon dispersion of planar-like silicene. The contributions of silicene and that of the ZrB$_2$ substrate are presented by *red* and *black circles*, respectively. The diameters of *circles* represent the absolute squares of the atomic components in the eigenvectors. **c** Brillouin zone of the silicene-($\sqrt{3} \times \sqrt{3}$) unit cell. **d** Top and side view of fully relaxed atomic positions in the domain structure. Protruding on-top Si atoms are *red*-colored for the sake of readability and the domain width is indicated. **a** © IOP Publishing. Reproduced with permission from [28]. All rights reserved. **b** Adapted with permission from [39]. Copyrighted by the American Physical Society

The calculated dispersions [28, 39] are in agreement with each other and with the experimental results except for the very soft mode found by Lee et al. [39] around the *M* point of the Brillouin zone. This acoustic branch purely originating from Si atoms approaches the "zero" frequency and thus represents a singular point of phonon instability that could cause a divergent response. A possible way to avoid this instability is to modify the periodicity of the system in order to reduce the size of the Brillouin zone in such a way that all *M* points of the Brillouin zone of the ($\sqrt{3} \times \sqrt{3}$) unit cell are excluded. The domain-structure of Fig. 12.7d, in good agreement with the large-scale STM image of Fig. 12.4a, complies with this criterion. It thus might be suggested that the periodic domain structure originates from a phonon instability of epitaxial silicene on ZrB$_2$(0001) [39].

12.2.6 Electronic Properties of Epitaxial Silicene on ZrB$_2$(0001)

12.2.6.1 Electronic Bands Structure

The valence band structure of epitaxial silicene was resolved by ARPES [25, 38]. The dispersion along the Γ-K_{Si}, where K_{Si} is the *K* point of the Brillouin zone of unreconstructed silicene (Fig. 12.8a, b), features bands labeled S$_1$ and S$_2$ which can

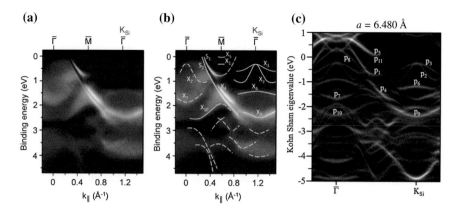

Fig. 12.8 Electronic band structure of epitaxial silicene on $ZrB_2(0001)$. **a** and **b** ARPES spectrum with and without guides for the eye. Photon energy of hv = 43 eV was used. **c** Computed band structure of the planar-like structure. The K point of the Brillouin zone of the unreconstructed unit cell (K_{Si}) and the Γ and M points of the Brillouin zone of the ($\sqrt{3} \times \sqrt{3}$) unit cell are indicated. Adapted with permission from [38]. Copyrighted by the American Physical Society

be identified as the surface states of bulk $ZrB_2(0001)$ [1–3] (see Fig. 12.1b). Several other features, not found in the band structure of bare $ZrB_2(0001)$ are thus attributed to the presence of silicene. The preeminent feature labeled X_2 centered on K_{Si} is reminiscent of the Dirac cone of free-standing silicene [32–34]. Due to back-folding of the bands into the first Brillouin zone of the ($\sqrt{3} \times \sqrt{3}$)-reconstructed unit cell, the band X_2 is mirrored as a feature labeled X'_2 centered on the Γ point.

The ARPES spectrum was compared with the band structure computed for the planar-like structure of epitaxial silicene on $ZrB_2(0001)$ [36, 38]. The translational symmetry is broken by the epitaxial conditions that give rise to the structure of the ($\sqrt{3} \times \sqrt{3}$) reconstruction. However the symmetry breaking might not be as strong as it would be in the results of the DFT calculations performed on a ($\sqrt{3} \times \sqrt{3}$) unit cell. In order to reflect the strength of the translational symmetry breaking as it really appears in ARPES spectra, the computed spectral weight, derived from the imaginary part of the one-particle Kohn-Sham Green's function, is unfolded to a larger Brillouin zone [3, 38, 40].

The best agreement between the ARPES spectrum and calculated band structure (Fig. 12.8c) was found for a lattice parameter of the planar-like structure of 6.48 Å (we note that the GGA functional is known to possibly give as much as up to 2 % error in the lattice constants). The prominent spectral features X_2 and X_3 centered on the K_{Si} point are particularly well reproduced by features p_2 and p_3. Similarly, the band X_5 may be related to the feature p_5 even though the latter is above the Fermi level. Features labeled p_{11}, p_4, p_6, p_7, p_1, p_8 and p_9 find their respective counterparts in the experimental spectrum as bands labeled X_1, X_4, X_6, X_7, S_1, S_2 and X_9.

Although the Γ-K_{Si} direction is perpendicular to the domain direction, it is noteworthy that the domain structure does not need to be taken into account to reproduce the experimentally measured band structure. It is thus concluded that the effect of the large-scale texturation on the band structure is negligible.

12.2.6.2 Orbital Character of the Electronic Bands

The good agreement between experimental and computed band structures allows for the analysis of the orbital character of the spectral features. The individual contribution of the Si $s + p_x + p_y$, Si p_z and outermost Zr d orbitals are shown in Fig. 12.9. Features labeled p_1 and p_8 and corresponding to the ZrB$_2$(0001) surface states S_1 and S_2 have no contribution from Si orbitals, giving confirmation that those bands are pure surface states of the substrate. In contrast, all the silicene-derived bands appear to be hybridized with Zr d orbitals, suggesting a non-negligible interaction between the silicene sheet and the Zr-terminated surface. Most importantly, features p_2 and p_3 in the vicinity of the K_{Si} point, have a strong contribution of the Si p_z orbitals giving evidence for the π-band character of X_2 and X_3. These bands are however hybridized with the Si $s + p_x + p_y$ orbitals in agreement with the expected intermediate sp^2/sp^3 hybridization of silicene. The analysis of the orbitals giving rise to features p_6 and p_7, suggests that the bands labeled X_6 and X_7 are also of π-character.

Due to the hybridization, the p_z states are spread out across a much wider binding energy range (approximately 5 eV) than the band width (3–3.5 eV) expected for the π-band of free-standing silicene [32–34].

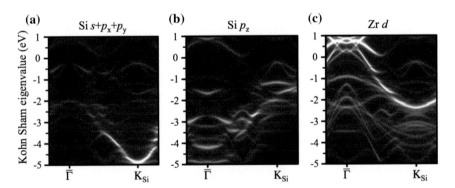

Fig. 12.9 Orbital character of the electronic bands of epitaxial silicene on ZrB$_2$(0001). Adapted with permission from [38]. Copyrighted by the American Physical Society

12.2.6.3   Microscopic Origin of the Silicene States

The mapping of the local density of states around the Fermi level by low-temperature scanning tunneling spectroscopy (LT-STS) [41] gave hints on the microscopic origin of the low-energy electronic states. The STS spectra (Fig. 12.10b) recorded in the center of the protrusions as shown in the STM image of Fig. 12.10a, indicate the existence of a 350 meV-wide band gap in the density of states. Its center is shifted by 60 meV below the Fermi level indicating the n-type character of epitaxial silicene on $ZrB_2(0001)$. The energies of the spectral features just below and above the gap are in good agreement with the band structure resolved by ARPES or calculated by DFT. In the filled states, the top of the X_2 and X_3 bands (p_2 and p_3 in the calculation), at 250 meV below the Fermi level, is at the same energy as the edge of the gap. The feature in the empty states and touching the Fermi level can be attributed to the band X_5 (p_5). The shift in energy from slightly below the Fermi level to above can be explained by a possible temperature dependence of the silicene structure and of its band structure. Assuming that the protrusions visible in the topographic STM image of Fig. 12.10a are centered on the protruding C-site atoms, the spectroscopic images (Fig. 12.10c, d) recorded at the energies of those features suggest that the valence and conduction states originate from the orbitals of B- and C-site atoms. However, the location of those states is different, which sheds light on the fact that they originate from different Si and Zr atoms.

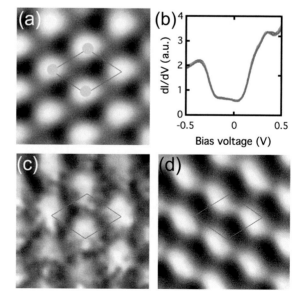

Fig. 12.10 Local density of states. **a** Topographic low-temperature STM image (2 nm × 2 nm, $V = -0.5$ V, and $I = 500$ pA). The rhombus emphasizes the $(\sqrt{3} \times \sqrt{3})$ unit cell. **b** STS spectra recorded on top of protrusion. **c** and **d** Spectroscopic images recorded at $V = -0.36$ eV and $V = 0.20$ eV. Modified from [41] (Color figure online)

12.2.6.4 Core-Level Excitations

In addition to giving hints on the chemical environment and the bonding configuration of the Si atoms, PES techniques can further provide a wealth of information
on how π-electronic states determine core-level excitations in epitaxial silicene.

The Si $2p$ spectrum recorded with a photon energy of $hv = 700$ eV (Fig. 12.11a)
differs from that recorded with $hv = 130$ eV (Fig. 12.3) by the presence of a
structured tail at the high binding energy side [26]. These features can be identified
as satellites replicating at binding energies that are 1.95 eV higher, the main four
peaks of the Si $2p$ core-level spectrum corresponding to the α and β components.
The energy difference between the main peaks and the satellites is comparable to
the energy of the prominent peak in the calculated optical absorbance of
free-standing silicene (1.7 eV), which is associated to the strong optical transition
between π-bands occurring at the M point of the Brillouin zone [42].

The Si $2p$ near-edge X-ray adsorption fine structure (NEXAFS) spectra were
recorded with different angles between the electric field of the light and the surface
normal (Fig. 12.11b) [26]. For the sake of comparison with bulk sp^3-hybridized
silicon, the NEXAFS spectrum of Si(111) is also displayed. The onset of the
adsorption for epitaxial silicene occurs at a photon energy 1 eV lower than that for
bulk silicon in agreement with the difference of binding energy of the Si $2p$
core-level between epitaxial silicene and bulk Si(111) [26].

The NEXAFS spectrum of silicene has a sharp onset at 98.7 eV followed by a
steady increase where several resonances can be distinguished. Features (1) to
(4) do not show any angular dependence and therefore can be attributed to π^*
resonances. The fact that those features are at the close proximity to the onset is in
agreement with the presence of states with a π-character just above the Fermi level.

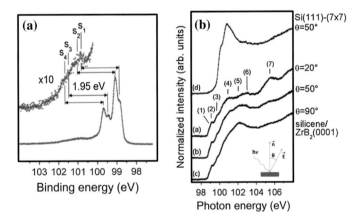

Fig. 12.11 Core-level excitations of epitaxial silicene on ZrB$_2$(0001). **a** Normal-emission Si $2p$
PES obtained at a photon energy of $hv = 700$ eV. Satellites features are labelled S$_1$–S$_4$. **b** Si $2p$
NEXAFS spectra recorded at different angle θ between the electric field and the normal. Spectral
features are labelled (1)–(7). The spectrum of the Si(111)-(7 × 7) surface is shown. Reprinted with
permission from [26]. Copyright 2014, AIP Publishing LLC (Color figure online)

12.2.7   Functionalization of Epitaxial Silicene by Adsorption of Atoms or Molecules

12.2.7.1   Tuning the Silicene Properties by Potassium Adsorption

As the ionic interaction of potassium atoms adsorbed on free-standing silicene was predicted to be stronger than with graphene [43, 44], the deposition of potassium on epitaxial silicene on $ZrB_2(0001)$ is expected to provide insightful information into the nature of the electronic states close to the Fermi level and in particular, to give hints on the interaction between silicene and the ZrB_2 surface. Nevertheless, the predicted strong electron donation of K adatoms to silicene [43–45] motivated the deposition of potassium as a straightforward and controllable method for tuning the electronic properties of silicene.

The deposition of 0.18 monolayer (ML) of potassium causes a drastic drop by 1.24 eV of the work function [46]. As shown on the low-energy electron diffraction (LEED) patterns of Fig. 12.12a, b, adsorption of K atoms does not affect the $(\sqrt{3} \times \sqrt{3})$ reconstruction of the pristine silicene, which suggests that K atoms are physisorbed rather than chemisorbed on silicene.

The investigation of the band structure by ARPES (Fig. 12.12c–f) [46] shows that upon adsorption of potassium, silicene-related bands undergo a rigid shift by

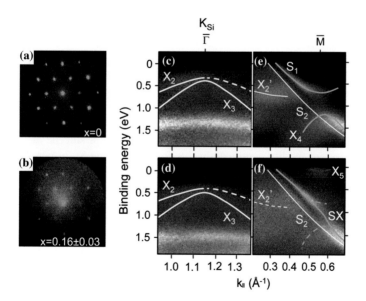

Fig. 12.12 Tuning of silicene band structure by adsorption of potassium. **a** and **b** μ-LEED patterns recorded at 30 eV before **a** and after **b** deposition of 0.16 ML potassium. **c–f** ARPES recorded with $h\nu = 43$ eV around the Γ point **c-d** and around the M point **e-f** of the Brillouin zone of the $(\sqrt{3} \times \sqrt{3})$ unit cell recorded before **c-e** and after **d-f** deposition of 0.18 ML of potassium. Reprinted with permission from [46]. Copyright (2013), AIP Publishing LLC (Color figure online)

0.15 eV towards higher binding energies. In particular, an electron pocket becomes visible at the M point of the Brillouin zone of the $(\sqrt{3} \times \sqrt{3})$ unit cell, which can be identified as the π-band X_5 (see Fig. 12.8) shifted further below the Fermi level. Since the $ZrB_2(0001)$ surface bands S_1 and S_2 do not experience any shift, it is concluded that charges are solely transferred from the K atoms to silicene.

Nevertheless, the appearance of a band denoted "SX" in Fig. 12.12f resulting from the partial hybridization of S_1 with the π-band X_4, suggests that electron donation to silicene enhances the hybridization with the ZrB_2 thin film resulting in a stronger interaction with the thin film surface. Since electron donation strengthens the interaction between silicene and the $ZrB_2(0001)$ surface, it can be foreseen that in contrast, the deposition of an electron acceptor may induce the decoupling of the silicene layer from the substrate.

12.2.7.2 O_2 Exposure and Effect of Al on Silicene Oxidation

Even though silicene is significantly less reactive than bulk silicon [47], as emphasized by the claimed stability of multilayer silicene in ambient air [48], the fact that it is not inert, like graphene, this remains a major hindrance to processing silicene. The study of the reactivity of silicene with O_2 is therefore of primary importance. The encapsulation of epitaxial silicene on Ag(111) by an Al_2O_3 thin film [49] that allows for the fabrication of a field effect transistor [50] after transferring to another substrate in air motivated the study of silicene reactivity with O_2 in the presence of a small amount of aluminium.

The low reactivity of silicene with O_2, already observed for epitaxial silicene on silver substrates [47, 48], was confirmed by the Si $2p$ spectrum recorded after the exposure of bare silicene on $ZrB_2(0001)$ to an O_2 dose of 4500 L. The conservation of the lineshape of the pristine silicene and the sole appearance of a small shoulder at higher binding energy (Fig. 12.13a) reflect the relative stability of silicene upon exposure to O_2.

The deposition of a fraction of a ML of aluminium on silicene at room temperature does not modify significantly the lineshape of the Si $2p$ core-level spectrum, suggesting that Al physisorbs on silicene. In contrast, aluminium has a terrible effect on the oxidation of silicene as evidenced by the Si $2p$ spectrum recorded after exposure to 4500 L O_2 in the presence of Al (Fig. 12.13b). The typical lineshape of the pristine silicene is still observed but is significantly less intense and an oxide peak arises at a 4 eV higher binding energy. The enhancement of silicene oxidation is attributed to Al-induced dissociative chemisorption of O_2 molecules yielding highly reactive atomic oxygen that is highly prone to oxidize silicene. The highest reactivity of silicene in the presence of a small amount of Al is in apparent disagreement with the successive encapsulation of silicene on Ag(111) by an Al_2O_3 thin film realized by the oxidation of a thin Al film grown on silicene [49]. It can, however, be explained by the fact that the amount of Al is much larger.

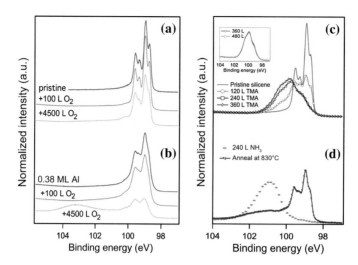

Fig. 12.13 Reactivity of epitaxial silicene on ZrB$_2$(0001). **a** and **b** Si *2p* spectra recorded after exposure to different O$_2$ doses on bare silicene (**a**) and after the deposition of 0.38 ML Al. **c** Evolution of the Si *2p* spectrum of pristine silicene upon exposure to TMA. **d** Si *2p* spectra after the subsequent exposure to NH$_3$ and annealing at 830 °C. For the sake of comparison, the spectrum of pristine silicene is shown. For all spectra, *hv* = 130 eV. **a** and **b** Reprinted with permission from [51]. Copyright (2014), AIP Publishing LLC. **c** and **d** Reprinted with permission from [53]. Copyright (2015), AIP Publishing LLC (Color figure online)

12.2.7.3 Tentative Encapsulation of Silicene by AlN Using an Atomic Layer Deposition Process

As the encapsulation of silicene, by sandwiching it between two single AlN honeycomb layers, was predicted to preserve its intrinsic properties [52], the passivation of silicene by an AlN single layer grown by atomic layer deposition (ALD) was attempted [53]. The ALD process consists of exposing silicene sequentially to the volatile precursors trimethylaluminium (Al(CH$_3$)$_3$, TMA) and ammonia (NH$_3$) [54].

As shown in Fig. 12.13c, the exposure of silicene to increasing TMA doses modifies dramatically the lineshape of the Si *2p* core-level spectrum. The appearance of new features indicates a chemical reaction between TMA and silicene [53] and the different distinguishable components correspond to different chemical interactions with adsorbed groups. For doses higher than 360 L of TMA no changes in the spectrum are found which indicates saturation of the surface.

The subsequent exposure to NH$_3$ done at 400 °C causes even stronger changes in the Si *2p* spectrum (Fig. 12.13d) since the Si *2p* peak is shifted by about 1.0 eV and the saturation is reached even faster (after exposure to 120 L).

The experimental evidence for the formation of complex compounds involving Si, Al, N, C and likely H atoms indicates that the proposed precursor does not allow for successful encapsulation of silicene on ZrB$_2$(0001) by ALD. Pristine silicene can be regenerated by annealing, but this regeneration is only partial.

12.3   Alternative Substrates for the Growth of Silicene

12.3.1   Silicene on ZrC(111)//NbC(111)

12.3.1.1   Sample Preparation

The spontaneous crystallization of silicene on $ZrB_2(0001)$ motivated the deposition of silicon on substrates with a similar surface with the aim of fabricating other forms of silicene. The (111)-oriented surface of zirconium carbide (ZrC) is also Zr-terminated. ZrC has a cubic structure made of the stacking of close-packed Zr and C layers along the $\langle 111 \rangle$ direction. As its lattice parameter (3.32 Å) is slightly larger than that of $ZrB_2(0001)$ (3.17 Å), ZrC(111) was thought to be an even better template for the growth of silicene. The lattice parameter of silicene is in epitaxy with ZrC(111) in such a way that the ($\sqrt{3} \times \sqrt{3}$) unit cell of silicene would be commensurate with the (2 × 2) unit cell of ZrC(111), is 3.83 Å and is in the range of the predicted values for free-standing silicene (3.8–3.9 Å) [32–34].

A ZrC thin film grown by molecular beam epitaxy (MBE) on a niobium carbide (NbC) (111) substrate with a cube-on-cube epitaxial arrangement was used as a substrate for the growth of silicon [55]. The surface structure resulting from silicon deposition at 527 °C on an oxide-free ZrC(111) surface bears striking resemblance to silicene on $ZrB_2(0001)$. The ZrC(111) surface is (2 × 2)-reconstructed and reversibly turns into (1 × 1) when the temperature is raised above 730–830 °C. The amount of silicon measured by Auger electron spectroscopy corresponding to completion of the (2 × 2)-reconstructed layer is close to that corresponding to silicene on $ZrB_2(0001)$. Nevertheless, similarly to other epitaxial forms of silicene, the Si-covered ZrC(111) surface is way less reactive than the bare surface.

In line with silicene on $ZrB_2(0001)$, a Si honeycomb structure whose ($\sqrt{3} \times \sqrt{3}$) unit cell is commensurate with the (2 × 2) unit cell of ZrC(111) was used as input for DFT calculations. The relaxed structure (Fig. 12.14a, b) resembles the planar-like structure found for silicene on ZrB_2 [36, 38]. Among the six atoms of the ($\sqrt{3} \times \sqrt{3}$) unit cell one is protruding at 3.74 Å above a Zr atom, three are sitting in bridge-site and two are sitting on fcc and hcp hollow sites. The rest of the atoms are almost all in the same plane at 3.3 Å above the Zr topmost atoms. The calculated valence charge density showed covalent character for the bonding within the silicon sheet.

12.3.1.2   Vibrational Properties of Epitaxial Silicene on ZrC(111)

The phonon dispersion (Fig. 12.14c) determined by HREELS [55] and by ab initio calculations, bears some resemblance to that of silicene on $ZrB_2(0001)$ [28]. Except for the soft acoustic surface phonon branch near the Γ point, the dispersion is small indicating that the Si–Si bonding is weaker than in graphene and no Kohn anomalies [56] were observed. The computed phonon dispersion indicates that the branches of the Si–Zr vibration modes are below 30 meV and that pure Si–Si

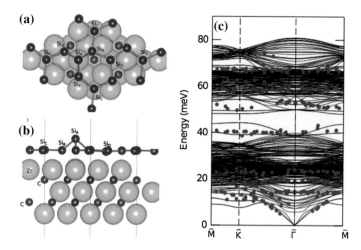

Fig. 12.14 Epitaxial silicene on ZrC(111): Structure and phonon dispersion. **a** and **b** Top and side views of the computed structure. **c** Phonon dispersion in the Brillouin zone of the ($\sqrt{3} \times \sqrt{3}$) unit cell determined experimentally (*red dots*) and calculated by DFT (*black lines*). Reprinted with permission from [55]. Copyright (2015) American Chemical Society

vibration modes, whose existence gives evidence for the formation of a continuous silicon layer, are found in the 40–60 meV region. The analysis of the vibration modes suggests that the silicon atoms are vibrating out-of-plane, in agreement with a buckled structure.

12.3.1.3 Electronic Band Structure of Silicene on ZrC(111)

In line with other epitaxial forms of silicene, the computed band structure (Fig. 12.15) does not exhibit Dirac cones [55]. One can note the striking similarity of the downward feature observed at the Γ point with that calculated for silicene on ZrB$_2$(0001) (Fig. 12.8c). Analysis of the orbital contribution of those bands confirms that they originate predominantly from Si atoms. The partial contribution of orbitals of the topmost Zr atoms to the Si states suggests hybridization of the silicene bands with the ZrC(111) surface states.

12.3.2 Epitaxial Silicene on Ir(111)

Iridium is a transition metal crystallizing in the fcc structure whose (111) surface has a lattice parameter of 2.71 Å. Silicon deposition on Ir(111) [57] substrates cleaned by several sputtering cycle gives rise to a ($\sqrt{7} \times \sqrt{7}$)-superstructure, which appears on STM images (Fig. 12.16a) in the form of a simple hexagonal lattice of protrusions. The periodicity (7.2 Å) is close to the expected value for the ($\sqrt{3} \times \sqrt{3}$)

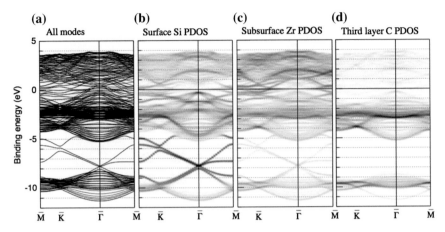

Fig. 12.15 **a** Electronic band structure of silicene on ZrC(111) in the Brillouin zone of the ($\sqrt{3} \times \sqrt{3}$) unit cell and contribution to the electronic bands of **b** the Si atoms, **c** the topmost Zr atoms, **d** the topmost C atoms. Reprinted with permission from [55]. Copyright (2015) American Chemical Society

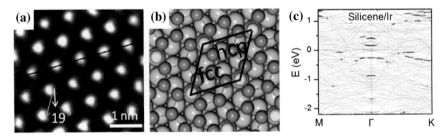

Fig. 12.16 Epitaxial silicene on Ir(111). **a** STM image (V = − 1.5 V, I = 0.05 nA). **b** Computed structure. Protruding Si atoms are *yellow-colored*. **c** Band structure represented in the Brillouin zone of the ($\sqrt{3} \times \sqrt{3}$) unit cell. The contribution of the Si atoms is *red-colored*. **a** and **b** Reprinted with permission from [57]. Copyright (2013) American Chemical Society. **c** Reprinted by permission from Macmillan Publishers Ltd from [58], Copyright (2014)

unit cell of free-standing silicene [32–34]. DFT calculations [56, 57] converged to the structure shown in Fig. 12.16b, are in agreement with the STM image of Fig. 12.16a. This silicon honeycomb structure is very similar to the planar-like structure assigned to epitaxial silicene on ZrB$_2$(0001) [36, 39], ZrC(111) [55] and Ag(111) [37]. Among the 6 atoms of the ($\sqrt{3} \times \sqrt{3}$) unit cell, one atom is protruding 2.83 Å directly above an Ir atom and the rest of the atoms are 2.0 Å above the Ir (111) surface. Two atoms are sitting respectively on the fcc and hcp hollow sites of the Ir(111) surface and three are sitting on bridge sites [57, 58].

The bonds between the silicon atoms have covalent character and the electrostatic interaction between the silicene layer and the iridium surface leads to a relatively high binding energy per Si atom of 1.6 eV [57, 58]. The computed band structure of epitaxial silicene on Ir(111) (Fig. 12.16c), shows a strong deviation from those for free-standing silicene and other forms of epitaxial silicene.

Considering the strong hybridization of the silicon layer with the Ir(111) substrate and the lack of features reminiscent of the Dirac cones in the band structure, it is questionable whether a π-electronic system exists in the epitaxial Si honeycomb structure on Ir(111) [58].

The hybridization of the electronic bands of silicon with those of the metal substrate was confirmed by the calculation of the work function of epitaxial silicene on Ir(111), which is 0.57 eV greater than that of free-standing silicene (4.48 eV). The loss of 0.13 electrons per Si atom, suggests a heavy p-type doping of silicene by the Ir substrate [58].

12.3.3 Self-assembled Silicon Nanoribbons on Au(110)

In a similar manner in which silicon nanoribbons can be grown on Ag(110) [59], the deposition of silicon at 400 °C on pristine Au(110)-(2 × 1), gives rise to well aligned asymmetric 1.6 nm wide nanoribbons (Fig. 12.17a) [60]. The Si $2p$

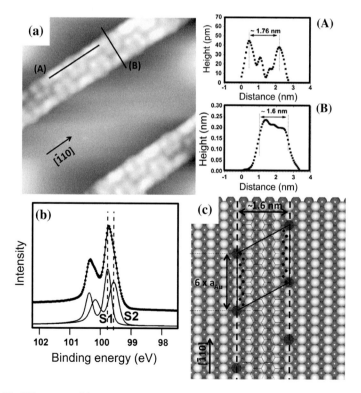

Fig. 12.17 Silicon nanoribbons on Au(110). **a** STM image (8 nm × 8 nm, V = −64 mV, and I = 3.5 nA) and profiles along the ribbon (**A**) and across (**B**). **b** Si $2p$ spectrum recorded with a photon energy of $h\nu = 147$ eV. Two components (S1) and (S2) are identified. **c** Proposed model for the Si nanoribbon. The Si honeycomb lattice is *black-colored*. The atoms of the topmost Au layer are in *light gray*. The Au atoms at the corners of the unit cell of the superstructure are in *dark gray* (in *blue* online). Reprinted with permission from [60]. Copyright (2013), AIP Publishing LLC (Color figure online)

spectrum (Fig. 12.17b) can be decomposed into two narrow spin–orbit split components that are respectively located at 99.76 and 99.58 eV, indicating well-defined environments of the Si atoms within the nanoribbons. The asymmetry of the peaks gives evidence for the metallicity of the nanoribbons. Coincidences between the lattice parameters of Au(110) and silicene lead to the model of Fig. 12.17c. Along the close packed [$\bar{1}$10] direction, 6 Au unit cells fit on 5 silicene unit cells with a lattice constant of 3.46 Å. Along the [100] direction, the ribbon width of about 4 Au unit cells fits well with 4 silicene rows. The weights of the two components (S1) and (S2) (0.52 and 0.48 respectively) are in qualitative agreement with the respective 0.6 and 0.4 fractions of Si atoms in the center and at the edge of the domains.

12.4 Stable Silicene Forms on Metals by Ab Initio Calculations

12.4.1 Silicene on Metal Versus Silicide

So far silicene, which can be defined as a silicon honeycomb structure possessing a π-electronic system, was found to be stable on a very limited number of substrates and the origin of the stability on those specific materials has not been clearly established yet. Therefore, ab initio calculations can be an insightful way to anticipate whether a given substrate is able, or not, to host a silicene structure without destroying its specific electronic properties. The sole experimental observation of silicene on conductive substrates may suggest that silicene is more prone to being stabilized on the surface of metallic materials. At first glance, the well-known strong reactivity of silicon with a wide variety of metals, which results in highly stable silicides [61, 62] suggests that the number of possible surfaces on which the electronic properties of silicene may survive, is limited. It should be noted that some silicides crystallize in the AlB$_2$ structure in such a way that buckled silicon honeycomb lattices are stacked along the c-axis together with close-packed metal interlayers [63–65]. Even though a strong hybridization between the metal and silicon layers is frequently observed [66], it was recently shown that Dirac cone-like bands exist in CaSi$_2$ [65].

12.4.2 Absence of Dirac Cones in Silicene on Metal Surfaces: A General Trend

The structure and electronic properties of silicon honeycomb layers put in contact with several (111)-oriented surfaces of metal substrates were computed [58, 67, 68]. Metals such as Mg, Au, Cu and Al were compared to metals for which experimental data is available, such as Ag [69], Ir [57] and Pt [70]. The lattice parameters of silicene sheets were adjusted to the metal surfaces in such ways that the silicene-($\sqrt{7} \times \sqrt{7}$) unit cell is commensurate with the Mg(111)-(2 × 2) unit cell and the silicene-($\sqrt{3} \times \sqrt{3}$) unit cell is commensurate with the ($\sqrt{7} \times \sqrt{7}$) unit cells of Cu(111), Pt(111) and Au(111) [58, 67, 68]. For silicene on Al(111), two epitaxial relationship were proposed: The silicene-($\sqrt{3} \times \sqrt{3}$) unit cell is commensurate with the Al (111)-($\sqrt{7} \times \sqrt{7}$) unit cell [67] and the silicene-(3 × 3) unit cell is commensurate with the Al(111)-(4 × 4) unit cell [69]. The latter allows for the comparison with the experimentally observed silicene-(3 × 3)/Ag(111)-(4 × 4) phase [67]. The silicene-metal interfacial structures can be classified into two categories. Substrates in the first category, including Mg and Cu give rise to structures similar to the so-called planar-like structure [57, 58]. The buckling of those forms of epitaxial silicene is high (respectively 1.52 and 1.25 Å for Mg and Cu) due to a protruding Si atom always located right above a metal atom [57, 58]. In contrast, the buckling of silicene on Au and Pt is less than that of free-standing silicene, likely in relation to the larger mismatch between silicene and the substrates [58]. A similar observation is made for the structure of silicene-($\sqrt{3} \times \sqrt{3}$)/Al(111)-($\sqrt{7} \times \sqrt{7}$) [58] whereas the buckling of the structure found for silicene-(3 × 3)/Al(111)-(4 × 4) [67], is 1.4 Å and thus larger than for silicene-(3 × 3)/Ag(111)-(4 × 4) [68].

Compared to Ir(111), silicene is relatively weakly bonded to the Al, Mg, Au, and Cu substrates, as the binding energies are 0.3, 0.43, 0.39, 0.63, and 0.86 eV/Si atom for silicene-($\sqrt{3} \times \sqrt{3}$)/Al(111)-($\sqrt{7} \times \sqrt{7}$), silicene-(3 × 3)/Al(111)-(4 × 4), silicene-($\sqrt{7} \times \sqrt{7}$)/Mg(111)-(2 × 2), silicene-($\sqrt{3} \times \sqrt{3}$)/Au(111)-($\sqrt{7} \times \sqrt{7}$) and silicene-($\sqrt{3} \times \sqrt{3}$)/Cu(111)-($\sqrt{7} \times \sqrt{7}$). It has to be noted that the interaction of silicene with metal substrates is much stronger than for graphene on metals (about 0.1–0.2 eV/C atom). The binding to the Pt(111) substrate (1.98 eV/Si atom) is particularly strong and may explain the formation of a surface silicide upon deposition of silicon on Pt (111) [69]. The strength of the binding of silicene with Pt(111) is reflected in the close distance between the silicon layer and the Pt(111) surface (Fig. 12.17c) and in the related band structure (Fig. 12.17d) in which, no features typical to silicene can be identified [58].

The work functions of epitaxial silicene on Au and Pt substrates are respectively 0.08, and 0.07 eV greater than that of free-standing silicene, and silicene loses 0.06, and 0.04 electrons per Si, respectively.

The work functions of epitaxial silicene on Mg, Al, and Cu substrates are 0.50, 0.11, 0.03 eV less than those of free-standing silicene, and as silicene receives 0.21, 0.06 and 0.04 electrons per Si atom it gives evidence for the heavy n-type doping of silicene by Mg and to the relatively low n-type doping of silicene by Cu and Al.

The calculated band structures of silicene on metals [58, 67] such as those for silicene on Ir(111) (Fig. 12.16c), on Al(111) and on Pt(111) (Fig. 12.18b, d), have in common the absence of features resembling Dirac cones. It is thus concluded that the more or less strong band hybridization between Si and bare metal substrates leading to the destruction of the Dirac cones cannot be easily avoided.

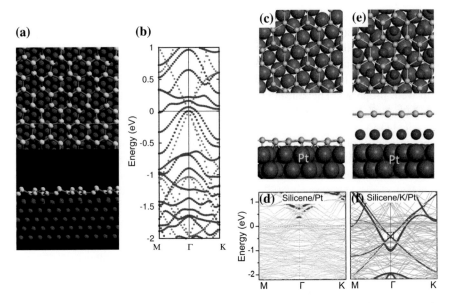

Fig. 12.18 Examples of computed structures and band structures of epitaxial silicene on metals. **a** and **b** Structure and band structure of silicene-(3 × 3)/Al(111)-(4 × 4) without the Al(111) substrate (*light gray/red* online). The band structure of free-standing silicene is also shown (*gray*). **c** and **d** Structure and band structure of silicene-($\sqrt{3}$ × $\sqrt{3}$)/Pt(111)-($\sqrt{7}$ × $\sqrt{7}$). **e** and **f** Structure and band structure of K-intercalated silicene-($\sqrt{3}$ × $\sqrt{3}$)/Pt(111)-($\sqrt{7}$ × $\sqrt{7}$). **d** and **f** The states contributed by the Si atoms are *dark gray-colored* (*red-colored* online). **a** and **b** Reprinted (adapted) with permission from [67] Copyright (2013) American Chemical Society. **c–f** Reprinted by permission from Macmillan Publishers Ltd from [58], Copyright (2014) (Color figure online)

12.4.3 Intercalation as a Way to Regenerate Dirac Cones

In order to verify whether Dirac cones destroyed by the hybridization between silicene and metal surfaces can be recovered by intercalation of potassium between silicene and metal substrates, structures and electronic band dispersions were computed when potassium atoms are inserted below the center of the Si honeycomb as illustrated in Fig. 12.18e [58]. As exemplified by the band structure of Fig. 12.18f, the band dispersion of intercalated silicene features Dirac cones for all the considered substrates. The recovered Dirac cone is located at 0.40–0.78 eV below the Fermi level, suggesting a n-type doping of silicene. However, for K-intercalated silicene on Ir(111), Mg(111), Cu(111) and Ag(111), a 0.15–0.40 eV gap is opened between the π and π^* bands, as a result of the breaking of the inversion symmetry between the two silicene sublattices and inter-valley interactions [58].

It was also theoretically demonstrated that the linear dispersion of silicene π-bands, which are destroyed when it is put in contact with a Cu(111) surface, can be restored by decoupling silicene from the conductive substrate by intercalating a single layer of hexagon boron nitride (h-BN) [68]. Dirac cones of silicene on h-BN-buffered Cu(111) are shifted below the Fermi level by 200 meV, an amount larger than that for silicene on free-standing h-BN (50 meV). This suggests that silicene is weakly coupled to the Cu(111) substrate. This is in agreement with the shift in the work function of silicene with respect to that of free-standing silicene that is related to a slight charge transfer from Cu to silicene.

12.4.4   Polygonal Silicene on Al(111)

By progressively tugging out one of the outermost Si atom of the silicene-(3 × 3)/Al (111)-(4 × 4) structure (Fig. 12.18a), the honeycomb lattice evolves into the two-dimensional Si structure of Fig. 12.19a, containing hexagonal, pentagonal and rectangular rings. Even though the honeycomb structure does not exist anymore, this new structure, called "polygonal silicene" remains periodic and is commensurate with the Al(111)-(4 × 4) unit cell. Except for one atom belonging to the central hexagon protruding at 2 Å above the other Si atoms, the structure of polygonal silicene is almost flat. The bond length associated with the rectangles and pentagons are in the 2.5–2.7 Å range, whereas those associated with the hexagons are shorter (approximately 2.4 Å). The polygonal silicene structure is found to be slightly less stable than the regular honeycomb structure (Fig. 12.18a). Analysis of

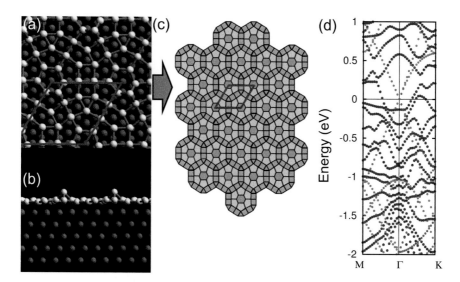

Fig. 12.19 Polygonal silicene on Al(111). **a** and **b** Top and side views of the epitaxial structure. **c** Color-coded schematic of the polygonal structure. *Blue, pink* and *green* (online) represent *hexagons*, *pentagons* and *rectangles* respectively. **d** Calculated band structures of polygonal silicene without the Al(111) substrate (*light gray/red* online) and for free-standing silicene (*gray*). Reprinted (adapted) with permission from [67] Copyright (2013) American Chemical Society (Color figure online)

the bonding in polygonal silicene shows that the covalent character of the bonds is less pronounced than in the regular hexagonal silicene on Al(111) due to a more pronounced electron transfer from Al to polygonal silicene. Despite the similarity in the lattice parameters of Ag(111) and Al(111), a polygonal silicene structure on Ag (111) could not be found [67].

The band structure of polygonal silicene (Fig. 12.19e) is significantly different to that of the regular honeycomb structure. The higher dispersion of the bands crossing the Fermi level suggests a higher conductivity in this form of silicene [67].

12.5 Summary

The existence of several epitaxial forms of silicene on substrates other than silver surfaces demonstrates that the stability of silicene is not specific to this metal. There are some common features to the different forms of silicene, such as the buckling of the epitaxial structure, the hybridization of silicene states with the substrate which result in a more or less strong deviation of the dispersion of the π-electronic states from the Dirac cones predicted for free-standing silicene. The number of epitaxial forms of silicene experimentally observed remains limited, most likely due to the high degree of hybridization between silicene and most of the conductive substrates. Among the few substrates on which silicene is observed, $ZrB_2(0001)$ is the one which has been the most studied experimentally and theoretically. Silicene on $ZrB_2(0001)$ thin films grown on Si(111) remains a unique example of the spontaneous formation of silicene. It is, therefore, a perfect benchmark for the study of the structural, electronic, mechanical and chemical properties of the graphene-like allotrope of silicon.

References

1. T. Aizawa, S. Suehara, S. Hishita, S. Otani, Phys. Rev. B **71**, 165405 (2005)
2. S. Kumashiro, H. Tanaka, Y. Kawamata, Y. Yanagisawa, K. Momose, G. Nakamura, C. Oshima, e-J. Surf. Sci. Nanotech. **4**, 100 (2006)
3. C.-C. Lee, Y. Yamada-Takamura, T. Ozaki, J. Phys.: Condens. Matter **25**, 345501 (2013)
4. H. Rosner, J.M. An, W.E. Pickett, S.-L. Drechsler, Phys. Rev. B **66**, 024521 (2002)
5. C. Jariwala, A. Chainani, S. Tsuda, T. Yokoya, S. Shin, Y. Takano, K. Togano, S. Otani, H. Kito, Phys. Rev. B **68**, 174506 (2003)
6. S. Otani, Y. Ishizawa, J. Cryst. Growth **165**, 319 (1996)
7. S. Otani, M.M. Korsukova, T. Mitsuhashi, J. Cryst. Growth **186**, 582 (1998)
8. H. Kinoshita, S. Otani, S. Kamiyama, H. Amano, I. Akasaki, J. Suda, H. Matsunami, Jpn. J. Appl. Phys. **40**, L1280 (2001)
9. T. Aizawa, W. Hayami, S. Otani, Phys. Rev. B **65**, 024303 (2001)
10. R. Armitage, J. Suda, T. Kimoto, Surf. Sci. **600**, 1439 (2006)
11. T. Aizawa, W. Hayami, S. Otani, Phys. Rev. B **65**, 024303 (2002)

12. N.L. Okamoto, M. Kusakari, K. Tanaka, H. Inui, M. Yamaguchi, S. Otani, J. Appl. Phys. **93**, 88 (2003)
13. J. Suda, H. Matsunami, J. Cryst. Growth **237–239**, 1114 (2002)
14. J. Tolle, R. Roucka, I.S.T. Tsong, C. Ritter, P.A. Crozier, A.V.G. Chizmeshya, J. Kouvetakis, Appl. Phys. Lett. **82**, 2398 (2003)
15. J. Tolle, J. Kouvetakis, D.-W. Kim, S. Mahajan, A. Bell, F.A. Ponce, I.S.T. Tsong, M.L. Kottke, Z.D. Chen, Appl. Phys. Lett. **84**, 3510 (2004)
16. R.A. Trivedi, J. Tolle, A.V.G. Chizmeshya, R. Roucka, C. Ritter, J. Kouvetakis, I.S.T. Tsong, Appl. Phys. Lett. **87**, 072107 (2005)
17. Y. Yamada-Takamura, Z.T. Wang, Y. Fujikawa, T. Sakurai, Q.K. Xue, J. Tolle, P.-L. Liu, A.V.G. Chizmeshya, J. Kouvetakis, I.S.T. Tsong, Phys. Rev. Lett. **95**, 266105 (2005)
18. S. Bera, Y. Sumiyoshi, Y. Yamada-Takamura, J. Appl. Phys. **106**, 063531 (2009)
19. A.L. Wayda, L.F. Schneemeyer, R.L. Opila, Appl. Phys. Lett. **53**, 361 (1988)
20. C.-W. Hu, A.V.G. Chizmeshya, J. Tolle, J. Kouvetakis, I.S.T. Tsong, J. Crystal Growth **267**, 554 (2004)
21. A. Fleurence, Y. Yamada-Takamura, Phys. Status Solidi (c) **8**, 779 (2011)
22. A. Fleurence, W. Zhang, C. Hubault, Y. Yamada-Takamura, Appl. Surf. Sci. **284**, 432 (2013)
23. R. Roucka, Y. An, A.V.G. Chizmeshya, J. Tolle, J. Kouvetakis, V.R. D'Costa, J. Menéndez, P. Crozier, Appl. Phys. Lett. **89**, 242110 (2006)
24. Y. Yamada-Takamura, F. Bussolotti, A. Fleurence, S. Bera, R. Friedlein, Appl. Phys. Lett. **97**, 073109 (2010)
25. A. Fleurence, R. Friedlein, T. Ozake, H. Kawai, Y. Wang, Y. Yamada-Takamura, Phys. Rev. Lett. **108**, 245501 (2012)
26. R. Friedlein, A. Fleurence, K. Aoyagi, M.P. de Jong, H. Van Bui, F.B. Wiggers, S. Yoshimoto, T. Koitaya, S. Shimizu, H. Noritake, K. Mukai, J. Yoshinobu, Y. Yamada-Takamura, J. Chem. Phys. **140**, 184704 (2014)
27. S. Kamiyama, S. Takanami, Y. Tomida, K. Iida, T. Kawashima, S. Fukui, M. Iwaya, H. Kinoshita, T. Matsuda, T. Yasuda, S. Otani, H. Amano, I. Akasaki, Phys. Status Solidi A **200**, 67 (2003)
28. T. Aizawa, S. Suehara, S. Otani, J. Phys.: Condens. Matter **27**, 305002 (2015)
29. C.-L. Lin, R. Arafune, K. Kawahara, N. Tsukahara, E. Minamitani, Y. Kim, N. Takagi, M. Kawai, Applied Physics Express **5**, 045802 (2012)
30. P. Moras, T.O. Mentes, P.M. Sheverdyaeva, A. Locatelli, C. Carbone, J. Phys.: Condens. Matter **26**, 185001 (2014)
31. D. Chiappe, C. Grazianetti, G. Tallarida, M. Fanciulli, A. Molle, Adv. Mater. **24**, 5088 (2012)
32. S. Cahangirov, M. Topsakal, E. Aktürk, H. Sahin, S. Ciraci, Phys. Rev. Lett. **102**, 236804 (2009)
33. G. Guzmán-Verri, L.C. Lew Yan Voon, Phys. Rev. B **76**, 075131 (2007)
34. C. Liu, W. Feng, Y. Yao, Phys. Rev. Lett. **107**, 076802 (2011)
35. T. Ozaki, Phys. Rev. B **67**, 155108 (2003)
36. C.-C. Lee, A. Fleurence, R. Friedlein, Y. Yamada-Takamura, T. Ozaki, Phys. Rev. B **88**, 165404 (2013)
37. L. Chen, H. Li, B. Feng, Z. Ding, J. Qiu, P. Cheng, K. Wu, S. Meng, Phys. Rev. Lett. **110**, 085504 (2013)
38. C.-C. Lee, A. Fleurence, Y. Yamada-Takamura, T. Ozaki, R. Friedlein, Phys. Rev. B **90**, 075422 (2014)
39. C.-C. Lee, A. Fleurence, R. Friedlein, Y. Yamada-Takamura, T. Ozaki, Phys. Rev. B **90**, 241402 (2014)
40. W. Ku, T. Berlijn, C.-C. Lee, Phys. Rev. Lett. **104**, 216401 (2010)
41. A. Fleurence, Y. Yoshida, C.-C. Lee, T. Ozaki, Y. Yamada-Takamura, Y. Hasegawa, Appl. Phys. Lett. **104**, 021605 (2014)
42. F. Bechstedt, L. Matthes, P. Gori, O. Pulci, Appl. Phys. Lett. **100**, 261906 (2012)
43. R. Quhe, R. Fei, Q. Liu, J. Zheng, H. Li, C. Xu, Z. Ni, Y. Wang, D. Yu, W. Gao, J. Lu, Sci. Rep. **2**, 853 (2012)

44. X. Lin, J. Ni, Phys. Rev. B **86**, 075440 (2012)
45. H. Sahin, F.M. Peeters, Phys. Rev. B **87**, 085423 (2013)
46. R. Friedlein, A. Fleurence, J.T. Sadowski, Y. Yamada-Takamura, Appl. Phys. Lett. **102**, 221603 (2013)
47. P. De Padova, C. Leandri, S. Vizzini, C. Quaresima, P. Perfetti, B. Olivieri, H. Oughaddou, B. Aufray, G. Le Lay, Nano Lett. **8**, 2299 (2008)
48. P. De Padova, C. Ottaviani, C. Quaresima, B. Olivieri, P. Imperatori, E. Salomon, T. Angot, L. Quagliano, C. Romano, A. Vona, M. Muniz-Miranda, A. Generosi, B. Paci, G. Le Lay, 2D Mater. **1**, 02103 (2014)
49. A. Molle, C. Grazianetti, D. Chiappe, E. Cinquanta, E. Cianci, G. Tallarida, M. Fanciulli, Adv. Funct. Mater. **23**, 4340 (2013)
50. L. Tao, E. Cinquanta, D. Chiappe, C. Grazianetti, M. Fanciulli, M. Dubey, A. Molle, D. Akinwande, Nature Nanotech. **10**, 227 (2015)
51. R. Friedlein, H. Van Bui, F.B. Wiggers, Y. Yamada-Takamura, A.Y. Kovalgin, M.P. de Jong, J. Chem. Phys. **140**, 204705 (2014)
52. M. Houssa, G. Pourtois, V.V. Afanasev, A. Stesmans, Appl. Phys. Lett. **97**, 112106 (2010)
53. H. Van Bui, F.B. Wiggers, R. Friedlein, Y. Yamada-Takamura, A.Y. Kovalgin, M.P. de Jong, J. Chem. Phys. **142**, 064702 (2015)
54. H. Van Bui, M.D. Nguyen, F.B. Wiggers, A.A.I. Aarnink, M.P. de Jong, A.Y. Kovalgin, ECS J. Solid State Sci. Technol. **3**, P101 (2014)
55. T. Aizawa, S. Suehara, S. Otani, J. Phys. Chem. C **118**, 23049 (2014)
56. S. Piscanec, M. Lazzeri, F. Mauri, A.C. Ferrari, J. Robertson, Phys. Rev. Lett. **93**, 185503 (2004)
57. L. Meng, Y. Wang, L. Zhang, S. Du, R. Wu, L. Li, Y. Zhang, G. Li, H. Zhou, W.A. Hofer, H.-J. Gao, Nano Lett. **13**, 685 (2013)
58. R. Quhe, Y. Yuan, J. Zheng, Y. Wang, Z. Ni, J. Shi, D. Yu, J. Yang, J. Lu, Sci. Rep. **4**, 5476 (2014)
59. P. De Padova, C. Quaresima, C. Ottaviani, P.M. Sheverdyaeva, P. Moras, C. Carbone, D. Topwal, B. Olivieri, A. Kara, H. Oughaddou, B. Aufray, G. Le Lay, Appl. Phys. Lett. **96**, 261905 (2010)
60. M.R. Tchalala, H. Enriquez, A.J. Mayne, A. Kara, S. Roth, M.G. Silly, A. Bendounan, F. Sirotti, T. Greber, B. Aufray, G. Dujardin, M. Ait Ali, H. Oughaddou, Appl. Phys. Lett. **102**, 083107 (2013)
61. A.H. Reader, A.H. van Ommen, P.J.W. Weijs, R.A.M. Wolters, D.J. Oostra, Rep. Prog. Phys. **56**, 1397 (1992)
62. J.A. Knapp, S.T. Picraux, Appl. Phys. Lett. **48**, 466 (1986)
63. P. Paki, U. Kafader, P. Wetzel, C. Pirri, J.C. Peruchetti, D. Bolmont, G. Gewinner, Phys. Rev. B **45**, 10555 (1992)
64. R. Baptist, S. Ferrer, G. Grenet, H.C. Poon, Phys. Rev. Lett. **64**, 311 (1990)
65. E. Noguchi, K. Sugawara, R. Yaokawa, T. Hitosugi, H. Nakano, T. Takahashi, Adv. Mater. **27**, 856 (2014)
66. L. Stauffer, A. Mharchi, C. Pirri, P. Wetzel, D. Bolmont, G. Gewinner, Phys. Rev. B **47**, 10555 (1993)
67. T. Morishita, M.J.S. Spencer, S. Kawamoto, I.K. Snook, J. Phys. Chem. C **117**, 22142 (2014)
68. M. Kanno, R. Arafune, C.L. Lin, E. Minamitani, M. Kawai, N. Takagi, New J. Phys. **16**, 105019 (2014)
69. P. Vogt, P. De Padova, C. Quaresima, J. Vila, E. Frantzeskakis, M.C. Asensio, A. Resta, B. Ealet, G. Le Lay, Phys. Rev. Lett. **108**, 155501 (2012)
70. M. Svec, P. Hapala, M. Ondracek, P. Merino, M. Blanco-Rey, P. Mutombo, M. Vondracek, Y. Polyak, V. Chab, J.A. Martin Gaco, P. Jelinek, Phys Rev B **89**, 201412 (2014)

Index

© Springer International Publishing Switzerland 2016
M.J.S. Spencer and T. Morishita (eds.), *Silicene*, Springer Series
in Materials Science 235, DOI 10.1007/978-3-319-28344-9

Printed in the United States
By Bookmasters